THE IMPLICATIONS OF CLIMATE AND SEA-LEVEL CHANGE FOR BANGLADESH

THE IMPLICATIONS OF CLIMATE AND SEA-LEVEL CHANGE FOR BANGLADESH

Edited by

R. A. WARRICK

Research Director
Centre for Environmental and Resource Studies (CEARS)
The University of Waikato
Hamilton
New Zealand

and

Q. K. AHMAD

Chairman
Bangladesh Unnayan Parishad (BUP)
(Centre for Research and Action on Environment and Development)
Dhaka
Bangladesh

KLUWER ACADEMIC PUBLISHERS

DORDRECHT / BOSTON / LONDON

A C.I.P. Catalogue record for this book is available from the Library of Congress

ISBN-13:978-94-010-6591-7 e-ISBN-13:978-94-009-0241-1
DOI:10.1007/978-94-009-0241-1

Published by Kluwer Academic Publishers,
P.O. Box 17, 3300 AA Dordrecht, The Netherlands.

Kluwer Academic Publishers incorporates
the publishing programmes of
D. Reidel, Martinus Nijhoff, Dr W. Junk and MTP Press.

Sold and distributed in the U.S.A. and Canada
by Kluwer Academic Publishers,
101 Philip Drive, Norwell, MA 02061, U.S.A.

In all other countries, sold and distributed
by Kluwer Academic Publishers Group,
P.O. Box 322, 3300 AH Dordrecht, The Netherlands.

Printed on acid-free paper

TABLE OF CONTENTS

Preface ix
Glossary of Bengali Terms xiii

CHAPTER 1: THE IMPLICATIONS OF CLIMATE CHANGE FOR BANGLADESH: A SYNTHESIS
Q.K. Ahmad, R.A. Warrick, N.J. Ericksen and M.Q. Mirza 1

1.1 Introduction 1
1.2 How might Climate Change? 1
1.3 How might Sea Level Change in the Bay of Bengal ? 7
1.4 What are the Possible Effects on Natural Resources? 12
1.5 What are the Possible Socio-Economic Effects? 17
1.6 What are the Legal Implications of Climate and Sea-Level Change? 22
1.7 How might the Coastal Zone of Bangladesh be Affected? 24
1.8 Policy Implications, Knowledge Gaps and Research Needs 28

CHAPTER 2: THE GREENHOUSE EFFECT AND CLIMATE CHANGE
R.A. Warrick, A.K. Azizul Hoq Bhuiya and M.Q. Mirza 35

2.1 Introduction 35
2.2 How has Climate Changed, and Why? 35
2.3 How is the Atmospheric Composition Changing? 40
2.4 Changing the Earth's Radiation Budget 50
2.5 Have Variations in Greenhouse Gases Caused Climate to Change? 54
2.6 How might Global Climate Change in the Future? 57
2.7 The Implications for Bangladesh 64
2.8 What are the Implications for Policy? 77
2.9 Summary and Research Directions 86

CHAPTER 3: SEA-LEVEL CHANGES IN THE BAY OF BENGAL
R.A. Warrick, A.K. Azizul Hoq Bhuiya, W.M. Mitchell, T.S. Murty and K.B.S. Rasheed 97

3.1 Introduction 97
3.2 What Causes Sea Level to Change? 98

3.3 Has Global Sea Level been Rising and will it Rise in the Future? 109
3.4 How are Relative Sea Levels Changing in the Bay of Bengal? 115
3.5 How do Cyclone-Induced Storm Surges Affect Sea Level? 123
3.6 Summary, Knowledge Gaps and Research Needs 129

CHAPTER 4: EFFECTS OF CLIMATE AND SEA-LEVEL CHANGES ON THE NATURAL RESOURCES OF BANGLADESH
H. Brammer, M. Asaduzzaman, and P. Sultana 143

4.1 Introduction 143
4.2 What is the Current Resource Setting of Bangladesh? 143
4.3 How might the Water Resources be Affected by Climate Warming? 156
4.4 How might Crop Agriculture be Affected by Changing Climate and Water Regimes? 168
4.5 How might Other Bio-Resource Sectors be Affected? 184
4.6 What Needs to be Done? 193

CHAPTER 5: SOCIO-ECONOMIC IMPLICATIONS OF CLIMATE CHANGE FOR BANGLADESH
N.J. Ericksen, Q.K. Ahmad and A.R. Chowdhury 205

5.1 Introduction 205
5.2 Climate and Society: an Interactive Perspective 206
5.3 How does Climate Currently Affect Bangladeshi Society? 208
5.4 What Societal Trends may Influence the Vulnerability of Bangladesh to Changes in Climate and Sea Level? 222
5.5 What are the Possible Socio-Economic Impacts of Climate Change on Bangladesh in the Future? 261
5.6 What Alternatives are there for Future Adjustment to Climate and Sea-Level Change? 271
5.7 Knowledge Gaps and Future Research Needs 275
5.8 What Should be Done? 277

CHAPTER 6: LEGAL IMPLICATIONS OF GLOBAL CLIMATE CHANGE FOR BANGLADESH
D. Freestone, M. Farooque and S.R. Jahan 289

6.1 Introduction 289
6.2 National Law Aspects 293
6.3 International Law Aspects 310

6.4 Looking Forward 327
6.5 Further Research 328

CHAPTER 7: CLIMATE CHANGE AND SEA-LEVEL RISE: THE CASE OF THE COAST
A. Kausher, R.C. Kay, M. Asaduzzaman and S. Paul 335

7.1 Introduction 335
7.2 The Coast of Bangladesh 336
7.3 Special Characteristics of the Coastal Zones 360
7.4 How might Climate and Sea Level Change in the Coastal Zone? 372
7.5 What are the Possible Effects on Coastal Water and Land? 377
7.6 What are the Possible Socio-Economic Consequences? 386
7.7 Environmental Policy for Coastal and Marine Areas 391
7.8 Knowledge Gaps and Future Research Needs 394
7.9 What Should be Done? 396

Index 407

PREFACE

Although the "greenhouse effect" and "global climate change" have been the subjects of scientific scrutiny for many decades, only recently have they received widespread public attention. Two major events helped generate this attention. First, in 1990 the Intergovernmental Panel on Climate Change (IPCC) published its findings on the science, impacts and policy implications of climate change. The findings of the IPCC, prepared and reviewed extensively by the world's leading experts in the field, confirmed that the increasing atmospheric concentrations of "greenhouse" gases, such as carbon dioxide, methane, nitrous oxide and the chlorofluorocarbons, could cause the world to warm and sea level to rise. Second, in 1992 the United Nations Conference on Environment and Development (UNCED) in Rio de Janeiro focussed the attention of the world's national governments, as well as organisations and individuals outside the governments, on the threat of global climate change. The Framework Convention on Climate Change (FCCC), signed by nations at UNCED, reflects both the concern about the effects of climate change and the urgent need for action to prevent or reduce its potential impacts, particularly with respect to the vulnerable developing countries of the world. Bangladesh ratified the FCCC on 15 April 1994.

The countries that have signed and ratified the FCCC are obligated to report to the Conference of the Parties (CoP) to the Convention on a number of inter-related issues. In particular, under Article 4 of the FCCC, these countries are required to prepare inventories of anthropogenic emissions by sources and removals by sinks of all greenhouse gases, to assess the possible future effects of climate change, and to develop strategies for mitigation or adaption. Efforts to lay the foundation for these reporting requirements are currently underway in a number of developing countries. This book, which focusses on Bangladesh, may provide an example of a national assessment — what is known, not known and needs to be known — of the effects of climate change that could well serve as a template for other developing countries to follow in carrying out their obligations to report to the CoP.

Bangladesh, one of the least developed nations of the world, may be one of the most vulnerable to climate change. The widespread flood in 1988 which submerged about two-thirds of the country, and the storm surge of April 1991 which resulted in an estimated death toll of 138,868 coastal inhabitants, are recent reminders of the degree to which the people of Bangladesh are subject to present-day variations in climate. The possibility of changes in climate and sea-level rise must be considered seriously in the context of the future development of Bangladesh.

In recognition of this fact, the Ford Foundation and the British Overseas Development Administration supported an investigation of the implications of climate change for the environment and people of Bangladesh, starting in 1992. The generous support provided is thankfully acknowledged. The study was carried out jointly by the Bangladesh Unnayan Parishad (BUP, Dhaka), the Centre for Environmental and Resource Studies (CEARS, University of Waikato, New Zealand) and the Climatic Research Unit (CRU, University of East Anglia, UK). The initial objective of the study was to review and assess the current state of knowledge concerning climate change and its implications for Bangladesh, and to identify how best to move forward with specific research projects at a later stage. The components of this interdisciplinary, collaborative study, as reflected in the chapters of this book, include the following: the greenhouse effect and climate change; sea-level changes in the Bay of Bengal; effects of climate and sea-level changes on the natural resources of Bangladesh; socio-economic implications of climate change for Bangladesh; legal implications of global climate change for Bangladesh; and climate change and sea-level rise - the case of the coast;

The analyses and specific findings reported in this book, tentative though they may be in certain cases, should provide a better understanding of the existing and emerging issues and, hence, be useful as inputs into the policy-making process. This assessment also suggests three main directions for further policy-relevant research in Bangladesh.

First, for some aspects of the problem (for example, climate and agriculture), sufficient data and models are available for conducting sensitivity analyses of the potential effects of climate and sea-level changes on Bangladesh. However, these data and models have not yet been combined in a way that would allow systematic analyses of the impacts of climate change and variability to be easily carried out. In these circumstances, the generation of new knowledge is perhaps less urgent than the integration of existing knowledge.

Second, for other aspects of the problem, basic data are not available and critical relationships between climate and environment are poorly understood (for example, subsidence, sedimentation and relative sea-level rise rates in the coastal zone; climate and vector-borne diseases). In such cases, the lack of basic knowledge precludes detailed analyses of the effects of climate change and variability. This lack of knowledge hinders comprehensive climate impact assessment for Bangladesh and the development and implementation of strategies for reducing adverse effects.

Third, it became apparent that there is little understanding of the full range of strategies by which the people and organisations in Bangladesh could, or would, adapt to climate and sea-level change (including fluctuations and extremes). It is this capacity for human response that largely determines the extent of vulnerability and resilience to environmental change. The lack of knowledge concerning human response represents a major gap in knowledge in attempting to assess the implications of global warming and sea-level rise for Bangladesh.

On behalf of all the authors who collaborated in the assessment, we should like to acknowledge the contributions of the following individuals: Raymond C. Offenheiser and Charles Bailey (The Ford Foundation) and Harry L. Potter (British ODA) for their support and advice; Hugh Brammer, Bo R. Döös, P. Michael Kelly, S.Z. Haider, and M. Maniruzzaman Miah (project Steering Group members), for their incisive comments and guidance; and all the participants of the International Workshop on Climate Change and its Implications for Bangladesh (held in Dhaka on June 10-11, 1993) for their helpful comments and suggestions for final revision of the various chapters. We are also grateful to Neil J. Ericksen, M.Q. Mirza and, especially, Claire Gibson for the additional hours spent editing. Finally, we wish to thank all others, particularly at BUP but also at CEARS, CRU and elsewhere, who have contributed in one way or another to this effort.

Q.K. Ahmad and R.A. Warrick
Editors

GLOSSARY OF BENGALI TERMS

aman — a type of rice crop which is grown in the *kharif* season. There are two types of *aman* (transplanted and broadcasted) and they are grown in the high lands and low lands, respectively.

aus — a rice crop which is planted during the *kharif*-I season in the low lands.

barsha — the normal rainy season.

bazar — a village market which opens at least twice in a week.

beel — natural depressions which are very rich in flood-plain fisheries. In Bangladesh the harvest (per hectare) from *beels* is the highest among all the inland-waterbodies.

bonna — flooding above the normal level.

boro — a rice crop usually planted in November-December and harvested in April-May. High yielding *boro* requires irrigation.

char — a sand bar which is formed within a river or estuary.

ghat — a small river terminal where people are transported from one bank to the other by boats.

gher — a form of fish trap made of bamboo sticks and tree branches to attract fish during the end of the monsoon.

hat — a village market which usually forms at least once a week. Villagers come to the *hat* with their products such as rice, fish, cloths, etc.

izardar — a kind of leasee. People who take lease of *ghats, hats* and *bazars* from the Government in exchange of an amount of money to collect taxes are known as *Izaradars*.

kaon — a type of crop which is grown in less fertile soil without any irrigation.

kharif — a cropping season (April-November)

khas — ownership of lands belonging to the Government. The Government can lease or donate these lands

kuchcha — house or shelter made of bamboos, trees and thatched.

mahals — a tract of government *khas* lands.

rabi — a cropping season (November- May). Usually vegetables, pulses and boro rice are grown during rabi season.

salami — an amount of money given to the government to take lease (*izara*) of *khas* lands, *ghats, hats, bazars*, etc.

salish — an occassional village court comprised of influencial elderly village people where small disputes are settled by mutual negotiation.

thana — the smallest administrative unit in Bangladesh.

taka — Bangladesh unit of currency.

upazila — in the mid 1980s all *thanas* were renamed *upazilas* (smaller than *zila* - i.e. district). They were renamed *thanas* in 1991.

zamindar — landlord. During Colonial period the British Government appointed many *zamindars* to collect land taxes. *Zamindari* was abolished after the partition of India.

THE IMPLICATIONS OF CLIMATE CHANGE FOR BANGLADESH: A SYNTHESIS

Q.K. Ahmad, R.A. Warrick, N.J. Ericksen and M.Q. Mirza*

1.1 INTRODUCTION

This chapter is a synthesis of the findings (reported in the following chapters) of an interdisciplinary, collaborative assessment of what is known, and needs to be known, about the possible effects of climate and sea-level change on Bangladesh. The assessment focussed on six major questions: How might climate change? How might sea level change in the Bay of Bengal? What are the possible effects on natural resources? What are the possible socio-economic effects? What are the legal implications of climate and sea-level change? How might the coast of Bangladesh be affected? In this chapter, the salient features relating to each question are summarised. For greater detail, including some references, the reader is directed to Chapters 2 to 7.

This chapter (as well as Chapters 2 and 3) begins by reviewing the scientific basis for the concern about global warming. While thorough scientific assessments, such as those of the Intergovernmental Panel on Climate Change, are widely available, a brief review is warranted here in order to provide the global context for understanding the possible regional changes in climate and sea level and their potential impacts on the resources and people of Bangladesh. We begin by focussing on global climate change.

1.2 HOW MIGHT CLIMATE CHANGE?

There is firm scientific evidence to show that the concentrations of "greenhouse"

* The authors are respectively: Chairman, Bangladesh Unnayan Parishad (BUP), Dhaka, Bangladesh; Research Director, CEARS, University of Waikato, Hamilton, New Zealand; CEARS Director; and post-graduate student at CEARS.

gases, such as carbon dioxide, methane and nitrous oxide, in the Earth's atmosphere have been steadily increasing. Furthermore, it is apparent that these increases are due largely to human activities like fossil fuel use, land use changes and agriculture. It is the current scientific consensus that such increases in greenhouse gas concentrations should cause global warming.

However, there are large uncertainties in attempting to predict the magnitudes, rates and regional patterns of future climate changes. Hence, the outlook for Bangladesh is unclear. In order to understand the key uncertainties and knowledge gaps in making predictions for the future, it is instructive to look first at climate change in the past.

1.2.1 Global Climate Changes in the Past

The past record shows that global climate varies naturally on all time-scales. During the Pleistocene (the last 2 million years) cold glacial epochs, associated with an expansion of continental ice sheets, were separated by relatively warmer interglacial epochs. The last glacial maximum occurred 18,000 years ago when the global temperature was 4°C to 5°C colder than today. Currently, the world is enjoying a warm, interglacial period. Because of such natural cycles in global climate, the world is headed toward another cold period over the next several thousand years.

On shorter time-scales, such as centuries to millennia, the global climate has also fluctuated between warm and cool periods. For example, during the Little Ice Age, which lasted several centuries and ended around the middle of the nineteenth century, the world was about 0.5°C cooler than today. Many such fluctuations have probably occurred throughout the last 10,000 years (the Holocene).

For the last 100 years, instrumental records of global climate changes are available. From these records it is evident that the world has warmed by about 0.3°C to 0.6°C, although large annual and decadal variations have occurred. It is not yet possible, however, to unequivocally attribute this past warming to greenhouse gases.

The causes of natural climate changes include variations in Sun-Earth orbits, solar output, volcanic activity and ocean circulation. Changes in the composition of the atmosphere, especially with respect to naturally-occurring concentrations of greenhouse gases such as water vapour, carbon dioxide, methane and nitrous oxide, can also cause climate to change. Evidence shows that the atmospheric composition has been changing markedly.

1.2.2 Changes in Atmospheric Composition

Over the last two hundred years, the atmospheric concentration of *carbon dioxide* has risen by about 27 per cent. The current increase is due primarily to fossil fuel (coal, oil, gas) consumption. Current emissions to the atmosphere from this source amount to about 6,000 million tonnes of carbon per year. A further 1,100 million tonnes per year may derive from the net effect of land-use changes, especially changes in forest cover. At present, the less-developed countries, with more than three-quarters of the world population, account for less than one-third of the total annual carbon emission. Bangladesh contributes a trivial 0.06 per cent of the global total.

The atmospheric concentration of *methane* has more than doubled over the last 200 years. Most of the increase has occurred in this century and has closely followed the growth of the global human population. The current growth rate in concentration is about 0.6 to 0.8 per cent per annum, based on observations over the last decade. The increase is believed to be the result of human activities such as: solid waste disposal; agriculture, particularly paddy rice production and animal husbandry; and fossil fuel-related leakages associated with gas pipelines and coal extraction. Bangladesh makes a contribution (if a very minor one) to the yearly emission of methane because of its extensive areas of seasonally or perennially flooded land and wetland paddy cultivation, as well as its animal husbandry.

The concentration of *nitrous oxide* has been increasing slowly. Although powerful greenhouse gases, the changes in concentrations have, so far, been small. Overall, nitrous oxides are a relatively minor contributor to the global warming problem.

Many of the *halocarbons*, especially those containing fluorine, bromine or chlorine like the widely-used CFC-11 and CFC-12, are very strong, long-lived greenhouse gases. They are not naturally-occurring, but rather were manufactured for use as refrigerants, foam blowing, spray-can propellants, and solvents, amongst many uses. Their atmospheric concentrations are very small, but grew rapidly over the last several decades at a rate of about 4 to 5 per cent per year. Because of their implication in the destruction of stratospheric ozone (the "ozone hole"), CFC production has been restricted through the provisions of the 1987 Montreal Protocol. The subsequent London and Copenhagen Amendments to the Protocol call for a complete phase-out of CFC production before the year 2000.

1.2.3 Effects on the Earth's Energy Budget

Greenhouse gases are transparent to incoming solar radiation, but readily absorb the infra-red radiation emitted by the Earth. In this way, naturally-occurring gases like water vapour and carbon dioxide "insulate" the Earth's surface and lower atmosphere, keeping them much warmer than would otherwise be the case — a natural greenhouse effect. Increasing the gas concentrations upsets the Earth's radiation balance, causing a change in the radiative forcing, or heating, of the climate system — an *enhanced* greenhouse effect. Carbon dioxide accounts for most of the estimated change in radiative forcing, and fossil fuel use for energy is the single largest source of anthropogenic emissions of carbon to the atmosphere. Bangladesh is responsible for perhaps 0.2 to 0.3 per cent of the current total radiative forcing change — a very minor contribution.

1.2.4 The Climate Sensitivity

From experiments using global climate models, it is estimated that a doubling of the atmospheric concentration of carbon dioxide (or the equivalent, taking into account other greenhouse gases) should eventually lead to an equilibrium warming of 1.5°C to 4.5°C, with a best estimate of 2.5°C. This equilibrium warming, conventionally called the *climate sensitivity*, can be thought of as the potential warming for a carbon dioxide doubling, without regard to the rate of change. The range of estimates for the value of the climate sensitivity derives from uncertainties in climate feedbacks — changes in water vapour, snow/ice extent, clouds — that act to enhance or lessen the initial radiative heating effects of greenhouse gases.

1.2.5 Projections of Global Warming

The current scientific consensus is that increasing atmospheric concentrations of greenhouse gases should cause the world to warm. The rate of warming, however, is very uncertain. In 1990, the Intergovernmental Panel on Climate Change (IPCC) estimated that with a "business-as-usual" (BAU) (i.e. with no policies to reduce emissions) scenario of greenhouse gas emissions, the world should be 3.3°C warmer by the end of the next century, with a range of uncertainty of 2.2°C to 4.9°C.

Subsequent analyses by IPCC and others using comparable emission scenarios suggest somewhat lower rates of warming, around 2.5°C. The revisions are due to improved predictions of future carbon dioxide concentrations, and, importantly, to recent studies showing that sulphate aerosols (associated largely with sulphur emissions from fossil fuel burning) and CFC-related ozone depletion should have a cooling effect, partially offsetting some of the global warming. Even so, the projected rates of global warming are greater than those which have occurred naturally over the last 10,000 years. Toward the end of the next century, the global temperature could well exceed that which has occurred over the last 150,000 years, at least.

1.2.6 Climate Change in Bangladesh

Over the past 100 years, the broad region encompassing Bangladesh has warmed by about 0.5°C (Figure 1.1). The warming trend is consistent with that of the Northern Hemisphere as a whole. As with the observed global warming, it is not yet possible to say unequivocally that the warming in the Bangladesh region has been due to greenhouse gases. There has been no discernible trend in average rainfall, although rainfall variability appears to have increased in recent decades.

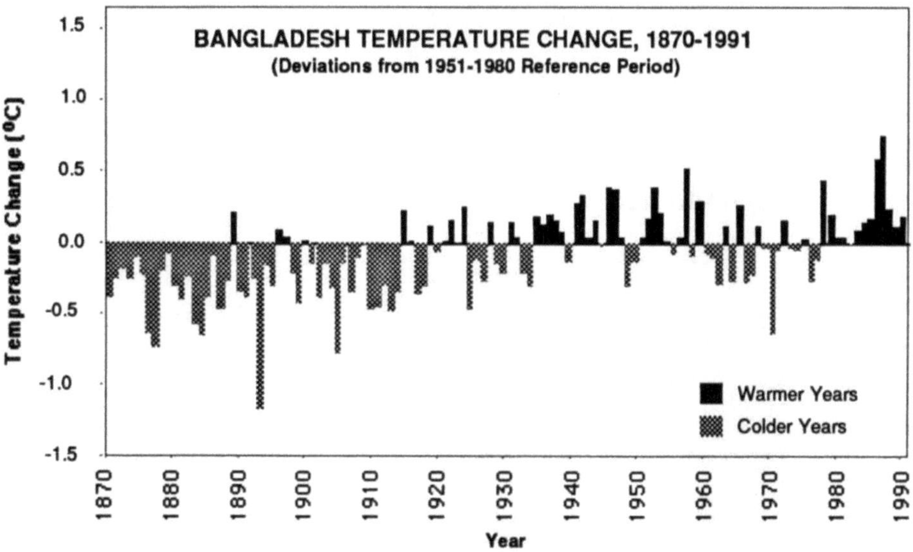

FIGURE 1.1: Bangladesh mean-annual temperature change, 1870-1991. The values are departures from the 1951-80 reference period

Source: Data provided by the Climatic Research Unit, University of East Anglia.

In the future, Bangladesh may get warmer and wetter. For the IPCC (1990) BAU emissions scenario, Bangladesh is projected to be 0.5°C to 2.0°C warmer than today by the year 2030, based on a range of global climate model results. Rainfall is more difficult to predict. However, climate models generally agree that regional monsoon rainfall should increase in a warmer world. The best estimate is a 10 to 15 per cent increase in average monsoon rainfall by the year 2030, although the uncertainties are very large (Figure 1.2). As yet, little can be said specifically regarding future changes in the frequency and intensity of cyclones in the Bay of Bengal.

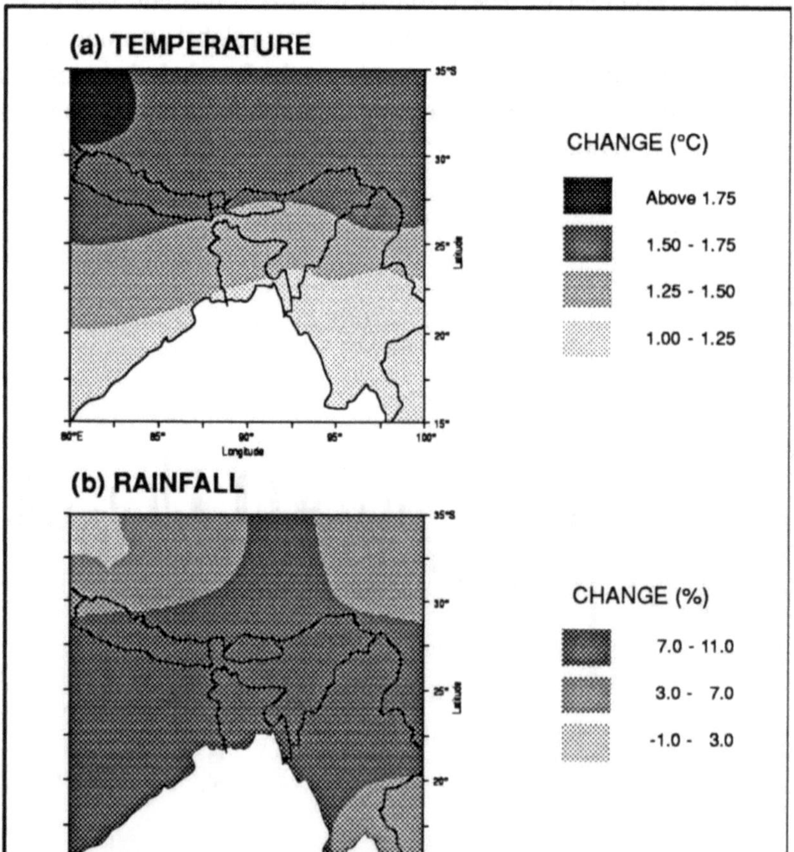

FIGURE 1.2: The projected best-estimate 1990-2030 climate changes for the monsoon (June to October): (a) temperature and (b) rainfall

1.3 HOW MIGHT SEA LEVEL CHANGE IN THE BAY OF BENGAL?

Scientific evidence strongly suggests that, on the decade-to-century time-scale, global warming should cause sea level to rise. The possibility of a future rise in sea level is of grave concern to Bangladesh, a developing country with a vast, low-lying, densely-populated deltaic coast. Sea-level rise threatens, potentially, to exacerbate problems already experienced in the coastal zone of Bangladesh, including erosion and inundation of land, salinisation of soil and water, and, especially, flooding from storm surges. However, such grim outcomes are by no means certain. There are large uncertainties involved in attempting to predict the effects of global warming on relative sea levels in the Bay of Bengal.

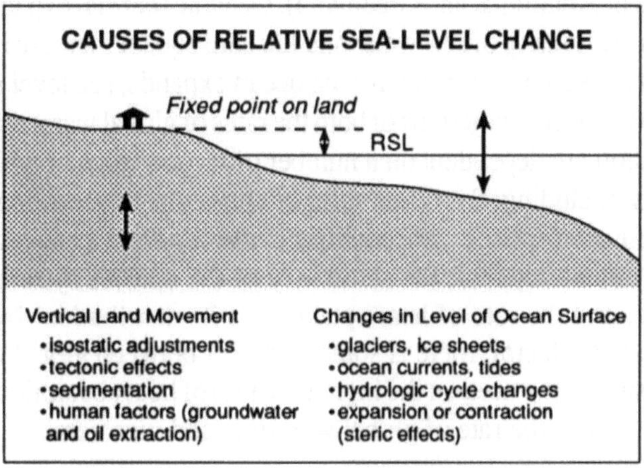

FIGURE 1.3: The causes of relative sea-level changes

Source: Warrick, 1993.

It is important to appreciate that what matters for the inhabitants of the coastal zone of Bangladesh is relative sea-level change — that is, the level of the sea in relation to the land. Relative sea level (as, say, recorded by a tide gauge) is influenced by vertical land movements, which occur locally and regionally from geological factors, as well as by changes in the level of the sea surface itself, which can be affected globally by changes in the volume of the world's ocean (Figure 1.3). Let us consider the global effects first.

1.3.1 Global Sea-Level Change

On geological time-scales, the evidence is clear that large changes in sea level have resulted from climate change, especially as a consequence of glacial-interglacial cycles. For example, at the height of the last ice age some 18,000 years ago, sea level was more than 100 metres lower than today. Over the past century, sea level has risen by 10cm to 20cm, most likely in response to a global warming of about 0.3°C to 0.6°C over the same period. Thus, based on the past, there is an apparent connection between changes in global climate and sea level which should not be ignored in projecting the future.

At the global scale, the enhanced greenhouse effect and global warming can change the ocean volume primarily in two ways: by *oceanic thermal expansion* and by *changes in the volume of land-based ice* (including mountain glaciers and the polar ice sheets of Antarctica and Greenland). Oceanic thermal expansion refers to the decrease in the density of sea-water and the consequent increase in the volume of the ocean as a result of warming. As the ocean expands, sea level rises (a steric rise in sea level). Future predictions of both the rates of global warming and oceanic thermal expansion are dependent on a number of factors, each of which is subject to uncertainties, including: the future rates of emissions of greenhouse gases and changes in their atmospheric concentrations; the resultant changes in radiative forcing of the climate system; the sensitivity of the climate system to radiative forcing change; and the rate of heat uptake (and its distribution) by the oceans. Because of the latter factor, there is some trade-off between global warming and thermal expansion of the oceans: the higher the rate of heat absorption and thermal expansion, the slower the rate of global warming, and vice versa.

Any shrinkage of the polar ice sheets (the grounded ice, not the floating sea ice) and the mountain glaciers and smaller ice caps would increase sea level. Whether a glacier or an ice sheet, the volume of ice is controlled by two opposing components of the *mass balance* — ice accumulation (e.g. through snowfall) and ablation (e.g. through melting/run-off, evaporation, or calving of icebergs). If the rate of accumulation exceeds that of ablation, the mass balance is positive and the glacier or ice sheet will grow, effectively lowering sea level. If, however, the ablation rate exceeds the accumulation rate, the mass balance is negative and ice volume will shrink, raising sea level. The scientific problem is to determine how possible future changes in climate will affect the mass balance components of the world's glaciers and ice sheets, and, hence, sea level.

With regard to the next century, the current scientific consensus is that global warming should increase the rate of sea-level rise. Most of the rise would probably be caused by thermal expansion of the ocean. The evidence suggests that global warming should also increase, on average, the net melting of mountain glaciers, raising sea level. With respect to the major ice sheets, the uncertainties are very large, but it appears that the Greenland ice sheet would also contribute positively to sea-level rise during the next century. In contrast, the evidence suggests that in Antarctica the accumulation rates would increase as the climate warms, thus lowering sea level and partially off-setting some of the projected rise.

While there is general agreement that sea level should rise, the projected rate of rise is very uncertain (Figure 1.4). In 1990, the Intergovernmental Panel on Climate Change (IPCC) estimated that with a BAU scenario of greenhouse gas emissions, global sea level would be 66 cm higher by the end of the next century, with a range of uncertainty of 31cm to 110cm. Subsequent projections suggest somewhat lower rates of rise, due to revisions of emission scenarios and models. Nonetheless, these estimates suggest a future rate of sea-level rise that is 2 to 6 times higher than that experienced over the last 100 years.

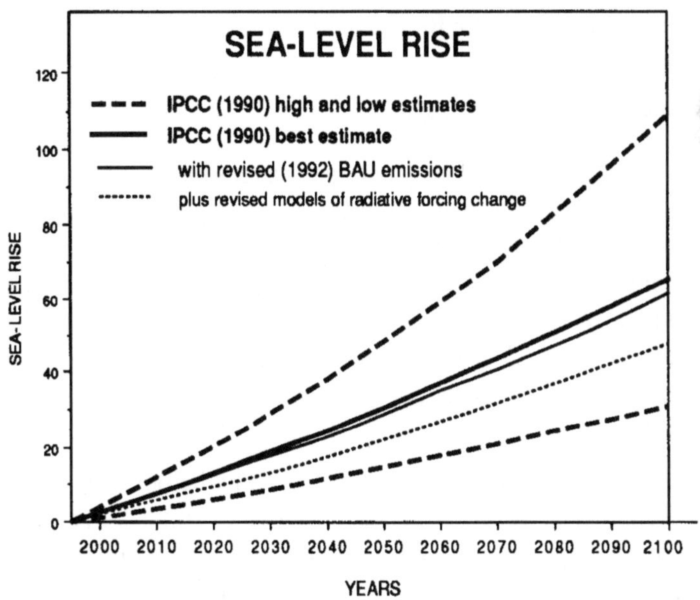

FIGURE 1.4: Projections of future sea-level rise

Source: IPCC projections from Warrick and Oerlemans, 1990; revised curves from Wigley and Raper, 1992.

1.3.2 Relative Sea-Level Change in the Bay of Bengal

For Bangladesh, relative sea-level change depends only partially on changes in the volume of the ocean. Geological factors, including tectonic movements, rates of subsidence/uplift and sedimentation rates, also play an important role because they affect the level of the land in relation to the sea. In general, the coastal area of Bangladesh is subsiding over the long term, due to the pressure on the Earth's crust from the thick layers of sediment accumulated over millions of years. In recent history, this subsidence apparently has, on average, been balanced by land accretion from sediment supply to the coast (which, itself, could be affected by climate change). This dynamic balance has precluded major advance or retreat of the coast of Bangladesh over the last 100 years.

However, locally and regionally, the rates of vertical land movements and their causes are not very well known. There is a network of tide-gauges in Bangladesh that could potentially be useful in addressing this issue, but the quality of the records from these gauges is questionable. Comprehensive field surveys using modern methods to determine subsidence and sedimentation rates have not been carried out. Better knowledge of the geological components is crucial for making future projections of sea-level change for Bangladesh.

1.3.3 Cyclones and Storm Surges

On short time-scales (minutes to days), large changes in sea level occur from storm surges. Such surges are caused by the change in atmospheric pressure and the wind stress (especially in the shallower water near the coast) associated with severe tropical cyclones. Storm surges have devastating consequences for Bangladesh. The last major storm surge in Bangladesh, in April 1991, left more than 138,000 dead in its wake (Figure 1.5). In recent history, the majority of global fatalities from surge events occurred in the countries bordering the Bay of Bengal and the adjoining Andaman Sea.

With respect to the future, the major issue is whether global warming will increase the frequency and/or intensity of severe tropical cyclones and their associated storm surges in the Bay of Bengal. While a minimum sea-surface temperature of about 27°C is required for the formation of tropical storms (and sea-surface temperatures are certain to rise with global warming), other conditions must be met before a storm intensifies into a severe tropical cyclone. At present, there is no firm evidence, either from observational data or model simulations, to conclude that global warming should worsen the problem of severe tropical cyclone and storm surge occurrence in the Bay of Bengal.

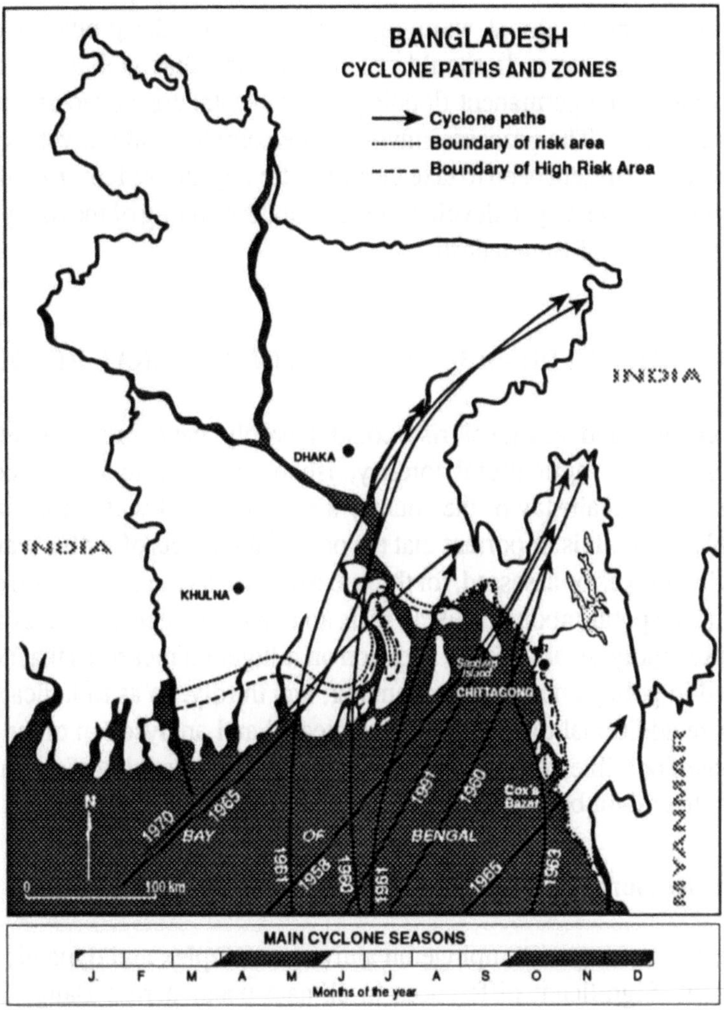

FIGURE 1.5: High risk cyclone hazard areas of Bangladesh and cyclone paths for the period 1958-1991

Source: FAP7, 1992; MCSP, 1993.

In the short term, the improvement of methods of surge prediction through the development of numerical models is warranted. At present, improvement of storm surge models is hampered by the lack of observational data required for model development and validation.

Overall, sea-level rise on various time-scales is a major problem confronting Bangladesh. Whether global warming will exacerbate the problems presently encountered is not clear. It should not be assumed that global warming will necessarily lead to a permanent flooding of the coastal zone or to an increase in storm surge events. The processes involved are complex and not yet well understood. There is an urgent need to take the precautionary steps of improving the data and knowledge necessary to develop a clearer understanding of the likely implications for the future of Bangladesh.

1.4 WHAT ARE THE POSSIBLE EFFECTS ON NATURAL RESOURCES?

Climate change and sea-level rise could possibly have profound effects on Bangladesh's water, agricultural, forestry, fisheries and livestock resources. These resources are the mainstay of the country's economy, and will remain so for the foreseeable future. It is important that the potential impacts of global warming on these vital resources be assessed, for three reasons: firstly, so as to inform decision-makers and the public about the likely scale and time-frame of the kinds of impacts foreseen; secondly, so as to focus attention on mitigation measures that need to be considered in policy and research planning; and thirdly, so as to indicate subject areas where additional data need to be collected and analysed in order to refine predictions of possible impacts, and thereby improve the planning of any mitigatory measures that might be required.

1.4.1 The Resource Setting

Greenhouse warming will impinge on a diverse, complex and dynamic physical environment. Significant differences in climate occur across Bangladesh, and annual variability causes periodic floods and droughts. Floodplains occupy 80 per cent of the country, hills 12 per cent and uplifted blocks (terraces) eight per cent (Figure 1.6). The characteristics of soils and seasonal flooding vary locally (even within villages) as well as regionally, and farmers' cropping patterns are closely adapted to these heterogeneous conditions. Dynamic changes in the environment and in land use are taking place under pressure of population growth and with the spread of irrigation and the provision of flood protection works.

Bangladesh's socio-economic environment is also diverse, complex and dynamic. Over 80 per cent of the about 115 million population are rural, mainly engaged in farming, fishing and trading. Land ownership is highly skewed. Over two-thirds of rural households are functionally landless. Poverty and malnutrition

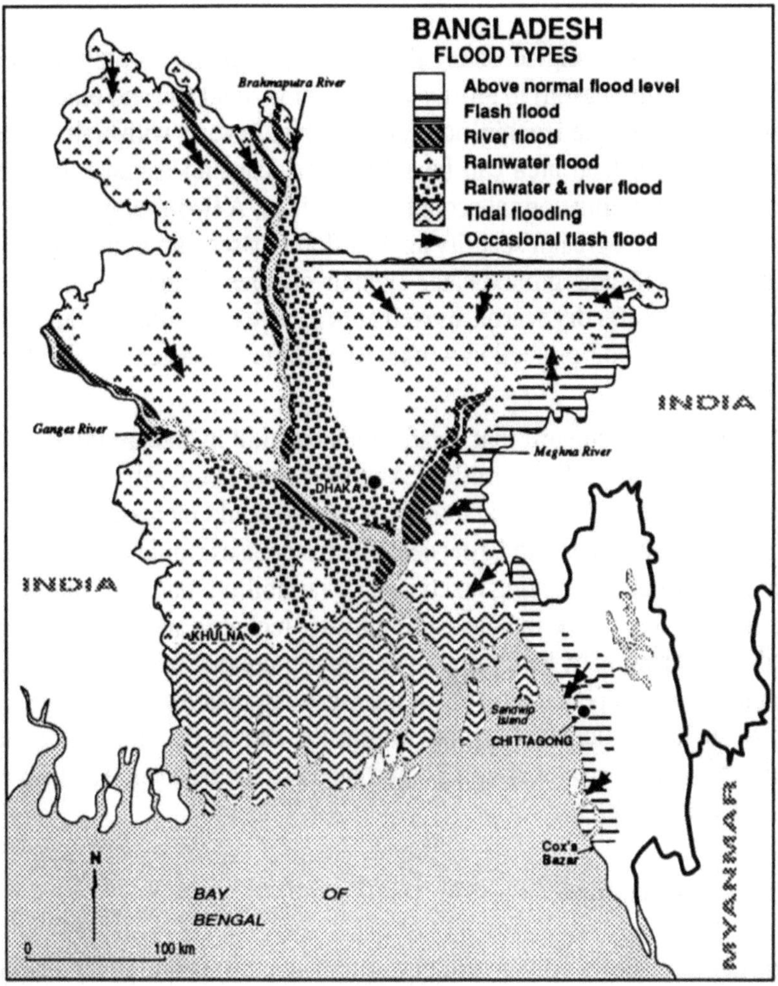

FIGURE 1.6: Flood types

Source: Brammer and Khan, 1991.

are widely prevalent. Rural population densities are high (>1000/km² in some parts) and rural-urban migration rates are accelerating. The country's population is projected to double by 2030, at which time the urban population could be one-third of the total. Thus, future greenhouse warming could affect a society that is much different from today's.

1.4.2 Water Resources

About 25 per cent of Bangladesh is submerged by river water or rainwater during the monsoon (rainy season) in most years. Housing, land use and other economic activities are well adapted to this 'normal' flooding. In years with high floods, over 60 per cent of the land area can be submerged, causing damage to crops and property, and disrupting normal economic activities. A two-pronged Government strategy is providing greater security against flood losses: promotion of irrigation, both from surface and groundwater (which enables farmers to increase rice production in the safer dry season, as well as to reduce exposure to drought losses); and provision of flood protection works, to exclude unwanted floodwater in the rainy season.

Climate models suggest that mean annual rainfall in and around Bangladesh could increase with global warming. As mentioned above, the "best-estimate" scenario for the year 2030 is that monsoon rainfall could increase by 10 to 15 per cent and winter rainfall by 5 to 10 per cent. Such increases would have both positive and negative effects. Assuming no change in variability, increased rainfall would reduce drought frequency, increase groundwater recharge and decrease irrigation demand. Furthermore, by increasing river flows, increased rainfall could also help to counteract salinity penetration in coastal areas associated with a rising sea level. On the other hand, increased rainfall within the catchment area of the Ganges-Brahmaputra-Meghna river systems could increase the frequency and severity of floods in Bangladesh.

1.4.3 Crop Agriculture

Farming in Bangladesh is diverse, intensive and dynamic (Figure 1.7). Farmers are flexible, adaptive and sensitive to both environmental variations and market prices. Over 100 different crops are grown, but rice occupies about 80 per cent of the cropped area (about 9 million ha). Two or three crops a year are grown in many areas, but average yields are low because of low levels of inputs such as fertiliser, in part caused by insecurity of production. The recent spread of irrigation has greatly increased rice production in the more secure dry season, but at the expense of pulses, oilseeds and grazing land. Flood protection works aim to increase security of crop production in the monsoon season. The projected doubling of food demand by 2030 will require further intensification and security of production, but will have to be obtained from a shrinking cropped area as urban and other land uses expand.

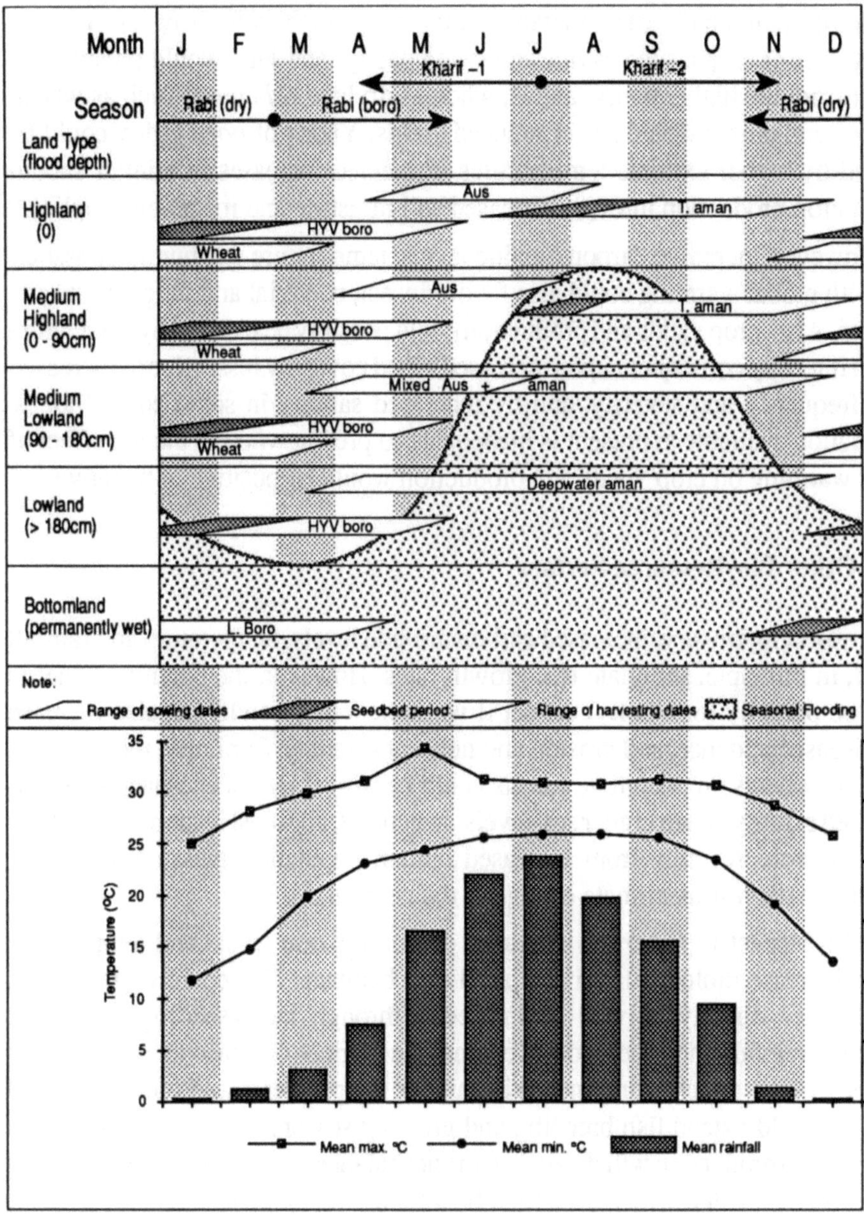

FIGURE 1.7: Rice and wheat crop calendar in relation to seasonal flooding, rainfall and temperature

wheat, jute and pulses would benefit more than maize and sugarcane.

The optimum range of temperatures for plant growth varies between crops. In general, *kharif* crops, such as *aus* and *aman* paddy, and jute, would presumably benefit from the higher temperatures, whereas dryland *rabi* crops such as wheat, potatoes and pulses would suffer reduced yields. Yields of *boro* paddy could be reduced by higher extreme temperatures. Projected increases in rainfall should benefit crop production through increased soil moisture and irrigation supplies.

However, increased carbon dioxide levels, temperature and humidity associated with global warming could also favour insect, bacterial and fungal pests and diseases. Also, crop yields and production could possibly be reduced by a reduction in light intensity consequent upon increased cloud cover, by losses due to increased flood frequency and severity, and by increased salinity in some coastal areas. Information available at present is inadequate to predict whether the net effect of global warming on crop yields and production would be positive or negative.

1.4.4 Other Bio-Resource Sectors

With respect to *forestry*, increased carbon dioxide levels, temperature and rainfall should, in principle, stimulate tree growth rates. However, the potential benefits could be partially (or wholly) off-set if warmer, wetter conditions also increase pest/disease incidence, soil erosion and nutrient leaching. Similarly, the potential beneficial effects on mangrove forests could be off-set by sea encroachment and increased salinity due to higher sea-levels. In general, all forest resources are likely to suffer more severely from increased human population pressure and over-exploitation than from climate change in the near future.

With respect to *fisheries*, increased carbon dioxide levels and temperature should increase biological activity in all fish habitats. This could increase fish growth rates directly as well as indirectly through increased food supplies. Projected rainfall increases could extend the period of seasonal flooding of floodplain land which, combined with higher temperatures and increased food supplies, could extend fish breeding and growing seasons. Impacts on brackish-water shrimp production will depend on the net balance, in different areas, between salinity reduction due to increased river flows and salinity increase due to sea-level rise. Quantitative assessment of potential changes in production due to global warming is highly uncertain because of the potentially greater impacts on fish habitats and stocks resulting from continuous increases in human population pressure and interventions in floodplain, estuarine and tidal fish habitats.

Higher summer temperatures could increase heat stress in *livestock*, but higher winter temperatures could reduce cold stress. Projected increases in rainfall and consequent flooding could increase the incidence of humidity-related diseases. In principle, higher carbon dioxide levels and temperature could increase plant food availability, but this benefit would be off-set by reduced grain and fodder availability due to more frequent flood losses, and loss of land in coastal areas from sea-level rise and salinisation. Such negative impacts would aggravate existing problems for livestock production caused by cropping and environmental changes induced by the increasing human population pressure on the country's land and water resources.

1.5 WHAT ARE THE POSSIBLE SOCIO-ECONOMIC EFFECTS?

Any changes in the natural resources of Bangladesh are bound to have social and economic effects. Some would be beneficial, others detrimental. For example, increased rainfall might increase the amount of water available for irrigation (a beneficial effect on agriculture), but increase the rate of soil erosion and leaching (a detrimental effect on agriculture). These impacts on agricultural resources (plant and soil) would affect the social and economic circumstances of farmers (and, thereby, other socio-economic sectors dependent upon their production) and, in turn, the ways in which the environment is modified by people — an interaction that ultimately affects human vulnerability to climate change.

1.5.1 Environment-Society Interaction

The socio-economic effects of climate change therefore arise from interactions between climate and society and how these affect both natural and managed environments. Traditionally, in Bangladesh, climatic variations have provided opportunities (resources) and imposed costs (hazards), depending on how society adapted to the environment. Thus, a bountiful floodplain rice-growing system, finely tuned to seasonal climate variations, is often disrupted by floods, droughts and cyclones. In the future, the extent to which Bangladesh will be affected (whether adversely or beneficially) will depend on the future technological, demographic and socio-economic trends and how they influence Bangladesh's ability to adapt in order to strike a new balance between resources and hazards.

Bangladesh is a developing country in transition from being a traditional rice-growing society. In the drive for modernisation, evolving technologies and economic and social structures alter existing systems and often make many sectors of, and groups in, society more vulnerable to extreme variations in climate and sea level.

For example, large-scale environmental interventions, such as flood control and irrigation, may buffer people from the ill-effects of recurrent floods and droughts and thereby enhance the resource base. On the other hand, changing land-use practices and intensification in the protected areas may increase vulnerability to the extreme, supra-design events (which may be made more likely under a changing climate), even though other benefits may accrue, such as improved crop production.

In the long run, Bangladesh's vulnerability may, however, depend more on the direction of technological, demographic, economic and social trends than on the rates of climate and sea-level change. This is because the pace of change in society is likely to be much more rapid than that of climate and sea-level change.

1.5.2 Population, Settlement and Climate Change

Since 1965, the population of Bangladesh rose steeply, doubling to about 110 million by 1991. The medium UN projection gives 235 million by the year 2025 and 305 million by 2050 (Figure 1.8). In rural areas the population density may increase by over half by the year 2025. The high-density rural areas will continue to supply migrants to low-density areas and to cities. The exposure of people to climatic extremes will persist and is likely to increase as more intense use is made of high-risk areas. The urban population is projected to grow at a faster rate, about five per cent per year. This increasing concentration of people in large urban areas could increase the risk of catastrophe from rare climatic events and is likely to create additional risks of climate impacts more akin to other urbanised countries, such as heat stress, urban flooding and urban drought. Overall, the trend of high population growth in Bangladesh should increase vulnerability to climate and sea-level change.

1.5.3 Migration, Employment and Climate Change

Escaping adversity due to lost land and employment lies behind most migration. Permanent movements are from the densely-settled core to less densely-settled periphery, and from rural to urban areas. Seasonal moves are an increasing trend. Exposure to natural disasters depreciates marginal landholdings and triggers many people to relocate. Limited opportunities mean many migrants relocate not only in areas at risk from climate extremes (drought-prone western districts, cyclone-prone coasts, and active floodplains), but also from adverse social and environmental

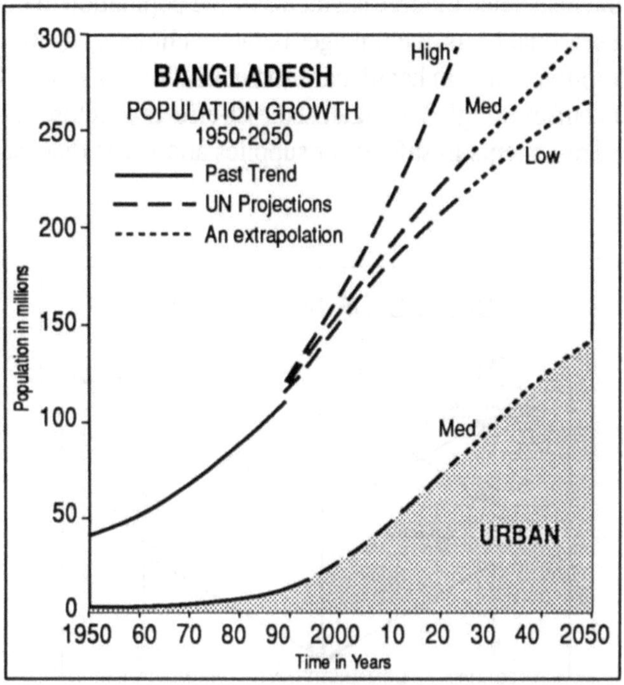

FIGURE 1.8: Growth in total population from 1950 to 1990; UN projections to 2025 and extrapolations to 2050 for low, medium and high variants of fertility). The proportion of urban population for the medium variant is included.

Source: From data in United Nations, 1989, 290-291.

conditions. In general, migrants are particularly susceptible to environmental disruptions, because they lack supportive infrastructure and employment. A continuation of high migration rates is likely to aggravate the potential socio-economic impacts of climate and sea-level changes in the future.

1.5.4 Health Care and Education

A population that is healthy and educated is better able to avoid poverty and the adverse effects of climate variations. While recent trends in improved health care and education in Bangladesh are encouraging, poverty and malnutrition remain rife, lowering the resistance of large segments of the population to disease (Figure 1.9). Even on a seasonal basis, the linkages between climate, nutrition and disease are apparent. Improvements in health care and education, as well as food production, would help buffer Bangladesh against the ill-effects of future climate change. So, too, will improvements in safe water supplies and waste disposal systems.

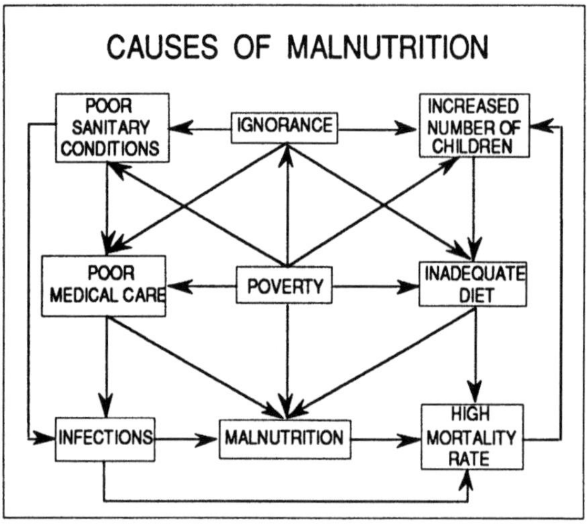

FIGURE 1.9: Socio-economic and environmental factors causing malnutrition in Bangladesh

Source: BRAC, 1979.

1.5.5 Water-Borne and Vector-Borne Diseases

Temperature, precipitation and humidity influence the incidence of water-borne (and air-borne) diseases. Bacteria, parasites and their vectors may breed faster and live longer in warmer, wetter conditions in Bangladesh. However, adverse climate is a necessary, but not sufficient, condition for these diseases. Sanitation problems, tied to poverty are the main conditions for diarrhoeal diseases (like cholera). Drought and flood facilitate their transmission. Rainfall and poorly maintained human settlements facilitate breeding of mosquitoes, and migrants from infected forest areas and/or migrants returning to the plains is a main reason for its resurgence on the lowlands (Figure 1.10). Climate change in the future could encourage such diseases, especially if economic development is impeded.

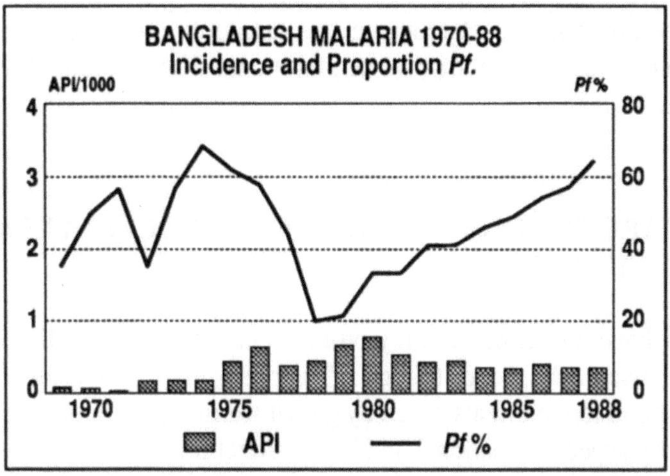

FIGURE 1.10: The incidence and proportion Pf of malaria in Bangladesh between 1979 and 1988. API is annual parasite incidence (the malaria positives per 1000 population under surveillance); Proportion Pf is the Pf (Plasmodium falciparum) infections per 100 malaria positive slides.

Source: International Assessment of the Malaria Programme, Bangladesh, 1989.

1.5.6 Future Scenarios

If the recent past becomes *business-as-usual* (BAU) in the future, then low economic growth, mass poverty and burgeoning unemployment are likely to persist. If accompanied by a pattern of extreme natural events and hazard adjustments similar to those of the recent past, loss of lives and property will escalate. Should adverse climate and sea-level changes occur under this BAU scenario, then the number of people at risk will increase, especially among the marginalised poor, and catastrophic losses will become more frequent.

Rather than focus on BAU investment and GNP growth targets, an *optimistic* scenario would emphasise productive employment targets aimed at releasing the creative energies of the country's poor people at the grassroots level. In this future of Bangladesh, a new market economy may evolve in which the poor are "mainstreamed" through an employment-based strategy anchored on basic education, skill training and organisational support at the local level. These social adjustments would be accompanied by an improved mix of structural and non-structural measures aimed at reducing the susceptibility of society to natural hazards — measures that would prove helpful should climate and sea levels change in the future.

1.6 WHAT ARE THE LEGAL IMPLICATIONS OF CLIMATE AND SEA-LEVEL CHANGE?

Legal issues have a bearing not only on the matter of responsibility for climate changes, but also on response strategies. The opportunities for effective national action to respond to the threats posed by global warming are both channelled and constrained by aspects of national and international law.

The negotiations that took place in preparation for the 1992 United Nations Conference on Environment and Development (UNCED) led to an acceptance of the general principle of "common but differentiated responsibilities" for climate change. This principle accepts that the developed countries have made a disproportionate contribution to global environmental degradation, many of the worst effects of which will be felt by developing states, such as Bangladesh. Because issues of proof are likely to remain highly problematic, the Framework Convention on Climate Change (FCCC), signed by nations at UNCED, is based on a *precautionary approach* — that action should be taken even before scientific proof becomes irrefutable. For countries like Bangladesh with much at stake, this approach appears eminently sensible. Any such strategies and policies to respond to the environmental impacts likely to be precipitated by climate change must be developed within the framework of national and international law.

1.6.1 National Law

The system of *property rights and land tenure* in Bangladesh will affect impacts of, and responses to, climate change. A number of impacts on the physical form of the country are envisaged through changing patterns of inundation and flooding, through possible increased incidence and severity of cyclones, and through sea-level rise. These changes will affect land tenure, probably depriving increasing numbers of people of their habitation or means of support. Such problems are not new for the Bangladesh legal system, a system which has developed its own practices of land holding to reflect the shifting fluvial soils of the delta region. Land will need to be available for flood and coastal defence works as well as for resettlement projects. The existing systems of allocation of government land to displaced persons provide some precedents but may not be adequate to deal with the sustained volume of population movements on the scales envisaged if the worst scenarios of climate and sea-level change eventuate.

Bangladesh is party to the main conventions dealing with atmospheric changes. With respect to *national environmental controls*, Bangladesh does have legislations to address some environmental issues. However, like most other countries, it has not yet developed national legislation to address specifically the factors which may limit its own contributions to the greenhouse effect.

Ideally, an integrated approach to resource management is most appropriate for dealing with the multifaceted problems of climate change. In Bangladesh there is a large *institutional administration framework* with a multiplicity of responsibilities that impinges upon the environmental and resource sectors potentially affected by climate change. Greater coordination and integration of institutional responsibilities may be required to minimise the cross-sectional impacts of climate change on Bangladesh in the future.

1.6.2 International Law

For Bangladesh, there are important issues concerning international *maritime zones*. Cyclone impacts and sea-level rise inevitably have impacts on the coastline — the line from which all the national maritime zones (territorial sea, contiguous zone, Exclusive Economic Zone (EEZ) and continental shelf) are defined. Despite the unusual method used for the 1974 Declaration of the Bangladesh coastal baseline, changes in the coastline might well cause jurisdictional problems and problems in delimiting borders with neighbours similar to those which currently exist over South Talpatty.

Participation in international *marine science research* projects, particularly those related to ocean changes, will be beneficial to Bangladesh. It seems advantageous to ensure that national procedures are in place to approve appropriate projects in Bangladesh waters.

Many environmental problems will be felt across national borders, and there are some clear principles for *international co-operation*. Not only do treaties and customary law impose obligations to inform neighbours and to co-operate in facing common hazards, but there is also an obligation not to take actions which will exacerbate the impacts of climate changes on other countries. Similarly there is an obligation not to use shared resources to the detriment of neighbours — as in the case of shared riverine resources.

Forests, including mangrove forests, and other fragile coastal and marine ecosystems, such as coral reefs, are important sources and sinks for greenhouse gases and provide natural protection and resources for coastal areas. With respect to the *protection of threatened ecosystems*, international treaty systems impose obligations to protect such areas and also provide channels for resources and management expertise.

The Vienna Convention on the Ozone Layer and its Montreal Protocol establish a system that may well be used as a model by the drafters of the Climate Convention protocols. It establishes a system of targets for the reduction and elimination of ozone damaging chemicals, the dates of which have been brought forward a number of times as the understanding of the threat has increased. It also recognises the special position of developing states by providing central funding to assist them to meet its obligations, and has an innovative monitoring and enforcement regime.

The Framework Convention on Climate Change establishes the framework for the negotiation of future protocols which will provide the detailed obligations for parties. The FCCC establishes a number of basic principles, including the differentiated responsibilities of countries and the particular vulnerability of certain groups of developing countries, which includes Bangladesh. It puts the onus on the developed countries to provide the financial and technical resources to assist developing countries to meet the general objective of stabilisation of greenhouse gas concentrations at levels which prevent dangerous anthropogenic interference with the climate system and which allow ecosystems to adapt naturally to climate change without threatening food production or economic development.

1.7 HOW MIGHT THE COASTAL ZONE OF BANGLADESH BE AFFECTED?

The sensitivity, vulnerability and resiliency of the coastal zone to future climate change and sea-level rise is of key concern to Bangladesh. This section first reviews the present status of the natural, social and economic systems of the coastal zone and how they interact, and then considers the potential impacts of future climate change and sea-level rise.

1.7.1 The Coastal Zone of Bangladesh

Depending on its character, the coastal zone can be divided into three main sections:

west, central and east coasts. The west and central coasts are part of the Ganges-Brahmaputra-Meghna delta systems, while the east coast is non-deltaic. The natural shape of the Bangladesh coastline is controlled by the forces of erosion and accretion involving tides, rivers and waves. Where wave forces are strong, straight coastlines of beaches and dunes are formed, as in the south-east. Where tidal forces dominate, complex patterns of tidal creeks result, as on the west coast. If the forces of rivers and the sediments that they carry dominate over tides or waves, the *chars* and estuaries of the central coast result.

The coastal mangrove forests are a multi-use natural resource of significance to the entire economy of Bangladesh. The Sundarban of the west coast is the largest natural single area of mangroves in the world, covering 587,000 hectares (Figure 1.11). The formerly extensive Chakaria Sundarban of the east coast now covers less than 1000 hectares due to felling for salt pans and shrimp ponds. Another 104 to 109

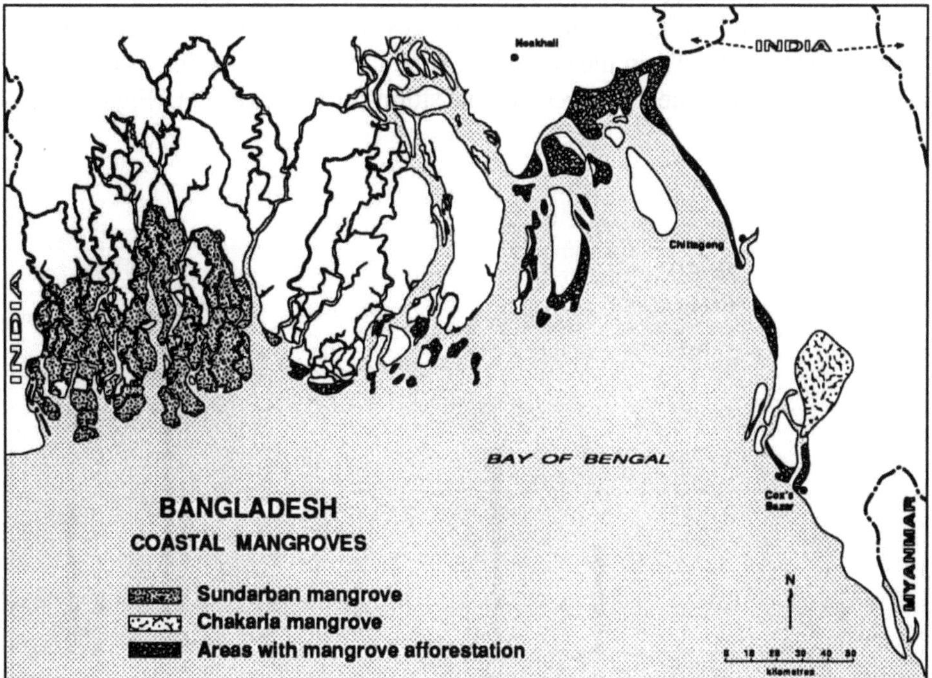

FIGURE 1.11: Coastal mangroves of Bangladesh

Source: ESCAP, 1987.

thousand hectares of coastal land were planted with mangroves under the Coastal Afforestation Project. The mangrove forests supply fuelwood, timber and a range of other products. The Sundarban also contains many endangered species of global importance, including the Royal Bengal Tiger.

Approximately 24 million people, or well over one-fifth of the total population of Bangladesh, live in the coastal zone. The coastal population is closely tied to the resources that land and water provide, either directly through fishing, agriculture and forestry, or indirectly through the processing of coastal fish, shrimp and timber. The two main urban centres in the coastal zone are Khulna and Chittagong, where most of the coastal industry is located. The people work within the seasonal moods of the coast: monsoon floods and rains, dry-season salinity and the seasonal threat of cyclones. Rapid population growth has pushed many people to the margin of survival in the coastal zone, where there may be as many as 45 to 50 million people by the year 2050. Such population growth could put further strain on coastal systems and resources; climate change and sea-level rise could then be a threat to a much higher number of people.

Cyclones and their associated storm surges are the major natural threat in the coastal zone, especially in the east and central coast. Over 600,000 people have been killed in Bangladesh by cyclones and storm surges since 1960 (Figure 1.12).

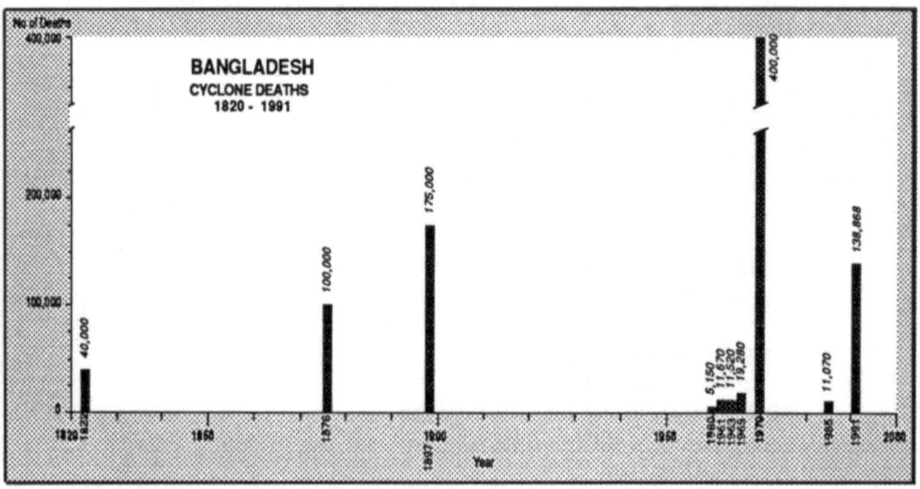

FIGURE 1.12: Number of cyclone deaths in the Bay of Bengal, 1820-1991

Source: Talukder, *et al.*, 1992.

The massive loss of life from cyclones is due to poorly constructed houses, lack of an adequate cyclone warning system and the extremely low-lying nature of much of the coastal zone. The worst cyclone since 1970 crossed the coast on 29 April 1991, claiming an estimated 138,868 lives.

1.7.2 Future Impacts on Coastal Resources

Being a low-lying country, the prospect of sea-level rise is cause for alarm. However, sea-level rise will not necessarily cause a permanent inundation of the present coastal zone of Bangladesh. The coastal zone exists in a dynamic balance between forces of erosion and accretion. Land subsidence and sea-level rise favour the loss of land to the sea, whereas supplies of freshwater and sediment to the coast promote the accretion of land. Whether climate change and sea-level rise will drown the coast depends critically on how the dynamic balance is maintained in the future. How the coast will respond to the cumulative effects of changes in sea level, waves, storms, river flow and sediment supply in the future is not clear.

It can be stated with greater certainty that the stresses currently being induced by population growth, over-exploitation of resources, and recurrent natural disasters are likely to be exacerbated by climate change and sea-level rise. One natural system at risk from such stresses is the mangrove forest.

The mangrove forests of the Sundarban, in their natural condition, are extremely resilient to fluctuations in mean sea level. However, over-exploitation of forest resources is decreasing their natural resilience, leaving the forests increasingly susceptible to possible climate and sea-level changes. Superimposed on these climate-induced changes will be modifications to the forest from human use. The pressure on forest resources will increase as the coastal population expands and if unsustainable forest use continues. The future of the forest under such multiple threats is uncertain. Sustainable management is vital to protect the forest resource now and to buffer the affects of climate change and sea-level rise in the future.

In the south-west, salinity levels are likely to change in the polders due to sea-level rise and increased rainfall. However, because sea-level rise will act to increase salinity, whereas increased rainfall will act to decrease salinity, the net effect on polders and other agricultural areas is uncertain.

1.7.3 Socio-economic Implications

In addition to the effects of climate and sea-level changes on agriculture, forestry and fisheries, there is a range of socio-economic effects in the coastal zone that warrants attention. These effects include population migration, changes in settlement patterns and health effects, particularly from possible changes in the supplies of potable water and the incidence of water-borne disease and malaria. It is not yet possible to predict such effects.

The impacts from climate change cannot be separated from the capacity of people to adapt. Can the coastal people of Bangladesh adapt to future climate and sea-level change? In the past, the coastal people have demonstrated a remarkable ability to adapt to changes in their environment. No doubt they will continue to do so in the future, although adaptation is likely to be more and more difficult. Environmental changes, including those arising from changes in climate and sea level, are expected to occur at the same time as the demands for food, fibre, shelter and energy are increasing due to the rapidly growing population. The critical issue, yet unresolved, is the rates, or thresholds, of change that will exceed the ability of the coastal people of Bangladesh to adapt.

Bangladesh, like other coastal countries of the world, must begin planning for integrated coastal zone management (ICZM) as stipulated in the Framework Convention on Climate Change. Policies and plans for ICZM must be based on a comprehensive understanding of the biophysical systems and their modification and use by human inhabitants. Such knowledge is vital to the development of policies to manage coastal resources sustainably and to avoid the adverse effects of future climate change and sea-level rise.

1.8 POLICY IMPLICATIONS, KNOWLEDGE GAPS AND RESEARCH NEEDS

Climate change is a global problem. It requires global solutions involving cooperative actions by all nations. The main obstruction to the implementation of measures to slow global warming is deeply rooted in the geo-politics of equity between the industrialised and developing world — between the poor and rich. In general, it appears that the developed nations should adopt measures to reduce greenhouse-gas emissions immediately, while extending economic and technological support to developing countries so that their future contribution to global warming is curtailed without impeding development. Equity will surely be a key component in any agreements on an emission reduction strategy involving international cooperation and participation of all nations. In the meantime, Bangladesh should consider additional strategies for adaptation to climate variability and change — for both the present and the future.

Traditionally, over many centuries, Bangladesh has learned to cope well with climate variations. Such traditional knowledge must be preserved for the future as Bangladesh undergoes development. Additional measures to reduce vulnerability to existing climatic extremes —floods, droughts, tropical cyclones — would provide immediate benefits while enhancing Bangladesh's future resiliency to such events should their frequency or severity increase as a result of global warming.

Climate change policy, planning and response in Bangladesh will be guided to some extent by the recent Framework Convention on Climate Change (FCCC). The FCCC was signed by 155 nations, including Bangladesh, at the United Nations Conference on Environment and Development (UNCED) in June, 1992. The ultimate objective of the Convention is the "...stabilization of greenhouse gas concentrations in the atmosphere...". For Bangladesh, the FCCC came into force on the ninetieth day subsequent to its ratification on 15 April 1994. Under Article 4 of the FCCC, Bangladesh has several obligations, including the development of a national inventory of anthropogenic sources and sinks of greenhouse gases, and the formulation of programmes containing measures to mitigate climate change and to facilitate adaptation. Currently, activities are underway, with support from the Global Environmental Facility (GEF), the U.S. Countries Studies programme and other agencies, to help enable Bangladesh to meet these obligations.

There are two major research directions that are needed to underpin the eventual formulation of policies for adaptive response and sustainable resource use in the face of climate and sea-level change. These were identified and elaborated by the participants of the International Workshop on Climate Change and its Implications for Bangladesh (held in Dhaka on June 10-11, 1993). They are:

- **to develop *integrated impact models* for Bangladesh that link existing climate, environmental and socio-economic models and data.** For Bangladesh, there is a need to develop methods for integrating existing models and data at different scales — global, regional, national, and local — in a way that allows analyses which are relevant to policy development and strategy implementation.

- **to determine the present vulnerability to environmental and socio-economic change, and to evaluate *strategies for preventing and mitigating the impacts* of possible climate change and sea-level rise in the future.** An important key to anticipating future impacts and guiding adaptive response is an understanding of the present degree of vulnerability to environmental fluctuations and the ways in which local inhabitants are able to cope with the changes. This research requires a local case-study focus.

In addition, there are more specific research needs identified by each of Chapters 2 to 7 of this series, as follows.

1.8.1 Climate and Sea-Level Change

In terms of research on climate change, the most immediate need is to carry out studies that would facilitate Bangladesh's task of complying with the requirements of Article 4 of the Framework Convention on Climate Change. These should include studies:

- to determine the magnitudes of the natural and anthropogenic sources, sinks and fluxes of greenhouse gases in Bangladesh and;

- to integrate climatic and environmental models and data for the purposes of examining future impacts and strategies for mitigation and adaptation in Bangladesh (as described above).

The issue of global sea-level change and its effects is one that involves scientific expertise from many nations of the world. Reducing the uncertainties in the projections of global sea-level rise is a continuing research endeavour, and Bangladesh could best contribute on a continuing basis as part of the international effort. There are, however, opportunities for research and monitoring that apply specifically to sea-level change in the Bay of Bengal and its implications for Bangladesh. These include:

- determination of the rates of subsidence within the coastal zone of Bangladesh, as well as the rates of sediment supply to the coast and its distribution;

- evaluation of existing tide-gauge data and correction of problems in instrumentation, recording and coverage;

- assessment of the feasibility of implementing the latest technologies for precision measurement of sea-level variations, with the aim of identifying the various contributions from the sea and land components of relative sea-level changes and;

- systematic collection of data on individual storms and surges for purposes of storm surge model development and validation.

1.8.2 Environment and Resources

The insidious impacts of projected changes in climate and sea level will impinge on environments already changing under pressure of population growth. Planning, research and environmental monitoring studies need to consider these factors together. In particular, research is required in the following areas:

- Hydrological modelling to assess the possible increased river discharges and internal flooding which could result from increased precipitation in and around Bangladesh. Such studies should indicate the implications for flood frequency and extent. The results should prove useful, for instance, in the design of flood mitigation measures.
- Agricultural and biological research to measure responses of plants, animals and fishes to higher temperatures and carbon dioxide levels, and to select biotypes tolerant of expected conditions. Existing models of Agroecological Zones (AEZ) could be used to estimate possible changes in moisture and thermal zones with climate change and consequent changes in crop suitability and potential yields.
- Forestry studies should focus on correlating individual mangrove species with salinity levels so as to improve predictions of probable changes with sea-level rise.

1.8.3 Socio-Economic Effects

What lack of knowledge impedes the ability of Bangladesh to better adapt to environmental change and vulnerability? First, in some areas there is a lack of fundamental knowledge concerning the relationship between climate variations and socio-economic effects. Second, there is uncertainty regarding the range of adaptive measures that are available for coping with environmental adversity, as noted above. Research is need to fill these gaps in knowledge. Specifically, studies need to be carried out with the following aims:

- to determine the extent to which traditional technologies are being adapted to changing socio-economic conditions;

- to examine how customary behaviour is being modified in response to changing social and environmental conditions;

- to increase understanding of the various forms of migration and resettlement of the landless, in order to anticipate the likely dimensions of problems that may arise if climate extremes worsen and sea-level rises and;

- to develop means of empowering the landless and poor with access to resources to ensure their resiliency in times of scarcity.

Fulfilling these needs requires a programme of interdisciplinary research (integrating social sciences and natural sciences) aimed at developing an optimum strategy for reducing vulnerability to climatic extremes. The outcomes of this research would provide decision makers with an indication of priorities for preparing against the adverse effects of climatic variation and change.

1.8.4 Legal Implications

The problems that climate change is likely to pose for Bangladesh are not necessarily new problems. The Bangladesh legal system has a wealth experience in dealing, for example, with inundation, flooding and population displacement. Rather, it is the scale of the problems that is likely to be different. In order to ensure effective implementation of policies and strategies related to climate change, more needs to be known about the ways in which legal provisions and institutions actually work in practice. At the national level, such research could include legal studies of:

- the actual operation of legal provisions and institutions on property rights;

- the establishment of wider systems of public land settlement;

- possible approaches to effective controls of pollution as well as of greenhouse gases and;

- effective, possibly community-based, protection of crucial life-supporting ecosystems.

At the interface between national and international law, work is needed on:

- the implementation of international rules at national level and;
- the way in which participation in treaty regimes may assist Bangladesh to achieve internally desirable objectives as well as to ensure coordination of its policies with international norms.

1.8.5 The Coastal Zone

There are several general areas in which research could improve the base of knowledge required to enhance adaptation and to guide policy in the coastal zone of Bangladesh, including studies of:

- the sensitivity of Sundarban mangrove species to changing salinity and other environmental conditions;
- the vulnerability of the Sundarban to increased future population pressure and resource use, within the context of potential changes in mangrove species and location due to climate change and sea-level rise;
- the sensitivity of coastal areas to potential changes in sediment supply and sea-level rise;
- relationships between climatic change, run-off and streamflow, and freshwater and sediment supply to the coast and;
- the vulnerability of coastal people and the possible adaptive strategies in the face of climate change and sea-level rise.

In the short term, the coordination and integration of existing coastal zone research programmes is needed in order to efficiently guide and carry out research.

REFERENCES FOR FIGURES

Bangladesh Rural Advancement Committee (BRAC) (1979) *Ten Diseases: A Treatment Guide for Medical Paraprofessionals.* BRAC, Dhaka.

Brammer, H. and Khan, H.R. (1991) Bangladesh country study. In Asia Development Bank, *Disaster Mitigation in Asia and the Pacific.* Asian Development Bank, Manila, pp. 72-149.

Economic and Social Commission for Asia and the Pacific (ESCAP)(1987) *Coastal Environmental Management Plan for Bangladesh.* Volume II- Main Report, ESCAP, Bangkok.

Flood Action Plan (FAP) 7 (1992) *Cyclone Protection Project II: Feasibility and Design Studies* (Main Report). Government of Bangladesh, Ministry of Irrigation, Water Development and Flood Control, Dhaka.

International Assessment of Malaria Programme, Bangladesh. Final Report, Extract, 1-28 October 1989.

Multi-purpose Cyclone Shelter Project (MCSP) (1993) Summary report. Bangladesh University of Engineering and Technology and Bangladesh Institute of Development Studies, Dhaka.

Talukder, J., Roy, G.D. and Ahmad, M. (1992) *Living with Cyclone: Study on Storm Surge Prediction and Disaster Preparedness.* Community Development Library, Dhaka.

United Nations (1989) *World Population Prospects, 1988.* United Nations, Department of International and Social Affairs, New York.

Warrick, R. A. (1993) Slowing global warming and sea-level rise: the rough road from Rio. *Transactions of the Institute of British Geographers,* 18, 140-148.

Warrick, R. A. and Oerlemans, J. (1990) Sea level rise. In J.T. Houghton, G.J. Jenkins and J.J. Ephraums (eds.) *Climate Change: The IPCC Scientific Assessment.* Cambridge University Press, Cambridge, pp. 257-281.

Wigley, T.M.L. and Raper, S.C.B. (1992) Implications for climate and sea level of revised IPCC emission scenarios. *Nature,* 357, 293-300.

THE GREENHOUSE EFFECT AND CLIMATE CHANGE

R.A. Warrick, A.K. Azizul Hoq Bhuiya and M.Q. Mirza*

2.1 INTRODUCTION

The threat of global climate change promises to be the single most pressing environmental issue as the world enters the 21st Century. Increasingly, it becomes incumbent upon decision-makers from a wide array of backgrounds to gain a sound understanding of the fundamentals of the climate change issue and how it relates to their areas of responsibility. The purpose of this chapter is to review the main elements of the climate change problem and the implications for Bangladesh in order to highlight the key scientific and policy issues involved. This is accomplished by addressing seven main questions: How has climate changed and why? How is the composition of the atmosphere changing? What are the effects on the Earth's radiation budget? Have variations in greenhouse gases caused climate to change? How might the global climate change in the future? How might the climate of Bangladesh change? What are the implications for policy? The chapter concludes with reflections on the next steps for action and research.

2.2 HOW HAS CLIMATE CHANGED, AND WHY?

The world's climate is partly governed by the composition of the atmosphere. The natural abundances of water vapour, carbon dioxide, methane and other so-called "greenhouse gases" keep the Earth's surface and lower atmosphere warm — a natural "greenhouse effect." Without them, the Earth would be a cold, inhospitable place, unable to support life as we know it. However, due to human activities largely related to energy use, the atmospheric concentrations of greenhouse gases have

* The authors are respectively: Research Director, CEARS, University of Waikato, Hamilton, New Zealand; Professor of Geography, Jahangirnagar University, Bangladesh and Research Scholar at BUP, Dhaka; and post-graduate student at CEARS.

recently been increasing — an "enhanced greenhouse effect" — with the threat of future global warming. It is the purpose of this chapter to present the scientific basis for the concern about global warming and its implications for Bangladesh.

It is essential that global warming and its associated impacts — whether in Bangladesh or elsewhere — be evaluated within the context of natural variability and change. In this way, we are better able to assess whether the inadvertent changes in climate and sea level caused by increases in the atmospheric concentrations of carbon dioxide and other "greenhouse gases" pose a serious threat to our environmental, social and economic systems. Thus, before reviewing the subject of the "greenhouse effect" and the prospect of global warming in the decades to come, it is instructive to examine the rates and magnitude of global climate changes in the distant past.

2.2.1 Climate Change in the Past — a Global Perspective

There is nothing new about "global environmental change". The environment of the planet Earth changes continually across a broad spectrum of time and space scales. Climate is no exception. Today's pre-occupation with warming brought about by human use of the world's natural resources often neglects this important fact.

Let us first consider the large swings in global temperature that began about two million years ago. At that time, after a period of relatively warm conditions lasting millions of years, the world began to experience glacial-interglacial cycles occurring roughly every 100,000 years. During this period of ice ages, termed the Pleistocene, the global temperature varied on average by about 4°C to 6°C as the world alternated between cold glacial epochs (when large continental glaciers covered vast portions of North America and Europe and, as a consequence, global sea level was more than 100 metres lower) and the warmer interglacial epochs. The temperature variations in the middle and high latitudes would have been greater than (perhaps double) the global average, while those of the lower latitudes (including Bangladesh) were probably less. The last glacial maximum occurred around 18,000 years ago. Today, the world is in an interglacial epoch. We are currently experiencing a global climate that is warmer than the preceding 100,000 years, as shown schematically in Figure 2.1a.

By 10,000 years ago, the last ice age had largely ended and the vast continental glaciers had retreated. Over most of the last 10,000 years (the Holocene), the global temperature probably fluctuated by no more than 2°C (Figure 2.1b). A notably warm period occurred during the mid-Holocene, reaching a maximum warmth

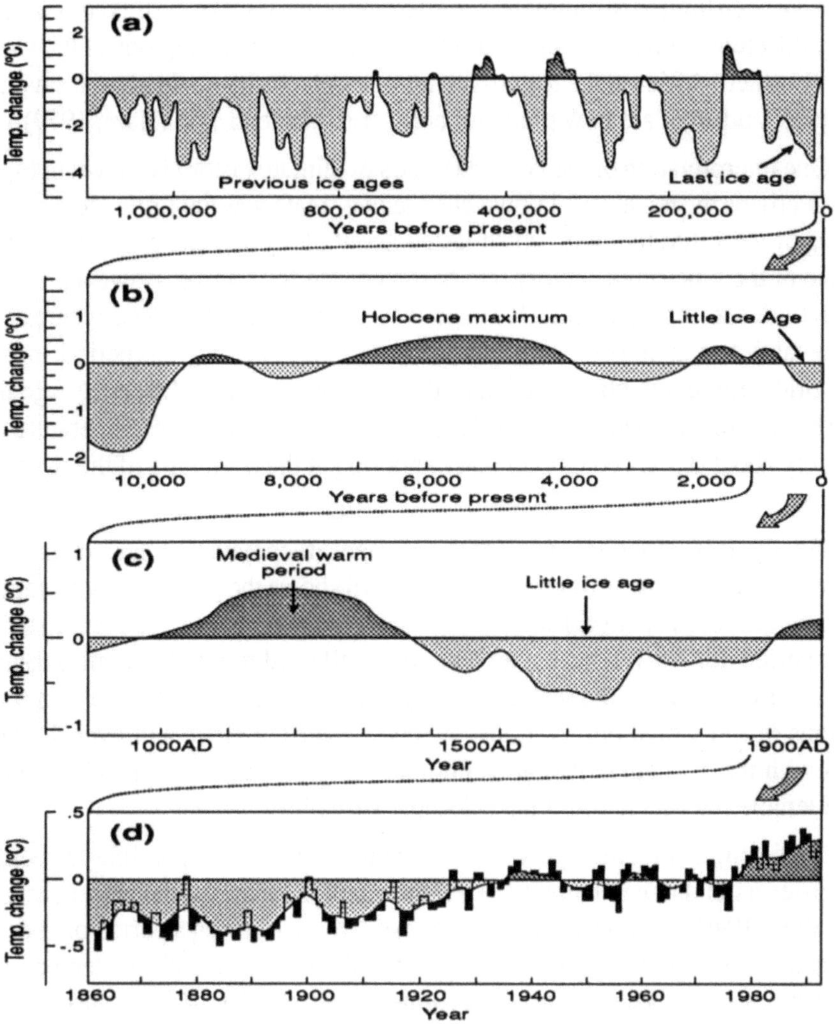

FIGURE 2.1a-d: Past changes in global temperature for (a) the last milion years; (b) the Holocene, the last 10,000 years; (c) the last 1000 years; and (d) the last 140 years of instrumental record. In panels a-c, the temperature changes are shown in relation to conditions nominally prevailing near the beginning of the twentieth century. In panel d, based on the instrumental record, the departures are shown in relation to the 1951-80 reference period; the smooth curve represents filtered values to show trends more clearly.

Sources: Panels a-c from Folland *et al.*, 1990; panel d courtesy of P.D. Jones, Climatic Research Unit, Univerity of East Anglia.

about 6000 to 5000 years ago, although recent evidence suggests that the warming may have been limited to summers in mid to high latitudes. During this period, some vastly different environmental conditions prevailed in many regions of the world. From Northern Africa to the Rajasthan, for example, lush grasslands and savannas occupied lands that are now predominantly arid and semi-arid (Hare, 1979).

One interpretation of the last 1000 years of climate history is shown in Figure 2.1c. Perhaps the most well-known change in climate was the Little Ice Age. The effects of the Little Ice Age, which lasted about 350 years and ended around the middle of the nineteenth century, are best recorded for Europe. However, there is some evidence that the Little Ice Age was global (Grove, 1988), whereas the Medieval Warm Epoch probably was not. For example, during this period, most of the world's mountain glaciers advanced; since the end of the Little Ice Age they have been retreating. During the Little Ice Age, the global average temperature was probably about 0.5°C cooler than that prevailing at the beginning of the twentieth century.

Finally, it is important consider the changes in climate that have taken place on the decade-to-century timescale. Figure 2.1d shows the global annual temperature changes for the period 1860-1991, as derived from instrumental records. This was a period during which weather data were collected routinely for most regions of the world, including the sea surface. The curve in Figure 2.1d represents many millions of individual observations, averaged over the globe and over the year. Even with the averaging, it is apparent that the global temperature still displays considerable inter-annual and inter-decadal variability.

By fitting a smoothed curve to the annual values, it is possible to identify longer-term trends (see Figure 2.1d). From just after the turn of the present century until about 1940, the global temperature rose. There followed a period of about 3-4 decades in which there was no obvious trend, and then the global temperature began to increase once again — this time rather sharply. With respect to this particular data set, the seven warmest years on record occurred after 1980; the warmest year was 1990, followed by 1991 and 1988. Overall, the global temperature has increased by about 0.3°C to 0.6°C over the last 100 years (Folland *et al.*, 1992).

2.2.2 Causes of Climate Change

On the 10,000 to 100,000 year timescale, changes in climate appear to be largely the result of orbital forcing changes. That is to say, factors such as the path of the Earth's orbit around the Sun, as well as the tilt of the Earth's axis in relation to the

Sun, influence the seasonal and latitudinal distribution of energy received at the Earth's surface from the Sun. The swings from glacial to interglacial epochs during the Pleistocene can be largely explained by this mechanism.

BOX 2.1: Climate changes in the past

The global climate varies naturally on all time scales. During the Pleistocene (the last 2 million years) cold glacial epochs, associated with an expansion of continental glaciers, were separated by relatively warmer interglacial epochs. The last glacial maximum occurred 18,000 years ago when the global temperature was 4°C to 5°C colder than today. Currently, the world is enjoying an interglacial period. On shorter timescales, the global climate has fluctuated between warmer and cooler periods, like the Little Ice Age. These have occurred throughout the last 10,000 years (the Holocene). Over the last 100 years, the world appears to have warmed by about 0.3°C to 0.6°C on average, though large annual and decadal variations are evident.

The causes of natural climate changes include variations in the Sun-Earth orbits, in solar output, in volcanic activity and in the ocean circulation, as well as in naturally-occurring greenhouse gas concentrations (particularly water vapour, carbon dioxide and methane).

For instance, during the last interglacial (the Eem; Figure 2.1a) the eccentricity of the Earth's orbit was considerably more than today, resulting in greater radiation receipt during the Northern Hemisphere summer, with temperatures that were perhaps 1°C warmer than the present climate. Based on theoretical understanding of such orbital forcing factors, climate changes at this timescale can be fairly well predicted. In the millennia to come, it is likely that the world should be headed for a relatively minor cool period and, in the longer term, another ice age. The next minimum cool period is expected about 5000 years from now — about 1 to 2°C cooler.

Superimposed on these long-term climate changes are shorter-term changes that are often much more difficult to attribute to specific causes. For instance, around 10,500 BP, during the period of Holocene warming as the world was emerging from the last glaciation, there occurred a cooling lasting about 500 years — a puzzling event known as the Younger Dryas. The causes of this short, abrupt cold period remain uncertain; they could have been related to changes in ocean circulation. The cause of the more recent Little Ice Age is also uncertain, although variation in solar output is a possibility.

On shorter timescales — years to decades — such factors as volcanic eruptions, variations in solar output, atmospheric/oceanic circulation changes, or atmospheric pollutants can influence the climate. For example, the 1991 eruption of Mount Pinatubo in the Philippines has slightly cooled the global surface temperature over several years, as other volcanic eruptions have done in the past. This is because the sulphur-rich materials of such volcanic eruptions have reached the stratosphere (upper atmosphere), causing it to warm as they absorb radiation, depriving the surface and troposphere (lower atmosphere) of heat. The occurrence of El Nino/Southern Oscillation (ENSO) events has profound effects on the climate of some regions of the world, and can explain a significant amount of the interannual variation in the global temperature record (for further explanation of ENSO, see below). Such natural factors will continue to exert an influence on climate in the future. It is quite possible that natural variations in the greenhouse gas content of the atmosphere have also contributed to past climate change. But the main concern now is that changes in greenhouse gas concentrations induced by human activity could well dominate natural mechanisms of climate change in decades to come.

2.3 HOW IS THE ATMOSPHERIC COMPOSITION CHANGING?

It is indisputable that the atmospheric concentrations of the major greenhouse gases — carbon dioxide (CO_2), methane (CH_4), nitrous oxide (N_2O) and the chlorofluorocarbons (CFCs) — have been increasing. Table 2.1 summarises the changes in concentrations of the major greenhouse gases. It is also clear, particularly in the case of CO_2, that human activities are to blame. What are the sources of emissions to the atmosphere? Who is responsible? At what rates are the gases being emitted? What are their "sinks" and the prospects for reducing the concentrations? Let us consider each greenhouse gas individually.

2.3.1 Carbon Dioxide

Precise measurements of the CO_2 concentration have been made since the 1950s. As shown in Figure 2.2, the concentration of CO_2 was 315 parts per million by volume (ppmv) in 1958 and rose to over 350 ppmv by 1990. Over the recent decade, the CO_2 concentration has been increasing at a rate of about 0.5 per cent per annum.

TABLE 2.1: The "pre-industrial" and 1990 concentrations of the major greenhouse gases and their recent annual growth rates

	Carbon Dioxide	Methane	Nitrous Oxide	CFCs
PRE-INDUSTRIAL CONCENTRATION	280 ppmv	800 ppbv	288 ppbv	nil
CONCENTRATION IN 1990	354 ppmv	1720 ppbv	310 ppbv	CFC11 = 280 pptv CFC12 = 484 pptv
RECENT ANNUAL GROWTH RATE	0.5 %	0.6 - 0.8%	0.2 - 0.3%	CFC11 = 4% CFC12 = 4%

Source: From data reported in Watson *et al.*, 1990; 1992.

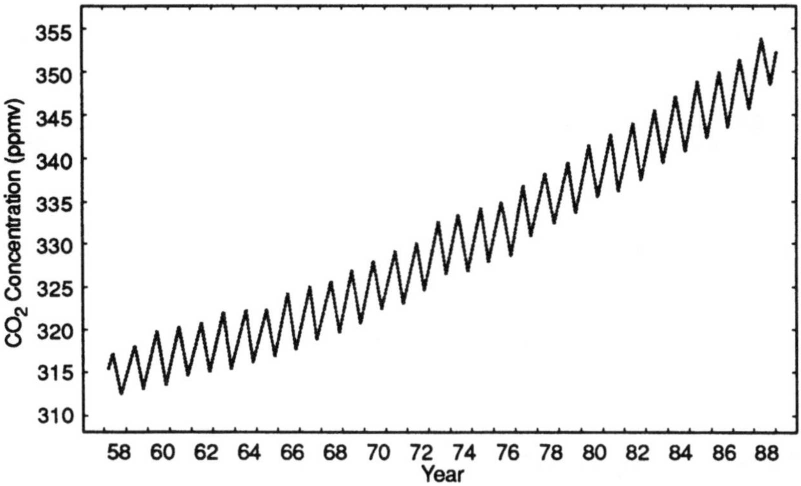

FIGURE 2.2: Monthly average CO_2 concentration (ppmv), observed continuously at Mauna Loa, Hawaii.

Source: Watson *et al.*, 1990.

The estimated 1993 concentration is 356 ppmv.

In order to estimate the CO_2 concentration for earlier periods, reliance must be placed on indirect measurements. One important method involves the analysis of samples of the atmospheric composition locked in ice cores extracted from the polar ice sheets. These samples indicate that the "pre-industrial" (nominally, the middle of the eighteenth century) concentration was about 280 ppmv (Figure 2.3). Thus, over the last two centuries, the CO_2 concentration has risen by about 27 per cent. The additional CO_2 is relatively quickly and evenly distributed around the world.

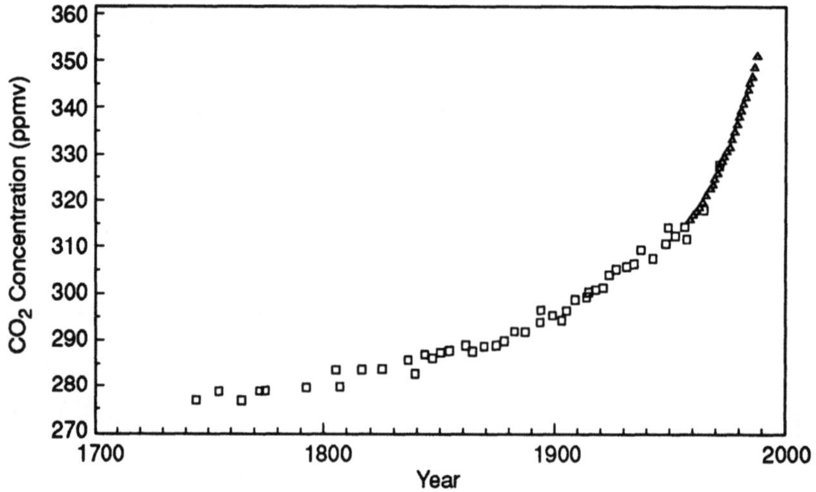

FIGURE 2.3: Atmospheric CO_2 increase during the past 250 years, as indicated by measurements of air trapped in ice cores.

Source: Watson *et al.*, 1990.

Why has the CO_2 concentration risen? To understand this, it is first necessary to know something about the global "carbon cycle". On the timescales that we are interested in here, carbon is cycled mainly between three reservoirs: the oceans, the atmosphere and the terrestrial biosphere. The ocean (especially the deeper layers) contains the largest store of carbon and, on an annual basis, is both a source and "sink" of atmospheric carbon depending on season and location. The exchange of carbon between the atmosphere and oceans amounts to about 90 GtC/yr (gigatonnes of carbon per year, where 1 Gt = 10^{15} grams = one billion tonnes). The flux between the ocean surface and the atmosphere depends, to a large degree, on the relative difference in carbon content between the two reservoirs (and other variables such as ocean temperature). Thus, if the atmospheric concentration is increased, some portion of the increase will be taken up by the oceans until a new equilibrium is reached.

The terrestrial biosphere also plays an important role in influencing the atmospheric CO_2 concentration. Through the process of photosynthesis, molecules of atmospheric carbon dioxide are assimilated by plants as they produce carbohydrates for growth. This carbon is temporarily stored in living biomass and non-living plant litter and soil humus (and, in the very long term, stored as "fossil" carbon—coal, oil and gas). The processes of respiration and decomposition release this carbon back to the atmosphere.

The effect of the biosphere on the atmospheric composition is illustrated by the seasonal cycle in atmospheric CO_2 concentration (with an amplitude of about 5 ppmv at Mauna Loa) evident in Figure 2.2. The cycle results from the draw down of atmospheric CO_2 in the spring and summer (as plants are vigorously photosynthesising) and the release of carbon to the atmosphere during the autumn and winter (when respiration dominates) by the terrestrial biosphere in the Northern Hemisphere (where most of the world's land mass is located). A store of carbon, amounting to about 2000 GtC, is maintained in the terrestrial biosphere (including soils) of the world from one year to the next. The tropical and boreal forest ecosystems, for example, contain a large portion of the total carbon stored in the world's terrestial ecosystems. Modifications of such ecosystems, either natural or human-induced, can change the biospheric carbon content and affect the fluxes between — and, consequently, the carbon contained within — the other two major reservoirs.

Over the last 200 years, two human activities have interfered with the carbon cycle and have led to an increase in atmospheric carbon dioxide: the extraction and combustion of fossil fuels; and changes in land use, primarily deforestation in the mid-latitudes and, more recently, the tropics. The Industrial Revolution in the West marked the beginning of the widespread use of fossil fuels, initially in the form of coal. In terms of carbon emitted per unit of energy available, coal is "dirtier" than either oil or, especially, natural gas. Later, after the middle of the twentieth century, the combined global consumption of oil and natural gas overtook that of coal. Around this time, the total emissions of carbon from fossil fuel burning (and cement manufacture, a minor contributor) accelerated. Over the last two decades, fossil fuel carbon emissions grew at an average rate of about 1.9 per cent per annum, from

4.1 GtC/yr in 1970 to 6.0 GtC/yr in 1990 (Marland, 1990; Watson *et al.*, 1992).

Global averages, however, mask the regional and national differences in emissions. In the case of fossil-fuel carbon emissions, the differences are substantial. For example, the developing countries (the world excluding all OECD countries, the former Soviet Union, and any remaining European countries), although containing nearly three-quarters of the world population, are responsible for less than one-quarter of the fossil-fuel carbon emissions. In comparison, the OECD countries, with about 15 per cent of the world's population, account for around 44 per cent of the total emission (Marland, 1990). One country, the United States, is solely responsible for 23 per cent of the total yearly fossil-fuel carbon emission to the atmosphere. In contrast, Bangladesh contributes a minuscule 0.06 per cent (based on WRI, 1990). In terms of per capita emissions, an average Bangladeshi is responsible for less than 1/100 the amount of an average American. The geographical disparities are marked.

The other main source of CO_2 emissions derives from changes in land use, principally through deforestation. For the 1980s, the best estimate is that the net carbon release from deforestation, mainly in the tropics — Brazil, South-East Asia, Central America, tropical Africa — amounted to about 1.6 GtC/yr (Bolin, 1986; Watson *et al.*, 1990, 1992). But unlike the fossil-fuel component, a very large range of uncertainty surrounds this estimate. The estimates range from near zero to as much as 3.6 GtC/yr (Houghton, 1991). Taking into account re-growth of forests in the mid-latitudes, the latest estimate of *net* carbon emission from land use changes is 1.1 GtC/yr (Schimel *et al.*, 1995).

Overall, the carbon emission from deforestation is currently relatively minor in comparison to the fossil-fuel emission. However, this was not always the case. Prior to the beginning of the twentieth century, widespread removal of temperate forest cover in the mid-latitudes of the Northern Hemisphere probably accounted for a large portion of the increases in atmospheric CO_2 that occurred during that time. Similarly, in Bangladesh, forests once covered a large portion of the country, but now only account for about 6 per cent of the land area, disappearing quickly over the last 20 years (Rashid, 1991). Over the last 200 years, the total cumulative carbon emission from the terrestrial biosphere may match that from fossil-fuel consumption (Warrick and Rahman, 1992).

The addition of over 7 GtC into the atmosphere each year through fossil-fuel burning and deforestation disrupts the carbon cycle. Where it all goes is not entirely known. Because of the higher atmospheric CO_2 concentration, the ocean is able to absorb more carbon and, on an annual basis, becomes a net sink for about 2.0 ± 0.8 GtC each year. About half of the total emission remains in the atmosphere each year. This leaves some portion unaccounted for — the so-called "missing sink". It is likely that the missing sink is the terrestrial biosphere (Tans *et al.*, 1990; Wigley and Raper, 1992). A possible mechanism for this is the "CO_2 fertilisation effect": higher CO_2 has been shown to stimulate photosynthesis and plant growth in controlled experimental conditions (although there is no direct evidence that this is occurring in natural ecosystems). In any event, the "sinks" are unable to absorb all of the annual anthropogenic emission, so the atmospheric CO_2 concentration increases as a consequence.

BOX 2.2: Carbon dioxide

Over the last two hundred years, the atmospheric concentration of carbon dioxide has risen by about 28 per cent. The current increase is due primarily to fossil fuel (coal, oil, gas) consumption. Current emissions to the atmosphere from this source amount to 6,000 million tonnes of carbon per year. A further 1,100 million tonnes per year may derive from changes in forest cover. At present, the less-developed countries, with more than three-quarters of the world population, account for less than one-third of the total annual carbon emission. Bangladesh contributes a trivial 0.06 per cent of the global total.

The prospect of stabilising the atmospheric CO_2 concentration in the short-term is remote. Because of the "lifetime" of CO_2 in the atmosphere (more properly, the time required for equilibrium to be re-established between the reservoirs of carbon), the response time between changes in emissions and concentrations is long. For example, if the world were somehow able to immediately stabilise global CO_2 emissions at current rates, the atmospheric concentration would continue to rise linearly throughout the twenty-first century. In order to stabilise the concentration at around the current level, CO_2 emissions would have to eventually drop by about 60 to 80 per cent (Watson *et al.*, 1990).

2.3.2 Methane

The concentration of methane has more than doubled since pre-industrial times, from an estimated globally-averaged value of 800 ppbv to 1720 ppbv (in 1990). Most of the increase has occurred since 1900, with a 30 per cent increase over the last four decades. The growth has been steady since 1978, when direct precision measurements began, although there is some evidence that the annual growth rate has slowed slightly in recent years (Steele *et al.*, 1992).

Methane is a product of anaerobic (oxygen deficient) processes. Thus, the sources of methane emissions to the atmosphere are highly varied and include a mix of natural and anthropogenic sources. Methane is emitted naturally from wetlands (including tundra, marshes, swamps), animals (from enteric fermentation), termites, oceans and freshwater bodies. However, it is likely that the observed *increases* in the CH_4 concentration are due to human-induced increases in emissions.

The human activity that appears to account for a large portion of the total anthropogenic CH_4 emissions is agriculture, particularly paddy rice production and animal husbandry (especially ruminants). The estimated leakage of natural gas from fossil fuel use is also large and results from gas drilling, venting, transmission and coal mining. Additional sources are landfills and biomass burning. The best current estimate is that these anthropogenic sources amount to about 360 Tg CH_4 per year (1 Tg = 10^{12} grams), but the uncertainties are very large indeed, about $\pm$50 per cent (Watson *et al.*, 1992).

It is difficult to quantify the CH_4 emitted from specific regions or nations. For example, one recent controversial report (WRI, 1990) estimated that most of the methane emissions originate in the developing world (according to this report, Bangladesh may account for as much as 1 to 2 per cent of the global total, primarily from rice production, but the data are very questionable). However, depending on how the data are interpreted, the opposite conclusion could easily be reached (e.g. see Agarwal and Narain, 1991).

One issue that pertains to Bangladesh is whether the natural ecosystems that cultivated rice replaced were large sources of methane. Wetlands are probably the largest source of global CH_4 emissions. Thus, it is likely that the extensive areas of swamps and mangroves in Bangladesh prior to agricultural encroachment naturally produced considerable amounts of methane. In such areas, agricultural development has probably simply meant trading one methane-producing land use with another. In drier forested areas, this was probably not the case and rice paddy production has meant a net increase in emissions. The recent expansion of irrigated wetland paddy (*boro*) cultivation on land that was previously used for dryland crops, or that lay fallow in the dry season, may have increased methane emissions. However, there are no reliable estimates of the net change in CH_4 emissions for Bangladesh.

The lifetime of CH_4 in the atmosphere is variable, depending on the interactions with other gases. But, in general, the lifetime (8 to 12 years) is considerably less than that of CO_2, so that the time between emission change and concentration response is comparatively short. For instance, a slow reduction of current CH_4 emissions by about 15 per cent over several decades would be sufficient to approximately stabilise the CH_4 concentration over the next century.

BOX 2.3: Methane

> The atmospheric concentration of methane has more than doubled over the last two hundred years. Most of the increase has occurred in this century and has closely followed the growth of the global human population. The current growth rate in concentration is about 0.6 to 0.8 per cent per annum. The increase is believed to be the result of human activities such as: solid waste disposal; agriculture, particularly paddy rice production and animal husbandry; and fossil fuel-related leakages associated with gas pipelines and coal extraction. Bangladesh makes a contribution (albeit a very minor one) to the yearly emission of methane because of its extensive areas of seasonally or perennially flooded land and wetland paddy cultivation, as well as its animal husbandry.

One of the results of the breakdown of methane is an increase in stratospheric water vapour. Because water vapour itself is a greenhouse gas, an additional radiative heating thus occurs as an indirect effect of increasing methane concentrations. (Globally, water vapour will, on average, increase as a consequence of warming, but this is considered a "climate feedback" and is dealt with in climate models — see below.) The magnitude of this additional heating may be equivalent to about one-third of the radiative forcing change due directly to methane itself.

2.3.3 Nitrous Oxide

The concentration of N_2O has not risen as quickly as the other gases. The 1990 concentration was 310 ppbv. According to the IPCC (Watson *et al.*, 1992), the pre-industrial concentration was about 288ppbv, although recent analysis (Leuenberger and Siegenthaler, 1992) suggest a lower value of 260 ppbv. Thus, the concentration of N_2O has risen by 8 to 19 per cent over the last two hundred years. Most of the increase has occurred in the present century, particularly since 1950. The recent annual rate of increase has been around 0.2 to 0.3 per cent (Watson *et al.*, 1992).

The sources of N_2O and their fluxes to the atmosphere are very uncertain. The natural sources are the oceans and soils. It remains unclear as to which anthropogenic sources are primarily responsible for the observed increases in atmospheric concentrations. It was once believed that fossil fuel and biomass burning played a large role, but it was later discovered that gas sampling procedures were faulty, so the earlier fossil fuel estimates have been revised downwards by an order of magnitude. The use of nitrate and ammonium fertilisers is a likely source, but estimates of the fluxes vary widely. The major sink for atmospheric N_2O (photolysis in the stratosphere) is known and fairly well quantified, and implies a long atmospheric lifetime, about 150 years.

Overall, after summing the sinks, sources and the atmospheric increase — the uncertainties included — the "budget" remains unbalanced. This implies a yet undiscovered source of N_2O (or an underestimate of the sinks). In large part, it is only because the increases have occurred in recent times and have tracked the growth of human population, agriculture and energy use that they are believed to be related to human activities.

2.3.4 Halocarbons (including CFCs)

The most important halocarbons in terms of contributing to the enhanced greenhouse effect have been the chlorofluorocarbons CFC-11 and CFC-12. Most halocarbons are not naturally occurring; rather they have been purposely manufactured for specific uses such as refrigerants, foam blowing, aerosol propellants (e.g. as used in spray cans), solvents and fire retardants — amongst a wide variety of uses. Thus, unlike methane and nitrous oxide, their sources are relatively few in number and thus clearly identifiable. More than 90 per cent of CFCs are currently produced in the industrialised countries. Bangladesh does not produce CFCs and is an insignificant consumer of CFCs.

The gases CFC-11 and CFC-12 (and other fully halogenated halocarbons) have long lifetimes in the atmosphere: 65 and 130 years, respectively. This is because there are no means by which they are effectively removed from the troposphere (lower atmosphere); it is only after they reach the stratosphere (upper atmosphere) that they are slowly removed by photodissociation (where they also are implicated in the destruction of the ozone layer).

The CFCs are very strong greenhouse gases. As a consequence, even though their atmospheric concentrations are relatively low — 280 and 484 parts per trillion by volume (pptv) in 1990 — their potential contribution to the greenhouse is large.

However, the effects of chlorine and bromine containing compounds is complicated by their destruction of ozone in the stratosphere, which has a cooling effect and partially offsets the greenhouse warming caused by halocarbons (see below). Over the last several decades, the atmospheric concentrations of CFC-11 and CFC-12 have grown at a rate of about 4 per cent per year, faster than any other of the major greenhouse gases.

However, controls on halocarbons were set in motion with the ratification of the "Montreal Protocol on Substances that Deplete the Ozone Layer" (UNEP, 1987) by the major CFC-producing nations (and many other nations, including developing countries). The driving force behind the Montreal Protocol was not the concern over global warming, but rather the threat to stratospheric ozone from the chlorine and bromine chemicals. However, the provisions for cuts in the production of halocarbons set out by the Montreal Protocol have important implications for future greenhouse warming as well.

BOX 2.4: Chlorofluorocarbons (CFCs)

> The CFCs, especially the widely-used CFC-11 and CFC-12, are very strong, long-lived greenhouse gases. They are not naturally-occurring, but rather were manufactured for use as refrigerants, foam blowing, spray-can propellants, and solvents, amongst many uses. Their atmospheric concentrations are very small, but grew rapidly over the last several decades at a rate of about 4 to 5 per cent per year. Because of their implication in the destruction of stratospheric ozone (the "ozone hole"), CFC production has been restricted through the provisions of the 1987 Montreal Protocol. Subsequent amendments to the Protocol call for a complete phase-out of CFC production before the year 2000.

Specifically, the Montreal Protocol aimed to achieve a 50 per cent cut in the 1986 level of production of the CFCs by the year 1998 (with some allowances for temporary growth in production in developing countries). Under this arrangement, even after allowing for the delayed releases and the lifetimes, both the concentrations of CFC-11 and CFC-12 would still continue to grow (but at a declining rate) to more than double their 1990 level by the end of the twenty-first century. Recent international agreements to strengthen the Montreal Protocol were made in London in 1990 which call for a complete phase-out of production of these CFCs by the turn of the century. If this occurs, the concentrations of CFC-11 and CFC-12 would fall to about one-third and one-half of their respective 1990 values by the end of the next century. Nonetheless, some important substitutes (e.g. HCFCs) would still grow in concentration and still contribute positively to the greenhouse effect (Mitchell and Gregory, 1992). Further recent amendments to the Protocol agreed upon in Copenhagen in 1992 seek to speed up the phase-out of CFC production. Overall, the role of halocarbons in future climate warming will probably be much less than previously thought.

2.4 CHANGING THE EARTH'S RADIATION BUDGET

Naturally-occurring concentrations of greenhouse gases, particularly carbon dioxide and water vapour, play an important role in influencing the climate system through their effects on the Earth's radiation budget. To illustrate, if the naturally-occurring concentrations of such greenhouse gases were absent from the atmosphere, the Earth's surface and lower atmosphere would be more than 30°C colder than today and a very inhospitable place in which to live (IPCC, 1990). Thus, there is a natural greenhouse effect that is beneficial for life on Earth.

The principal issue of concern to us, however, is the prospect of an enhanced greenhouse effect. This is the situation the world now faces in which human-induced increases in the atmospheric concentrations of greenhouse gases threaten to cause major changes in the global climate.

2.4.1 The Effect of Greenhouse Gases

The climate of the Earth is controlled by the difference between the incoming solar radiation from the Sun and the outgoing infrared radiation emitted by the Earth-atmosphere system — the radiation balance. The Earth's radiation balance can be upset by changes in either of the two components. For example, changes in solar irradiance or Earth-Sun orbital parameters (mentioned above) influence the amount and distribution of incoming solar radiation. The major greenhouse gases influence the outgoing infrared radiation.

Greenhouse gases, although relatively transparent to incoming short-wave radiation, are quite able to absorb and emit the infrared radiation, some of which is re-directed back to the Earth's surface. An increase in greenhouse gas concentrations initially causes a reduction in the infrared radiation leaving the Earth-atmosphere system. This results in an imbalance in the radiation budget — *a change in radiative forcing*. To re-establish a balanced, or equilibrium, condition, the Earth's surface and troposphere (the lower 8km to 16km of the atmosphere) must warm in order to increase the radiation emitted to space. In short, increasing the concentrations of greenhouse gases should cause global warming.

2.4.2 Past Changes

From knowledge of the radiative properties of each gas, the change in radiative forcing can be calculated. In 1990, the IPCC estimated that the overall cumulative change in radiative forcing that has occurred since pre-industrial times (mid-eighteenth century) is about 2.5 Wm^{-2} (Watts per square metre, averaged over the year and the globe) — approximately equivalent to "turning up" the Sun by 1 per cent (Shine *et al.*, 1990). Nearly one-quarter of the total radiative forcing change since the eighteenth century has occurred over the last decade, which reflects the recent high growth rates of greenhouse gas emissions.

The importance of CO_2 to such recent changes in radiative forcing is shown in Figure 2.4. Because CO_2 dominates the change in radiative forcing, because fossil fuel use accounts for most of the CO_2 emissions (as well as being an important source of methane), and because global consumption of fossil fuels has been growing rapidly, *energy use* is the key factor in the enhanced greenhouse effect and future climate change. Assuming that CFC production will be cut drastically — as per the Montreal Protocol and the tightening of its provisions — fossil fuel use increasingly becomes the dominant issue to address in considering strategies to slow global climate change.

2.4.3 Effects of Sulphate Aerosols and Ozone Depletion

Two recent scientific findings have led to a downward revision of estimates of greenhouse-gas-induced radiative forcing changes. First, the effect of CFCs may be much less than previously thought because of their destruction of ozone in the lower stratosphere. The depletion of ozone is calculated to have a cooling effect on the Earth's surface and lower atmosphere (WMO, 1992). It has been calculated that this *ozone-depletion feedback* effect of CFCs largely offsets their direct contribution to greenhouse warming (Ramaswamy *et al.*, 1992; Isaksen *et al.*, 1992; Wigley and Raper, 1992).

Second, previous estimates of radiative forcing changes did not take into account the *cooling effect of sulphate aerosols*. Besides emitting carbon dioxide,the burning of fossil fuels also creates sulphur dioxide, which leads to the formation of sulphate aerosols in the atmosphere. Sulphate aerosols have a cooling effect that should partly offset the CO_2-induced warming (Isaksen *et al.*, 1992; Jonas *et al.*,

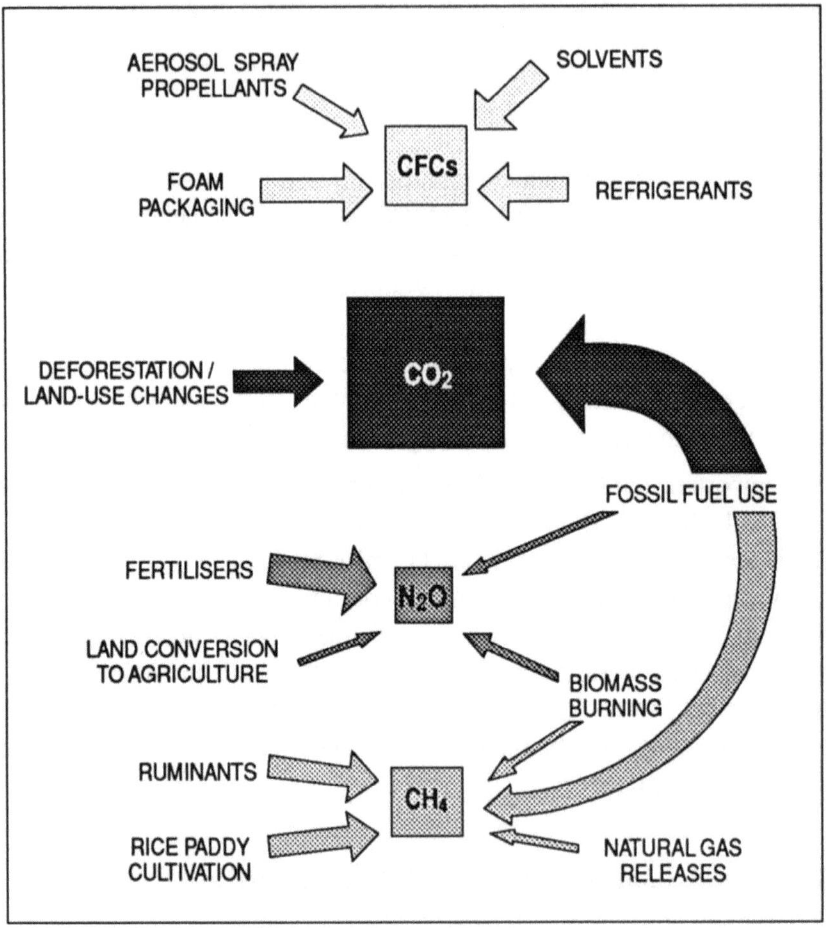

FIGURE 2.4: The main greenhouse gases and their anthropogenic sources. The sizes of the boxes are proportional to the contribution to recent radiative forcing changes, whereas, for each gas, the sizes of the arrows indicate the relative importance of each source (excludes the effects of sulphate aerosol cooling and ozone-depletion feedback).

Source: Warrick *et al.*, 1990.

1995). Taking into account both the sulphate aerosol effect and the ozone-depletion feedback should reduce estimates of net radiative forcing changes (as noted above) and thus lower past and future projections of global warming and sea-level rise (Houghton *et al.*, 1992, 1995; Wigley and Raper, 1992; also see Chapter 3).

BOX 2.5: Changes in radiative forcing

> Greenhouse gases are transparent to incoming solar radiation, but readily absorb the infrared radiation emitted by the Earth. In this way, naturally-occurring gases like water vapour and carbon dioxide "insulate" the Earth's surface and lower atmosphere, keeping them much warmer than would otherwise be the case — a natural greenhouse effect. Increasing the gas concentrations upsets the Earth's radiation balance, causing a change in the *radiative forcing*, or heating, of the climate system — an *enhanced* greenhouse effect. Nearly one-quarter of the past change in radiative forcing may have occurred over the last decade alone. Carbon dioxide accounts for most of the change. Fossil fuel use for energy is the single largest source of the enhanced greenhouse effect. Bangladesh is responsible for perhaps 0.2 to 0.3 per cent of the current total radiative change, a very minor contribution.

2.4.4 Who's to Blame?

Geographically, it is difficult to say precisely which countries are responsible for the changes in radiative forcing. There are large uncertainties, for example, in the sources and sinks for methane and nitrous oxide, and there is no universally-agreed procedure for attributing responsibility. However, some general patterns are evident. First, the industrialised countries of the world (particularly the OECD countries, Eastern Europe and the former Soviet Union) are clearly responsible for the large part of the changes in radiative forcing that have so far taken place. Second, on a per capita basis, there is a very big discrepancy between the less-developed countries and the rest of the world.

These points are illustrated in Figure 2.5. The less developed countries, in which about 75 per cent of the world's human population resides, probably account for somewhere between 27 per cent and 35 per cent of the enhanced greenhouse effect (Warrick and Rahman, 1992; excludes consideration of sulphate aerosol effect and ozone-depletion feedback). A major portion of their contribution stems from methane associated with agriculture and from CO_2 associated with deforestation. An inhabitant of a less developed country contributes, on average, 6 to 8 times less to the greenhouse effect than a citizen of a developed country. The average contribution from a Bangladeshi citizen is far less.

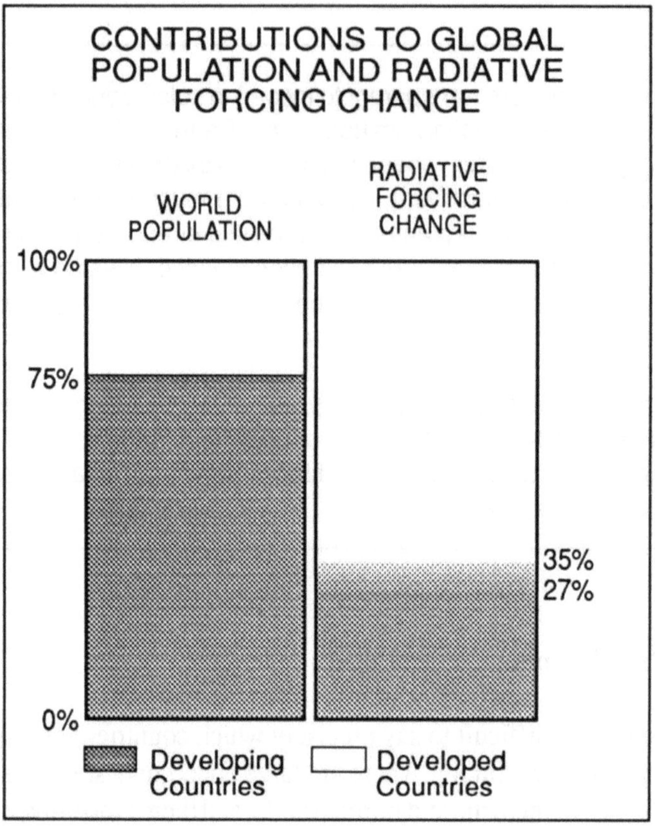

FIGURE 2.5: The relative contributions of developed and developing countries to radiative forcing change due to total world population (left bar) and to increases in greenhouse gas concentrations (right bar), excluding effects of sulphate aerosols and ozone depletion feedbacks. The developed world, with only one-quarter of the population, accounts for about two-thirds of the enhanced greenhouse effect.

2.5 HAVE VARIATIONS IN GREENHOUSE GASES CAUSED CLIMATE TO CHANGE?

2.5.1 The Geological Record

On the 10,000 to 100,000-year timescale, there is strong evidence that greenhouse gas concentrations and global temperature are associated. For example, Figure 2.6 shows the curves for temperature and for carbon dioxide and methane concentrations for the Antarctic region over the last 160,000 years (inferred from deuterium and oxygen isotope changes as derived from analyses of cores extracted from the ice sheet). There is a striking correspondence between the temperature curve and the curves of CO_2 and CH_4 concentrations. This is strongly suggestive of a positive causal relationship.

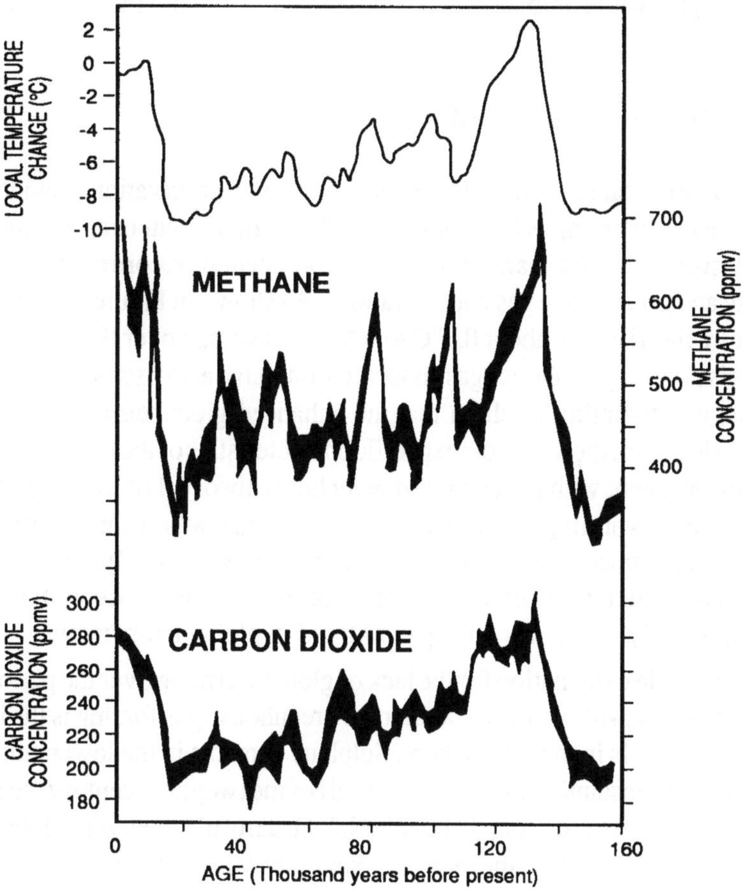

FIGURE 2.6: Carbon dioxide (bottom) and methane (middle) concentrations during the past 160,000 years as determined on the ice core from Vostok, Antarctica. Temperature changes (top) are estimated, based the measured deuterium concentrations.

Source: Houghton *et al.*, 1990, as adapted from Barnola *et al.*, 1987 and Chappellaz *et al.*, 1990.

However, due to the temporal resolution of the data on this timescale, it remains uncertain as to whether the greenhouse gases are the cause or the effect of the climate changes. At this timescale, they are probably both. Regardless, the change in CO_2 and CH_4 (and N_2O) concentrations between the glacial and interglacial epochs could partly explain the 4°C to 5°C global temperature decrease experienced during the last glacial maximum — a decrease only partly explicable by changes in Earth-Sun orbital parameters. Thus, at least on this timescale, greenhouse gases appear to have been influential in affecting the Earth's temperature.

2.5.2 The Instrumental Record

On shorter timescales — decade to century — the observational evidence for greenhouse gas warming is less convincing. This might seem contradictory, since over the last one hundred years we have a far more accurate record of both variations in climate and greenhouse gas concentrations. As shown in Figure 2.1d, the global temperature has risen by about 0.3°C to 0.6°C on average over the last 100 years. During this period, greenhouse gas concentrations have also risen substantially. On this basis, it is tempting to claim that the enhanced greenhouse effect has been detected. However, there are some specific characteristics of the temperature curve that are inconsistent with the enhanced greenhouse theory. For instance, there is a period of rapid warming up to about 1940, but that was a time during which greenhouse gas concentrations were increasing relatively slowly. Then, for about three decades after mid-century, when greenhouse gas concentrations began increasing rapidly, there was no apparent trend in global temperature.

One possible explanation for the lack of global warming over the period 1940-1970 which would still allow one to involve greenhouse gas forcing is the increase in sulphate aerosols in the atmosphere. Sulphate aerosols in the lower atmosphere have increased substantially in the second half of the twentieth century, due largely to fossil fuel use. As mentioned above, sulphate aerosols tend to cool the Earth's surface. Since mid-century, the increase, particularly in the Northern Hemisphere, may have been sufficient to offset a large part of the greenhouse-gas-induced warming (Charlson *et al.*, 1992; Wigley 1989; Wigley and Raper, 1992; Isaksen *et al.*, 1992).

The instrumental record is a very "noisy" one. There is considerable year-to-year and decade-to-decade variation caused by a variety of factors (a good deal of which may be attributable to natural "random internal variation" of the climate system). Science is not yet able to fully understand and predict these variations. It remains uncertain as to how much of the observed global warming is due to the increasing greenhouse gases. Nevertheless, the enhanced greenhouse effect remains the prime suspect for the global warming that the world, on average, has been experiencing. Monitoring climate change is essential if we are eventually going to be able to claim that the enhanced greenhouse effect has been detected (Wigley and Barnett, 1990; Wigley and Raper, 1991).

2.6 HOW MIGHT GLOBAL CLIMATE CHANGE IN THE FUTURE?

How much is the world expected to warm? The calculation would be relatively straightforward if one were only concerned with the direct change in global temperature from an instantaneous change in concentration. From theory it can be readily determined, for example, that a doubling of the concentration of carbon dioxide should cause the world to warm by 1.2°C (Cubasch and Cess, 1990).

But the real world is not so simple. There are secondary, feedback effects arising from the climate system itself (which includes the atmosphere, hydrosphere, biosphere and cryosphere) that can affect both the magnitude and rate of warming. The principal feedbacks result from changes in atmospheric water vapour content, clouds, ice and snow cover, and land surface characteristics. These feedbacks can amplify or diminish the direct warming effect from increases in greenhouse gas concentrations. It is the uncertainty in the direction and strength of these climate feedbacks that present major difficulties in predicting future global warming. The principal way in which the sensitivity of global climate to increased greenhouse gas forcing has been explored is through the use of complex computer-based climate models — general circulation models, or GCMs.

2.6.1 Equilibrium Changes in Climate from GCMs

General circulation models are three-dimensional, physical models (based on the fundamental laws of physics) that attempt to simulate the world's climate. Because of their complexity, GCMs require the use of the world's most sophisticated supercomputers. In concept, GCMs provide the only available means of realistically examining climate feedbacks, their interactions and the implications for regional climates. In practice, however, there are many deficiencies in the use of GCMs for this purpose.

For example, one important deficiency is the inability to properly model cloud processes, which tend to operate at finer spatial scales than GCMs are currently able to resolve. This, in turn, creates inaccuracies and uncertainties in both the direction and magnitude of the feedbacks resulting from radiative forcing changes, since clouds can act either as positive or negative feedback, depending on such details as water content, droplet size, optical depth, etc. Deficiencies in cloud modelling also create problems in correctly predicting the temporal and spatial patterns of precipitation. Thus, in simulations of the present climate, GCMs often produce grossly inaccurate depictions of precipitation in specific regions in attempts to simulate current climate patterns (Gates *et al.*, 1992).

Another major deficiency has been the rudimentary treatment of the oceans. The oceans are critical in influencing both the regional changes in climate and the rate of change (see below). Most GCMs have included the oceans only in simplistic, static fashion, which has precluded, for instance, the possibility of determining the influence of ocean current changes or detailed re-distribution of heat within the oceans. Several of the world's major climate modelling centres have recently been addressing this problem by linking together dynamic ocean and atmospheric GCMs into a fully coupled system — CGCMs.

Despite such deficiencies, however, GCMs and CGCMs perform reasonably well at simulating the major features of the circulation (such as the high and low pressure systems) and the seasonal and spatial temperature differences. For this reason, there is cautious confidence in the use of these climate models for examining the global and continental scale effects of greenhouse forcing. However, for the regional effects (including Bangladesh and the surrounding region) such model results have to be viewed with a good deal of caution — as scenarios of what *could* happen, not as predictions of what *will* happen.

Most of the experiments that have been conducted using GCMs to estimate the effects of increased greenhouse gas concentrations have been to determine the equilibrium response of the climate system. Typically, in such an experiment the GCM is first run with a base level of CO_2 concentration (say, 300 ppmv) until the climate approaches a steady-state (equilibrium) condition. Then, the GCM is re-run with double the concentration (600 ppmv) and allowed once again to approach equilibrium conditions. The difference between the two climates is taken to be the equilibrium change for a doubling of CO_2 (equivalent to a change in radiative forcing of 4.4 Wm^{-2} and a surrogate for other greenhouse gases as well). The conventional parameter used to express this equilibrium response is the change in global-mean surface air temperature, often referred to as the climate sensitivity. Not surprisingly, different GCMs often produce different estimates of the climate sensitivity because of the differences in the ways in which various processes (e.g. clouds) are modelled.

Upon reviewing the results of GCM experiments, the Intergovernmental Panel on Climate Change scientific assessment (IPCC, 1990; Houghton *et al.*, 1990;

1992) concluded that the value of the climate sensitivity for a CO_2 doubling falls within the range 1.5°C to 4.5°C, with a best-estimate of 2.5°C. Thus, even though the climate sensitivity values differ, GCMs consistently find that the net effect of climate feedbacks is positive, thereby enhancing the direct greenhouse-gas-induced heating effect of a CO_2 doubling (1.2°C, noted above). The range is due to uncertainties in the overall magnitude of the positive climate feedback effect.

One of the strongest feedbacks is that of water vapour. It is well understood that, as the world warms, the hydrological cycle is speeded up and the atmosphere is able to hold, on average, more moisture. Since water vapour itself is an important greenhouse gas, the direct warming is magnified by about a factor of 1.7 due to its increase.

Sea ice and snow cover feedbacks are also important, particularly for the higher latitudes in the Northern Hemisphere winter. Firstly, the initial warming results in a thinning of the sea ice cover (which acts as an insulating blanket), thus increasing the loss of heat from the underlying sea surface to the atmosphere during the winter. Secondly, warming causes a decrease in the sea ice extent, which allows more radiation to be absorbed by the sea instead of reflected back to space. Finally, a similar effect results from the decrease in the duration and areal extent of snow cover over land, which increases the absorption of heat by the land surface. Partly for these reasons, GCMs consistently find an enhanced high latitude warming in the winter half of the year (in the order of twice the global-mean value) (Mitchell *et al.*, 1990; Gates *et al.*, 1992).

Perhaps the climate feedback with the greatest uncertainty is that of cloud changes. Depending on the circumstances, clouds can have a cooling or warming effect. The issue is, which effect will dominate as the world warms? Depending on how GCMs model the clouds, the answer is different. In the most extreme simulations, a strong negative cloud feedback would be sufficient to counteract the other positive feedbacks; in contrast, a strong positive cloud feedback would push the climate sensitivity above a value of 5°C. Adequately resolving cloud processes is thus a major goal of climate modellers.

In summary, despite the uncertainties, there is general agreement that, on average, an increase in the atmospheric concentrations of CO_2 and other greenhouse gases should cause the world to warm. No major climate model experiments show global cooling.

Temperature is only one aspect of climate change. Regarding other climate elements, there are a few general findings about which there is high confidence. For example, there is agreement that the hydrological cycle will "speed up" as a consequence of warming, and that global average precipitation will increase, in the order of 2 to 3 per cent per degree Celsius. However, in general, less reliance can be placed on predictions of changes in other climate variables such as regional precipitation, radiation, windiness and severe storm frequencies (Mitchell *et al.*, 1990; Gates *et al.*, 1992).

2.6.2 Time-Dependent Changes in Climate

So far, we have been discussing the possible *equilibrium* changes in climate that might occur as a consequence of an instantaneous increase in greenhouse gas concentrations, without regard to the timing of such changes. But what is more relevant for purposes of assessing impacts and response are the rates of change — that is, projections of the *time-dependent* (or transient) changes in climate.

In addition to the climate sensitivity discussed above, there are two other major factors that will influence the rate of climate change. The first is the future emissions of greenhouse gases and the consequent changes in concentrations and radiative forcing. (Past radiative forcing changes are important, too, as we shall discuss below). Unlike most GCM equilibrium experiments in which CO_2 is instantaneously doubled, in the real world greenhouse gas emissions and concentrations change year by year. The uncertainties in the future rates of change are very large, but at least this is one crucial factor that is within the capacity of society to alter. As mentioned above, the critical issue is fossil-fuel combustion. Fossil-fuel carbon emissions in "business-as-usual" (i.e. do nothing) scenarios typically increase by more than a factor of three by the end of the twenty-first century.

The second major factor is the rate of oceanic heat uptake. As the mixed layer of the ocean surface warms in response to a change in radiative forcing, heat penetrates slowly to the deeper layers of the ocean. This means that, for an instantaneous change in radiative forcing, the surface and lower atmosphere would not reach their potential temperature change until many decades later. In effect, this thermal inertia of the oceans introduces a delay in the response of the climate to greenhouse gas increases.

Thus, in order to make realistic projections of transient changes in climate, oceanic and atmospheric GCMs must be coupled (CGCMs) and must be run with gradual, year-by-year changes in greenhouse forcing in order to take into account the thermal inertia of the oceans. This poses problems for these complex models because of their very large computer requirements. Thus, relatively few GCM transient climate experiments have been conducted using fully-coupled ocean-atmosphere models (e.g. Washington and Meehl, 1989; Stouffer *et al.*, 1989; Cubasch *et al.*, 1992; Murphy, 1992).

Furthermore, the large computational requirements of GCMs create difficulties in making multiple runs to examine, for instance, the climate effects of various policy options for reducing greenhouse gas emissions. For this purpose, it is more practical to use simpler models that parameterise many of the complicated processes incorporated in GCMs. This tactic has been adopted by the IPCC in making time-dependent projections of temperature change (and sea-level change — see Chapter 3) at the global scale. Fortunately, for similar sets of assumptions, the results of simple models compare favourably with GCM results. Various projections of global temperature change using such models have been made (e.g. Jaeger, 1988; Bretherton *et al.*, 1990; Mitchell and Gregory, 1992; Wigley and Raper, 1992, 1993), all of which differ somewhat because of different assumptions and different forcing scenarios. For our purposes here, we shall focus on the 1990 IPCC projections, as they were the product of the last full international scientific assessment of the greenhouse effect.

BOX 2.6: The sensitivity of climate

From experiments using global climate models, it is estimated that a doubling of the atmospheric concentration of carbon dioxide (or the equivalent) should eventually cause an equilibrium warming of 1.5 to 4.5°C, with a best estimate of 2.5°C. This equilibrium warming, conventionally called the *climate sensitivity*, can be thought of as the potential warming for CO_2 doubling, without regard to the rate of change. The range of estimates for the value of the climate sensitivity derives from uncertainties in climate feedbacks — changes in water vapour, snow/ice extent, clouds — that act to enhance or lessen the initial radiative heating effects of greenhouse gases.

In 1990 the IPCC estimated that the global-mean temperature could increase at an average rate of about 0.3°C per decade with a "business-as-usual" (i.e. a "do nothing") scenario of greenhouse gas emissions, reaching 3.3°C by the year 2100 (within a range of uncertainty of 2.2°C to 4.9°C) (Bretherton *et al.*, 1990). The 1990 IPCC projections are shown in Figure 2.7 and Table 2.2. The high and low temperature change curves reflect the uncertainty in the value of the climate sensitivity (in essence, the climate feedbacks).

The next full IPCC Scientific Assessment will be published in 1996. Scientific developments since 1990 suggest that the projections of global warming could be revised downwards, for several reasons. First, emissions projections may be lowered. For example, the IPCC, in its supplemental report (Houghton *et al.*,1992) revised its emission scenarios, inferring a lower business-as-usual case (Leggett *et al.*, 1992). By itself, this implies a slightly lower rate of warming (as indicated by the thin solid curve in Figure 2.7, from Wigley and Raper, 1992). Second, there has been progress in quantifying the effects of three factors that reduce somewhat the projections of future radiative forcing changes, namely, the terrestrial biospheric sink for carbon dioxide, the cooling effect of sulphate aerosols, and the stratospheric-ozone-depletion feedback — all of which tend to partially off-set greenhouse warming (see above). One analysis (Wigley and Raper, 1992; also see Mitchell and Gregory, 1992) suggests that when these factors are taken into account as well, the best-estimate projection of global warming may be about one-quarter lower than that of IPCC 1990 (as shown by the dotted curve in Figure 2.7) mostly due to sulphate aerosol effects.

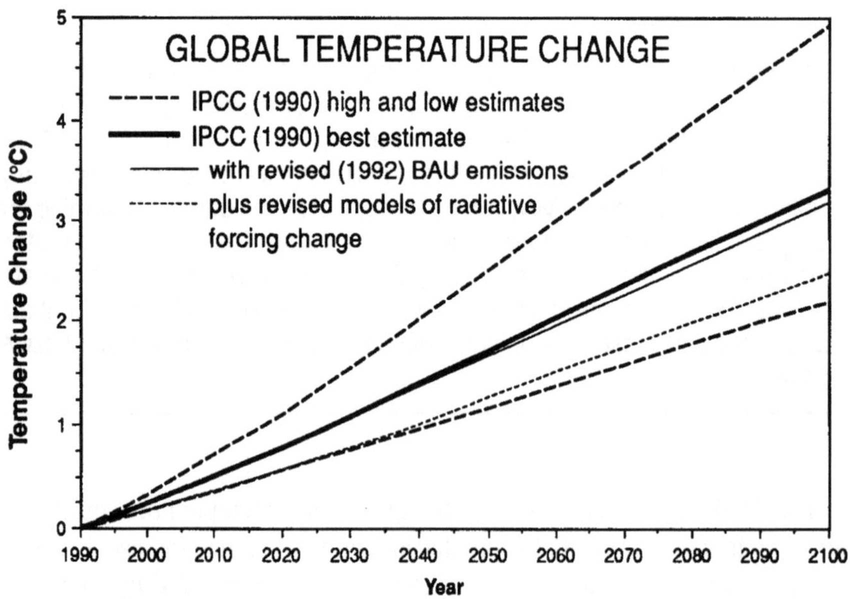

FIGURE 2.7: Projections of future global warming

Source: IPCC projections from Bretherton *et al.*, 1990; revised curves from Wigley and Raper, 1992.

TABLE 2.2: The IPCC (1990) projections of future global-mean temperature change for the business-as-usual emissions scenario.

Model Assumptions	Global warming scenario by year (°C) 2010	2030	2050	2100
Low	0.3	0.7	1.2	2.2
Best-estimate	0.5	1.1	1.7	3.3
High	0.7	1.5	2.5	4.9

Source: IPCC, 1990; Bretherton *et al.*, 1990.

Although the range of estimates of future global warming for business-as-usual (and other) scenarios is large and subject to revision, there are two general points that should not be overlooked. First, despite the large uncertainties, it is generally agreed that increasing concentrations of greenhouse gases should cause global warming. Second, even the implied lower projections would still represent rates of global warming that are higher than those experienced over the last 10,000 years of the Holocene. It is in this global context that the specific implications for climate change and its effects on Bangladesh must be considered.

BOX 2.7: Projections of future global warming

The current scientific consensus is that increasing atmospheric concentrations of greenhouse gases should cause the world to warm. The rate of warming, however, is very uncertain. In 1990, the Intergovernmental Panel on Climate Change (IPCC) estimated that with a "business-as-usual" scenario of greenhouse gas emissions, the world should be 3.3°C warmer by the end of the next century, with a range of uncertainty of 2.2°C to 4.9°C. Subsequent analyses by IPCC and others suggest somewhat lower rates of warming if factors such as sulphate aerosols are taken into account. Nevertheless, such rates of global temperature change are greater than those which have occurred naturally over the last 10,000 years. By the second half of the next century, the global temperature could well exceed that which has occurred over the last 150,000 years, at least.

2.6.3 Are the Projected Changes Large or Small?

One can obtain a sense of the importance of the projected changes in global temperature by referring back to Figure 2.1. If the world warms according to the best estimates for the year 2050, we would have to search back to the Eem interglacial, approximately 120,000 years ago, to find a period of comparable warmth. During the depths of the last glacial epoch, 18,000 years ago, the average global temperature was only about 4°C to 5°C cooler than today. Yet, a global temperature change of this magnitude is within the realm of possibility for greenhouse warming — but occurring over a century, not thousands of years. In short, in terms of climate history, the estimates of future greenhouse warming represent very significant changes in the planetary environment.

2.7 THE IMPLICATIONS FOR BANGLADESH

In order to assess the meaning of global climate change for Bangladesh, it is first necessary to understand the present climate, and then ask how that climate could be altered through greenhouse gas forcing.

2.7.1 Present Climate and its Variability

Bangladesh enjoys a tropical climate, although the Tropic of Cancer (23.5°N) passes through the middle of the country, dividing Bangladesh almost into two equal halves. The climate is characterised by high temperatures, heavy rainfall, often excessive humidity and fairly marked seasonal variations throughout the year. The country is well guarded by the Himalayas on the north, and is within the domain of a very active monsoon with a reversal of seasonal wind.

The region used to be known as the "land of six seasons" with regard to crop sowing and harvesting times, and cultural festivities. The notion lost some of its relevance with the initiation of various development programmes, particularly in the water resource sector during the early 1960s. Now, the seasonal cycle is often divided into three main periods, namely, the summer (March to May), the monsoon (June to October), and the winter (November to February).[1] Other important aspects of climate in Bangladesh include weather-related extreme events, e.g. tropical cyclones, tornadoes, nor'westers and droughts.

The weather records from the Department of Meteorology, Government of Bangladesh, provide the basis for examining the climatology of Bangladesh. Although some long-term records of temperature and precipitation exist for the larger region, modern meteorological measurements in Bangladesh itself began with four stations in 1948. Another 16 stations were added in 1958. A total of 31 stations operated by the Department are currently functioning, of which at least eight were installed after 1971. Weather records for 15 stations for 30 years (1958-87) are chosen to represent the climatology on the basis of their location, period of record and extent of missing data. The seasonal and annual temperature and rainfall means and coefficients of variation (CVs) are shown in Table 2.3.

The calculated mean winter temperature for 30 years is 21.0°C with a 1.7 per cent coefficient of variation. A steady fall in maximum and minimum temperatures from November marks the beginning of winter season, and by the end of December the north-western corner of the country records a mean minimum of 9°C. The same trend is also visible in the Sylhet hills at the north-east corner. At that time, Cox's Bazar experiences a mean minimum of 15.5°C. January is the coldest month in Bangladesh with a minimum of temperatures recorded in northern Dinajpur and southern Sylhet. The isotherms of mean January temperatures extend east-west, running almost parallel. The 16°C line passes just north of Dinajpur town, and the 19.5°C line below Barisal and Maijdi Court and then turns to south along the Chittagong coast.

TABLE 2.3: Bangladesh (area-averaged) mean annual and seasonal temperatures and rainfall, 1958-1987

	Winter		Summer		Monsoon		Annual	
	Mean (°C or mm)	CV* (%)	Mean (°C or mm)	CV (%)	Mean (°C or mm)	CV (%)	Mean (°C or mm)	CV (%)
Temperature	21.0	1.7	27.5	1.9	28.2	0.8	25.7	0.9
Rainfall	64	53.4	410	33.2	1795	11.3	2202	13.1

*CV = Coefficient of Variation = 100 (standard deviation /mean)

Source: Data from Department of Meteorology, Government of Bangladesh.

In the summer, the yearly maximum temperatures are reached in the month of April, when the western part of the country experiences temperatures that are sometimes over 40°C. The mean maximum temperature ranges from 30.4°C in Cox's Bazar to 36°C in Rajshahi. A hot dry westerly wind springs up almost every day and blows hard for at least five hours in the western part of the country. Infrequent thunderstorms bring some relief. In the eastern part, there are more frequent thunderstorms (nor'westers) and the intense heat of the west is rare. The 30-year mean summer temperature is 27.5°C.

The south-west monsoon normally reaches the north-west Bay of Bengal (20°N) and the Myanmar coast by June 1. It enters Bangladesh by the end of first week of June at the earliest and withdraws by the start of the fourth week in October at the latest (Table 2.4). On the average, the monsoon in Bangladesh lasts for 122 days within a range of 112 to 139 days (Table 2.4). The monsoon begins with a marked fall in temperature, because of the rainfall. Throughout the season the mean maximum temperature over most of the country remains around 31°C and the mean minimum about 6°C less.

TABLE 2.4: Onset and withdrawal of south-west monsoon

Year	ONSET	WITHDRAWAL	No.of Days
1979	14 June	06 Oct.	115
1980	06June	22 Oct.	139
1981	07June	17 Oct	133
1982	17 June	06 Oct.	120
1983	09 June	17 Oct.	123
1984	06 June	25 Sept	112
1985	07 June	03 Oct.	119
1986	19 June	10 Oct.	114
1987	10 June	04 Oct.	117
1988	12 June	08 Oct.	119
1989	13 June	14 Oct.	124
1990	14 June	12 Oct.	121
1991	10 June	12 Oct.	125
			Average = 121.6

Source: Data obtained from Department of Meteorology, Government of Bangladesh, Dhaka.

During the monsoon season, the mean temperature for the 30-year period is around 28°C and the variability is 0.8 per cent (Table 2.3). The mean temperature is relatively higher during the monsoon than during summer. This is because the range between maximum and minimum temperatures is very high during summer; the minimum temperature during summer is relatively very low.

As shown in Table 2.3, Bangladesh receives heavy rainfall. Mean annual rainfall increases rapidly towards the north-east to an average of about 4060mm at Sylhet. Rainfall also increases towards the south-east to a mean of about 3600mm near Cox's Bazar. Lower rainfall occurs in the west with an average annual of 1415mm at Rajshahi. The distribution of the average annual rainfall makes the isohyets swing round in a curve from the south toward the east and then towards the north-east. The mean annual rainfall for 30 years is over 2200mm.

About 82 per cent of annual rainfall occurs in the five months of the monsoon period, June through October. The 30-year mean monsoon rainfall amounts to about 1800mm, with variability of around 11 per cent, which is less than the annual. This fact indicates that monsoon rainfall is relatively stable with variability ranging from around 8 per cent to 13 per cent between the 10-year time periods. Winter is the driest season in Bangladesh. The 30-year mean-winter rainfall amounts to about 64mm with a variability of around 53 per cent.

What influences the monsoon rainfall amounts? Elsewhere in the Asian sub-continent (particularly India) and southeast Asia, monsoon rainfall has been shown to be weakly related to the El Nino/Southern Oscillation (ENSO) phenomenon (Mooley and Pathasarathy, 1984; Ropelewski and Halpert, 1987,1989). Historically, the El Nino events refer to years of anomalously warm sea surface temperatures in the eastern Pacific Ocean off the coast of Peru. Such events are associated with a weakening of the high pressure system over the southeastern Pacific and the winds that maintain the upwelling of nutrient-rich, colder waters off the South American coast. These events are linked to the Southern Oscillation, which refers to the out-of-phase relationship between the atmospheric pressure over the south-eastern Pacific and over the Indonesian region; the two pressure systems are highly negatively correlated. ENSO events, in turn, show a negative correlation with Indian monsoon rainfall — that is to say, Indian monsoon droughts are more likely to occur during strong ENSO events. This suggests that there may also be some connection between Bangladesh's monsoon rainfall characteristics and ENSO. However, the specific relationships between ENSO and Bangladesh's climate have not yet been thoroughly investigated, so the question remains open.

Of the weather-related extreme events, tropical cyclones occur frequently in Bangladesh and cause severe human casualties, widespread damage to property and constant set-backs to economic development. During the period 1976-1988, the frequency of cyclones (wind speed greater than 35 knots) was 4.0 per month, as compared to 2.7 per month during 1891-1960. Bangladesh is also subjected to devastating tornadoes and nor' westers every year during summer, but little is known about their frequency of occurrence. Droughts are usually considered climatically-controlled extreme events in Bangladesh. During 1950-1979, severe drought conditions affecting at least 20 per cent of the country's area prevailed in seven years; very severe droughts occurred in 1957 (47 per cent), 1972 (42 per cent) and 1979 (42 per cent) (GOB, 1989).

2.7.2 Has Climate Changed in the Bangladesh Region?

For some climate variables, like temperature, changes tend to be rather uniform over large areas. In such cases, climate data from surrounding areas can be used to extend climate records and to develop region-wide trends. Other variables, like rainfall, exhibit large variations from one place to another. For these, area-wide averages (using appropriate weighting schemes) can help to "smooth out" the local variations in order to identify regionally-coherent trends. For Bangladesh, this has been done using region-wide data from the global climate data set maintained by the Climatic Research Unit at the University of East Anglia, England.

Mean-annual regional temperatures, expressed as departures from the reference period 1951-1980, are shown in Figure 2.8. It is evident that, on this timescale, the Bangladesh region has been getting warmer. From the latter part of the last century, there has been, on average, an overall warming of about 0.5°C, comparable in magnitude to the observed global warming. Interestingly, the shape of curve also follows closely that of the Northern Hemisphere land temperature curve: a warming trend during 1910-1940, a slight cooling trend until the mid-1970s, and resumed warming thereafter (Folland *et al.*, 1990, 1992). For the period 1979-1991, 12 out of the 13 years were warmer than the reference period. As with the global temperature change, however, it is not yet possible to attribute this warming in Bangladesh to the enhanced greenhouse effect.

The annual rainfall for the Bangladesh region for the period 1870-1991 is shown in Figure 2.9. Unlike the temperature curve, there is no discernible long-term trend in mean-annual rainfall. What is striking, however, is the natural year-to-year variation in rainfall, which is quite large. There has also been an apparent increase in rainfall variability during recent decades (which has been noted by others, e.g. Meehl and Washington, 1993).

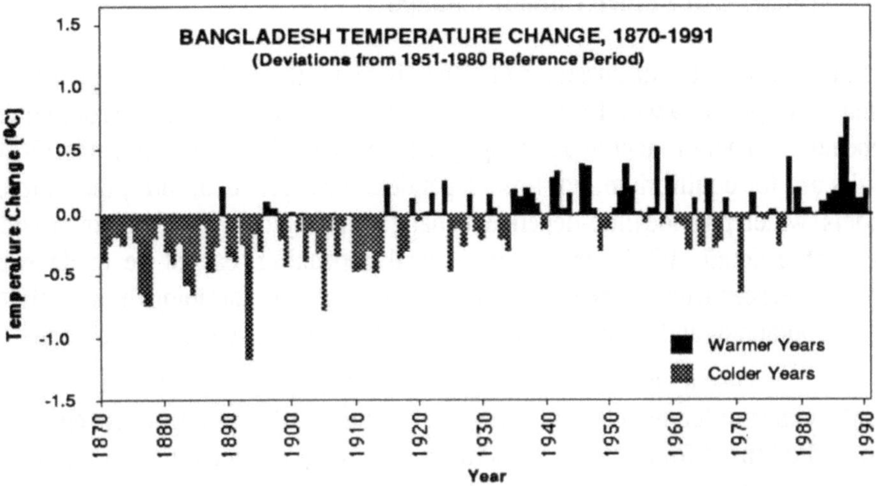

FIGURE 2.8: Bangladesh mean-annual temperature change, 1870-1991. The values are departures from the 1951-1980 reference period.

Source: Data provided by the Climatic Research Unit, University of East Anglia.

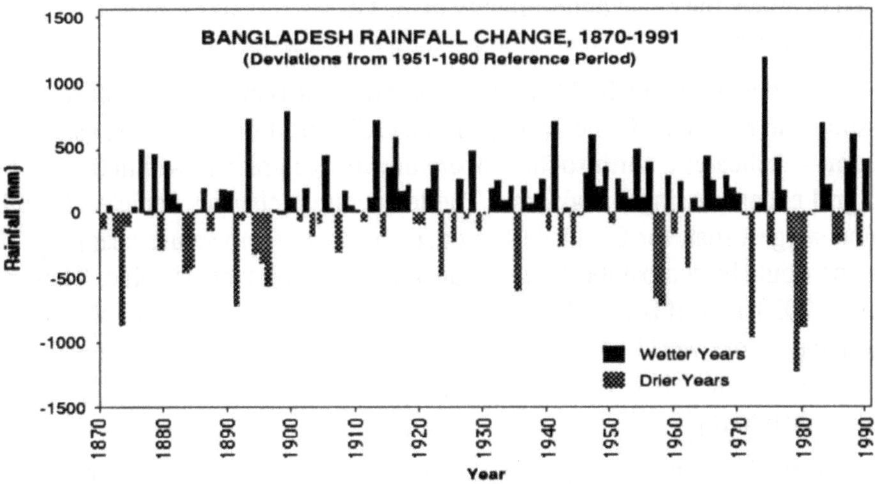

FIGURE 2.9: Bangladesh regional annual rainfall, 1870-1990. The values are departures from the 1951-1980 reference period.

Source: Data provided by the Climatic Research Unit, University of East Anglia.

2.7.3 Scenarios of Future Climate Change

It is evident from the discussion so far that there are three obstacles to predicting climate changes for a specific region like Bangladesh: first, there is low confidence in the regional details of climate change predicted from GCMs; second, most GCM results are for equilibrium, not time-dependent, changes; and, third, the simpler models (which provide time-dependent changes at the global scale) do not provide geographic detail. While predictions, strictly speaking, cannot be made with confidence, scenarios can be constructed for purposes of examining the sensitivity of environmental and socio-economic systems to possible changes in climate.

Here, we use a method of scenario construction that combines the results of GCMs with simpler models to give regional scenarios of time-dependent changes. The method (used for Europe and elsewhere) involves the derivation of "composite" patterns of temperature and precipitation change for the Bangladesh region by averaging together the results of seven major GCM experiments. These are standardised to give the pattern of change per degree of global warming. The patterns are then scaled by the time-dependent projections of global warming given by simpler models (here, we use one projection only, the "business-as-usual" middle projection of IPCC 1990 — refer to Figure 2.7). Because there is a sample of seven GCMs, not only can the average be used (which might be considered the "best-estimate" pattern), but the "high" and "low" patterns can also be derived and scaled to reflect the range of uncertainty in the GCMs. Let us examine the results for Bangladesh.

For temperature, all GCMs indicate warming for Bangladesh in all seasons. As shown in Figure 2.10, the average of the GCM results — the "best-estimate" pattern — indicates a north-south temperature change gradient within the 20° by 20° grid encompassing Bangladesh. The temperature changes depicted for this region suggest that, for Bangladesh proper, the mean temperature change in all seasons would be comparable to that of the global annual-average value. Thus, by the year 2030 most of Bangladesh would be approximately 1.0°C to 1.5°C warmer than today if the world continues on a "business-as-usual" path of greenhouse emissions (Figure 2.10). By 2050, Bangladesh is projected to be 1.5°C to 2.0°C warmer than today.

It should be borne in mind that the individual GCMs differ in the projected degree of warming. The uncertainty in the GCM results is of the order of ±50 per cent. Thus, by the year 2030, the range of uncertainty for the "business-as-usual" warming scenario is about 0.5°C to 2.0°C.

With respect to rainfall, the GCMs differ widely in both the direction and magnitude of predicted changes and for this reason must be viewed as very unreliable. For instance, although the best-estimate GCM pattern shows slightly wetter winter conditions by the year 2030 (Figure 2.11a), the range, based on inter-GCM differences, varies from about 5 per cent drier to more than 19 per cent wetter.

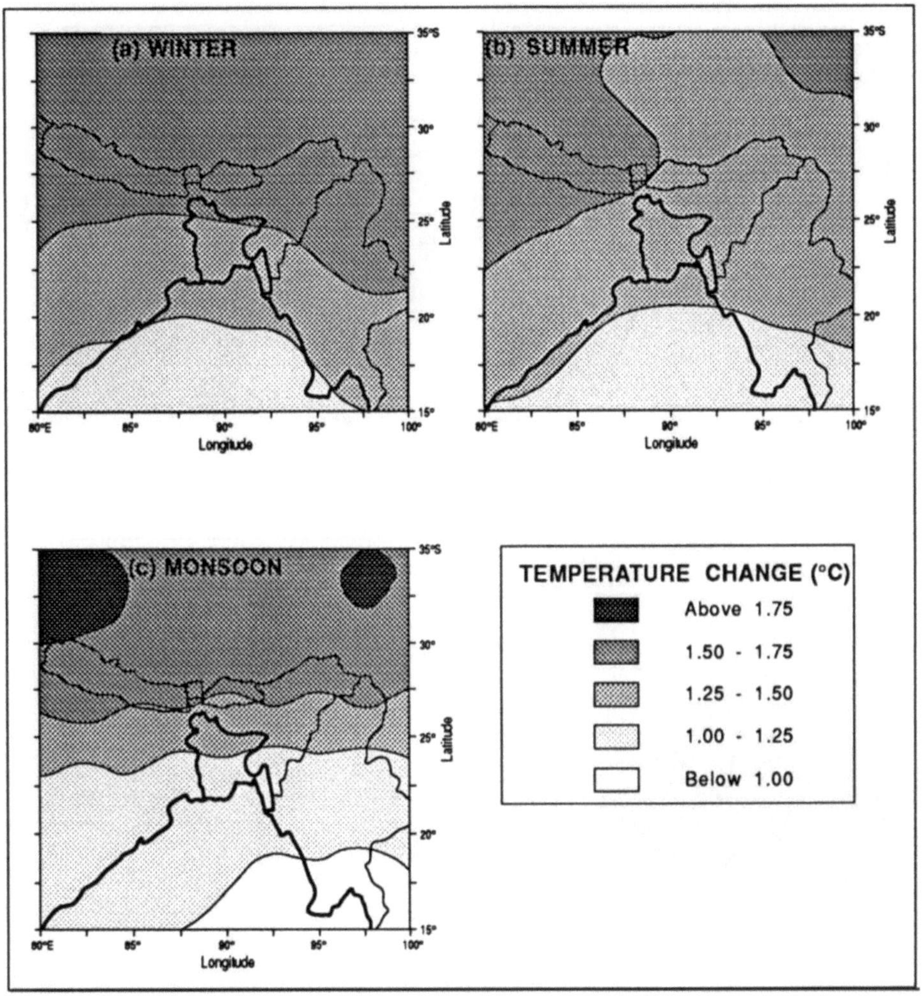

FIGURE 2.10a-c: The projected best-estimate 1990-2030 changes in average seasonal temperature (°C) in the Bangladesh region: (a) Winter (November-February); (b) Summer (March-May); and (c) Monsoon (June-October).

The rainfall changes for winter, however, may not have much practical significance. Winter rainfall is so low and so variable (in relative terms) that farmers do not rely on rainfall. Rather, crops depend on residual soil moisture or irrigation; any rainfall that occurs is a bonus, provided that it is not too heavy. For farmers, the

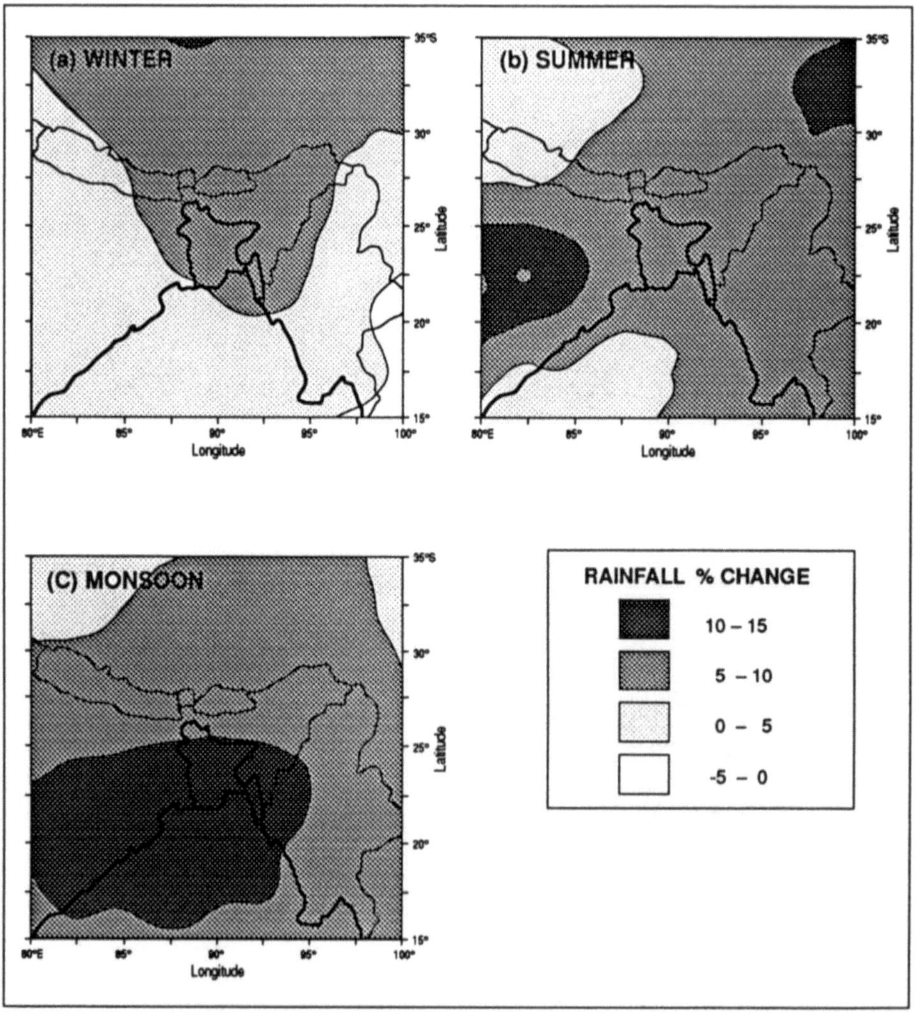

FIGURE 2.11a-c: The projected best-estimate 1990-2030 changes in seasonal total rainfall (in %) in the Bangladesh region: (a) Winter (November-February); (b) Summer (March-May); and (c) Monsoon (June-October).

change in summer pre-monsoon rainfall is more critical. The best estimate from the GCMs is that summer rainfall will increase 5 to 10 per cent by 2030 (Figure 2.11b), but, again, this is subject to very large uncertainties.

For Bangladesh, one potentially important aspect of climate change is how the monsoon season will be affected. There are some physical grounds for believing that monsoon rainfall affecting Bangladesh and the surrounding region will increase — namely, that the temperature gradient between the sea and the Asian continent (the driving force behind the monsoon) will steepen with enhanced mid-continental warming. In general, GCMs agree that the monsoon season should be wetter in a warmer world (Gates *et al.*, 1992). A few GCMs show very sharp increases, while others project no change at all. As shown in Figure 2.11c, the best-estimate scenario is a 10 to 15 per cent increase in average rainfall during the monsoon period by the year 2030 (with a range of 0 to 20 per cent based on inter-GCM comparisons). By the year 2050, rainfall during the monsoonal period is projected to increase by about 15 to 20 per cent (with a range of 0 to 35 per cent). Such increases in the strength of the monsoon could well be accompanied by a lengthening of the monsoon season. Again, the model uncertainties at this regional scale are very large.

BOX 2.8: **Bangladesh regional climate change - past and future**

Over the past 100 years, the broad region encompassing Bangladesh has warmed by about 0.5°C. The warming trend is consistent with that of the Northern Hemisphere as a whole. There has been no discernible trend in average rainfall, although rainfall variability appears to have increased in recent decades.

In the future, Bangladesh may get warmer and wetter. For the IPCC (1990) "business-as-usual" emissions scenario, Bangladesh is projected to be 0.5°C to 2.0°C warmer than today by the year 2030, based on a range of GCM results. Rainfall is more difficult to predict. However, climate models generally agree that regional monsoon rainfall should increase in a warmer world. The best estimate is a 10 to 15 per cent increase in average monsoon rainfall by the year 2030, although the uncertainties are very large.

2.7.4 Possible Changes in Climate Variability and Extremes

An important issue is whether the *variability* of climate, as well as the mean, will change as a consequence of global warming. With the advent of global coupled ocean-atmosphere models, GCMs are improving in their ability to simulate aspects of climate variability on a range of time-scales. For example, one climate model indicates an increase in variability of monsoon rainfall with doubled atmospheric CO_2 concentration (Meehl and Washington, 1993). However, more research is required to confirm such findings. There is not, as yet, any firm, reliable evidence to suggest future changes in the variability of climate due to greenhouse gas forcing, particularly at the Bangladesh regional scale. In the words of the recent (1992) IPCC assessment,

> Model experiments with doubled CO_2 give no clear indication of a systematic change in the variability of temperature on daily to interannual time-scales, while the changes of variability for other climate features appear to be regionally (and possibly model) dependent (Gates *et al.*, 1992).

A related issue concerns the possible change in the frequency of *climatic extremes*. Such changes can occur even if the variability (or shape of the frequency distribution) does not itself change. This can be illustrated with the frequency distribution of monsoon (June-October) rainfall totals, as shown in Figure 2.12a. This distribution, approximately normal in shape, is based on values for the period 1870-1990, representing an area-wide average for a 5° latitude by 5° longitude "grid-box" centred on Bangladesh.[2] The distribution indicates that the mean monsoon rainfall is 2367 mm, with a standard deviation of 327 mm. From the set of values, extreme seasons can be identified with respect to the distribution and the likelihood of their occurrence simply estimated. A few recent years are noted on the curve for the sake of illustration.

For example, the 1974 season was the wettest on record. As indicated in the top panel in Figure 2.12a, a season like 1974 has a probability of occurring (or being exceeded) of 0.0017 in any given year. Put in other terms, such a wet season occurs (or is exceeded) about once every 600 years, on average. At the other "tail" of the distribution, a very dry season like that of 1979 occurs only once every 400 years, on average. How might the likelihood of such events change with global warming?

The lower panel in Figure 2.12b attempts to illustrate how the probabilities of such climate extremes might shift with a projection of a 12 per cent increase in average monsoon rainfall by the year 2030. Let us simply assume that the mean rainfall increases (from 2367mm to 2651mm) and that the variability about the

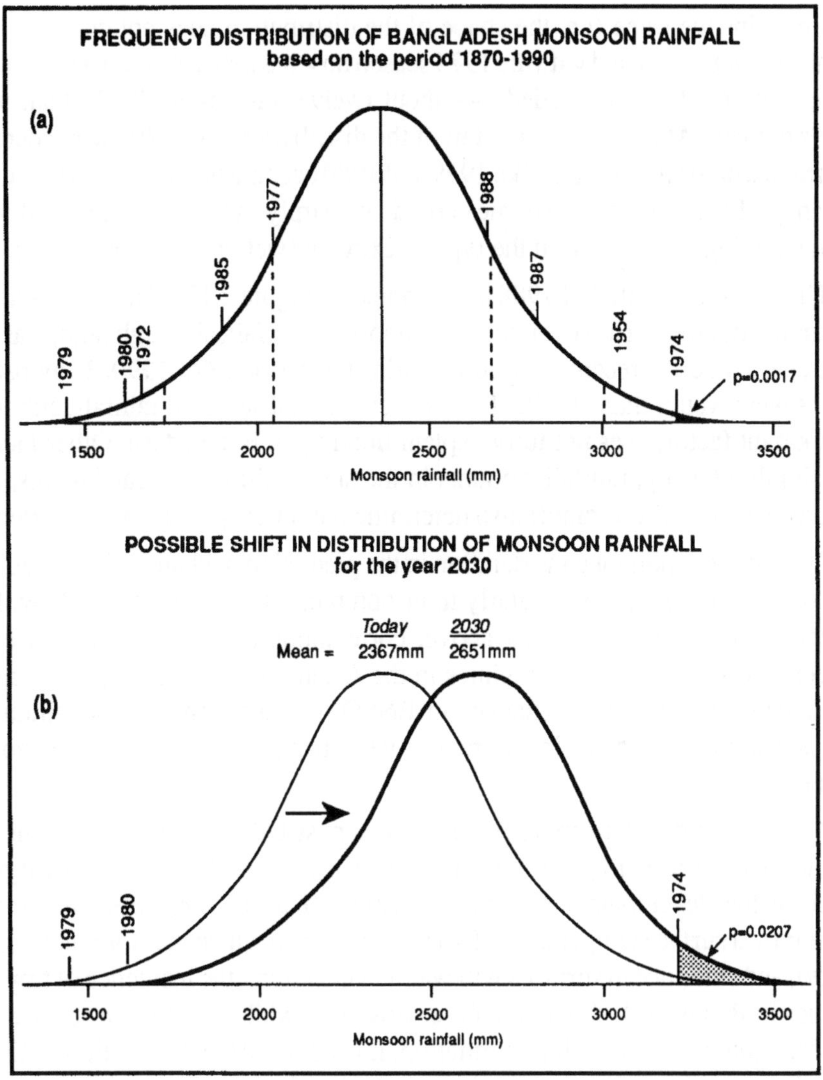

FIGURE 2.12a-b: Frequency distribution of Bangladesh monsoon rainfall(June-Oct). (a) Present distribution based on the period 1870-1990, with selected extreme wet and dry seasons noted. The shaded area denotes the probability of the extremely wet monsoon of 1974 being equalled (or exceeded) in any given year. (b) The shift in distribution for a 12 per cent increase in mean rainfall projected for the year 2030, assuming no change in variability. The stippled area under the curve indicates the increase in probability of occurence of an extremely wet monsoon like 1974.

Source: Based on data provided by the Climatic Research Unit, University of East Anglia.

mean remains the same (i.e. the shape of the distribution does not change). The rainfall amount associated with the 1974 season now has a sharply increased chance of occurring (or being exceeded) — about twelve times more likely than with today's climate. At the other "dry" tail of the distribution, by 2030 the chances of a recurrence of the dry seasons like 1979 and 1980 become negligible. In short, the frequency of such climatic extremes could be sharply increased or reduced with global warming, depending on the type of extreme (wet or dry respectively).

The monsoon rainfall of 1988 is also noted on Figure 2.12a. This year is noted for extreme flooding. However, the total monsoon rainfall, although high, was not as extreme as in other recent monsoons (during the period 1870-1990, 22 monsoon seasons were wetter than 1988). This is because total seasonal rainfall, although an important factor, does not fully explain flooding in Bangladesh. Other factors like rainfall intensity, rainfall amounts in the surrounding river catchments, and Himalayan snowfall and runoff also determine the severity of Bangladesh floods.

A possibly important consideration in the prediction of changes in variability and extremes in rainfall, particularly monsoon rainfall, is how the ENSO will be affected in the future. As noted above, correlations have been found to exist between ENSO and monsoon regimes in the Asian-Pacific region. Until climate models can reliably predict changes in ENSO, the question of how rainfall in Bangladesh and elsewhere in the region will change will not be satisfactorily answered.

Another topic of concern to Bangladesh is possible changes in the frequency and intensity of *severe tropical storms (tropical cyclones)*. It is sometimes argued that since the development and intensification of tropical cyclones requires a minimum sea surface temperature of about 27°C (from which the storm derives its energy), and since ocean surfaces will warm in the future, the incidence of tropical cyclones in Bangladesh should increase. However, while a warm sea surface is a necessary condition for cyclone formation, it is not a sufficient condition. Other atmospheric conditions, such as weak vertical wind shear over the storm centre and strong low-level cyclonic vorticity with high mid-tropospheric humidity, appear instrumental in the intensification of a tropical disturbance into a severe storm (Raper, 1993). Unfortunately, the horizontal resolution of GCMs used for climate studies is generally too coarse to model such factors explicitly (Mitchell *et al.*, 1990). In short, little can be said specifically regarding future changes in cyclone frequency and intensity in the Bay of Bengal (Gates *et al.*, 1992; Raper, 1993; Lighthill *et al.*, 1994. Also see Chapter 3).

2.8 WHAT ARE THE IMPLICATIONS FOR POLICY?

It is generally acknowledged that increasing greenhouse gas concentrations will cause changes in the global climate, although the rate and regional details of such changes are not clearly understood. There are large uncertainties and gaps in scientific understanding. But policy-action cannot afford to wait until all the uncertainties and gaps in knowledge are resolved. For Bangladesh, this sense of urgency is heightened since Bangladesh may be especially vulnerable to the consequences of the enhanced greenhouse effect to which it has contributed practically nothing. What are the choices?

2.8.1 General Response Strategies

Figure 2.13 presents a generalised picture for the greenhouse effect showing various components, feedbacks and policy options available to combat the greenhouse effect. There appear to be three possible categories of response: passive adaptation, active adaptation, and reduction of emissions.

Passive Adaptation

Passive adaptation implies a strategy of non-anticipatory response — that is, letting society adapt to changes as they occur, while research continues to narrow the scientific uncertainties. There are drawbacks associated with this passive approach. First, if the rate of global warming is as fast as suggested in our scenarios above, both the rate and the degree of climate change and other consequences will far exceed recent human experience. Second, because of the time lag in climatic and other responses, the longer society avoids actions to slow global warming, the larger are the environmental changes to which we are eventually committed. Finally, while some countries would be able to enjoy the luxury of passive adaptation, Bangladesh and other developing nations are not so fortunate. It is true that, in the past, passive adaptation has allowed Bangladesh to survive many catastrophic events. But, the projected impacts of climatic changes on the environment could be long-lasting and outside the range of human experience, so that historic wisdom cannot necessarily be relied upon to provide the means of adaptation.

Active Adaptation

Active adaptation implies a strategy of response in advance of any changes. Actions could be directed towards either reducing vulnerability or distributing the impacts.

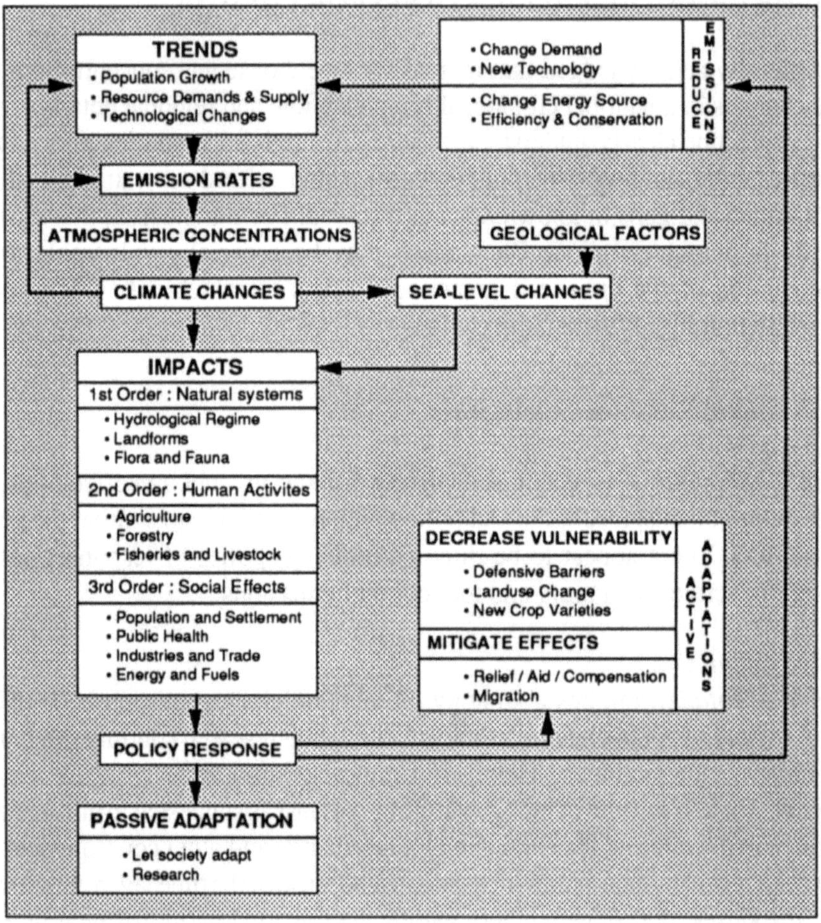

FIGURE 2.13: Flowchart for the greenhouse effect showing various components, feedbacks and the policy options to combat the greenhouse effect

Source: Modified from Warrick *et al.*, 1990.

One method for *spreading or distributing the impacts* is to compensate those who suffer ill-fortune, as through relief, rehabilitation and government compensation schemes at the national or international levels. Such an approach is widely adopted as a means of coping with natural hazards. For Bangladesh, however, it would be difficult to sustain a relief programme in the face of persistent, widespread disruption from climate change. Bangladesh is already hard-pressed in maintaining its present level of production and consumption with an existing two billion US dollar external assistance programme annually. It would be difficult for Bangladesh to divert resources away from existing investment and consumption programmes to develop new, additional measures for relief and rehabilitation that have value only in the event of future climate change. Rather, it would be advisable in the short term to strengthen the relief programmes already in place as we move into an uncertain world of changing climate.

An active adaptation strategy also includes measures to *reduce or limit vulnerability*. Climate "impacts" result from populations, settlements and economic activities being in inappropriate places at inappropriate times as the climate changes. Anticipating these changes and adjusting accordingly can prevent or reduce the potential adverse effects (see Chapter 5). For example, from international experience, impacts associated with current sea-level rise have been minimised by encouraging alternative uses of land in coastal zones subject to salinization or erosion, or by the construction of elaborate systems of barriers to prevent inundation by the sea (CZMS, 1990). Agricultural impacts could be reduced by developing crop varieties that are heat-tolerant or salt-tolerant and take advantage of a high CO_2 concentration (Tegart *et al.*, 1990. Also see Chapter 4).

BOX 2.9: Adapting to climate change

Adaptation strategies are required to decrease vulnerability to climate change. Traditionally, over many centuries, Bangladesh has learned to cope well with climate variations. Such traditional knowledge must be preserved for the future as Bangladesh undergoes development. Additional measures to reduce vulnerability to existing climatic extremes — floods, droughts, tropical cyclones — would provide immediate benefits while enhancing Bangladesh's future resiliency to such events should their frequency or severity increase as a result of global warming.

For Bangladesh, however, measures for reducing vulnerability that require large technological inputs may not be feasible in the context of climate change. Such measures often require long lead times and large expenditure. For example, the technical feasibility of constructing coastal defences is questionable because of the geology, geography and hydro-meteorological conditions of Bangladesh. The costs of coastal engineering structures could easily run into many billions of US dollars, which far exceeds the capability of Bangladesh to manage by itself. Importantly, such measures also demand information about how regional climates and sea level will change, which is presently highly uncertain.

On the other hand, investment in strategies to reduce Bangladesh's vulnerability to *existing* climatic extremes (floods, droughts, cyclones) may well serve the double function of alleviating suffering today while building the capacity to buffer society from a possible increase in the frequency or severity of such events in the future. For instance, such measures might include the enhancement of flood warning systems or increasing the number of reinforced storm shelters.

An important element in reducing vulnerability may be in maintaining the large array of measures that have, traditionally, allowed local populations to cope with climate variations over many centuries (e.g. building houses on raised mounds; changing seed varieties, planting dates or crop mix). Steps should be taken to preserve such knowledge in the process of social and economic development.

Reduce Emissions

The third strategy involves slowing the rate of climate change by *reducing greenhouse gas emissions*. There is wide-based political support for measures to reduce emissions, as evidenced by the recent United Nations Conference on Environment and Development and the Framework Convention on Climate Change (see below). The strategy requires tackling the problem at its primary source — putting an acceptable brake on the burning of fossil fuels and its energy-related activities.

The principal obstacle that appears to stand in the way of reducing CO_2 emissions is the slow advancement of agendas for action by developing and developed nations. The great bulk of past and current emissions of greenhouse gases have come from the highly industrialised nations; therefore, the first and largest response to the global warming threat should presumably come from these same nations. Commendably, strong actions have been taken to curtail the production of CFCs. Similarly, actions are now required by these nations to first stabilise, and then reduce, CO_2 emissions, as through increasing energy efficiency and reducing wasteful fossil fuel use.

Although the industrialised countries currently have the immediate responsibility for reducing fossil-fuel CO_2 emissions, the long-term goal of stabilising atmospheric concentrations of greenhouse gases will require global cooperation. In the developing countries (containing 75 per cent of the world's population), the CO_2 emissions are still relatively small (less than 20 per cent of the global total).

However, in comparison to the industrialised countries, the current *growth rate* of CO_2 emissions in developing countries is large, about 4 to 5 per cent per year on average. This means that, even if the highly industrialised nations were to reduce emissions substantially, the gains would be offset in a matter of decades by burgeoning energy use in the developing world. In short, while the industrialised countries are undeniably responsible for getting the world into the present mess, the goal of effectively slowing future global warming is not attainable without participation by the developing countries — as unfair as this might seem (Warrick, 1993). Thus, the problem of who should accept responsibility for reducing greenhouse gas emissions is not easily resolved. This problem lies at the heart of most substantive issues raised at the international negotiating table, as discussed below.

2.8.2 Framework Convention on Climate Change

In late 1990, the Ministerial Declaration of the Second World Climate Conference called for the creation of a *Framework Convention on Climate Change* (FCCC) and recommended that consideration be given to the need for funding facilities, like the then-proposed *Global Environment Facility* (GEF), to assist developing countries in implementing the Convention (Jaeger and Ferguson, 1991). The Convention would outline a set of principles and obligations to help subsequent negotiations to produce specific targets and quantitative reductions over an agreed time-frame, which could be added as 'protocols' at a later stage. An Intergovernmental Negotiating Committee (INC) was constituted and given the task of formulating the FCCC for signing at the United Nations Conference on Environment and Development (UNCED) in Rio de Janeiro, Brazil in June 1992. Bangladesh was one of the 155 signatory countries to the FCCC, which came into force on 21 March 1994 after ratification by 50 countries (it was ratified by Bangladesh on 15 April 1994).

Implications for Bangladesh and Other Developing Countries

The ultimate objective of the FCCC "is to achieve...stabilization of greenhouse gas concentrations in the atmosphere at a level that would prevent dangerous anthropogenic interference with the climate system". The final version of the FCCC admits that "... the largest share of historical and current global emissions of greenhouse gases has originated in developed countries..." but stops short of specifying which countries should assume responsibility for emissions reductions. This is less specific than an earlier draft (28 February 1992) which recognised that "Greenhouse gas emissions have come primarily from developed countries, and those countries have the main responsibility for combating climate change."

Country responsibilities for emissions reductions remains a key issue in international negotiations.

For developing countries, one major dilemma relating to the FCCC is perceived to be its incompatibility with their goals of economic development. In many developing countries, economic development through widespread industrialisation has emerged as a high priority aim, to the detriment of other sectors like agriculture. In Bangladesh, for example, the contribution of the agricultural sector to the Gross Domestic Product (GDP) has dwindled from 59 per cent in 1973-1974 to 37 per cent in 1991-1992 (GOB, 1973, 1990). Like Bangladesh, many developing countries have turned their focus from agriculture to industrialisation for future economic development. And industrialisation requires energy, which, in most instances, necessitates the use of fossil fuels. Thus, reductions in the growth of carbon dioxide emissions from fossil fuel use could well be perceived as a constraint to development.

The FCCC recognises this possible dilemma. It admits that in comparison to the industrialised countries, the per capita emission rates in developing countries are relatively low, and agrees that "the share of global emissions originating in developing countries will grow to meet their social and development needs." Furthermore, the FCCC affirms "that responses to climate change should be coordinated with social and economic development in an integrated manner with a view to avoiding adverse impacts on the latter, taking into full account the legitimate priority needs of developing countries for the achievement of sustained economic growth and the eradication of poverty." However, there is no indication regarding the time-span over which high rates of greenhouse gas emissions should be allowed to continue and/or the maximum levels of emissions from developing countries that could be sustained. Again, such issues are left for subsequent negotiation. Undoubtedly, the most difficult task for developing countries in future negotiations will be to strike an acceptable balance between their contribution to the international goal of stabilising atmospheric concentrations of greenhouse gases and their pressing needs for socio-economic development.

Commitments and Issues of Implementation

Article 4 of the FCCC specifies the commitments of all Parties to the Convention "...taking into account their common but differentiated responsibilities and their specific national and regional development priorities, objectives and circumstances... ." These are presented, in paraphased form, in Table 2.5. The first two, relating to greenhouse gas inventories and programmes for mitigation and adaptation measures, present clear obligations to all parties, while the remainder lack theprecision to make them enforceable (see Chapter 6).

TABLE 2.5: Commitments by all parties under the Framework Convention on climate change (paraphrased)

Develop national inventories of the sources and sinks of greenhouse gases.
Formulate programmes containing measures to mitigate climate change and to facilitate adaptation.
Promote and cooperate in the development of practices and processes that control, reduce or prevent greenhouse gas emissions.
Promote sustainable management, conservation and enhancement, as appropriate, of sinks and reservoirs of greenhouse gases.
Prepare for adaptation to the impacts of climate change; develop integrated plans for coastal zone management, water resources and agriculture.
Take climate change into account in social, economic and environmental policies and actions.
Promote and cooperate in scientific and other research related to the climate system.
Promote and cooperate in education, training and public awareness related to climate change.
Promote and cooperate in the exchange of relevant information related to the climate system and climate change and to the socio-economic consequences of response strategies.
Communicate to the Conference of the Parties information related to implementation.

Source: U.N. Framework Convention on Climate Change, Article 4.

The Convention states clearly that it is the obligation of the developed countries to provide the necessary financial and technical support to developing countries for implementation (Art. 4):

3. The developed country Parties and other developed Parties...shall

provide new and additional financial resources to meet the agreed full costs incurred by developing country Parties in complying with their obligations under Article 12, Para 1 [developing national inventories of anthropogenic emissions by sources and removals by sinks of all greenhouse gases not controlled by the Montreal Protocol]. They shall provide such financial resources, including the transfer of technology, needed by developing country Parties to meet agreed full incremental costs of implementing measures that are covered by paragraph 1 of this Article [see Table 2.6] and that are agreed between a developing country Party and the international entity or entities referred to in Article 11 [Financial Mechanism].

4. The developed country Parties and other developed Parties included in annex II shall also assist the developing country Parties that are particularly vulnerable to the adverse effects of climate change in meeting costs of adaptation to those adverse effects.

With respect to the "particularly vulnerable" countries, the FCCC identifies the following categories in relation to implementing actions concerned with funding, insurance and transfer of technology (Art. 4 [8]): (a) small island countries; (b) countries with low-lying coastal areas; (c) countries with arid and semi-arid areas, forested areas and areas liable to forest decay; (d) countries with areas prone to natural disasters; (e) countries with areas liable to drought and desertification; (f) countries with areas of high urban atmospheric pollution; (g) countries with areas with fragile ecosystems, including mountains ecosystems; (h) countries whose economies are highly dependent on income generated from the production, processing and export, and/or on consumption of fossil fuels and associated energy-intensive products; and (i) land-locked and transit countries. Bangladesh falls into both categories (b) and (d) and is presumably given priority status, although developing countries have not yet been ranked according to their levels of vulnerability, economic capability and technological base.

One of the main obstacles to implementing the FCCC, particularly for developing countries, concerns the provision of financial and technical support to meet their commitments. The criteria for the provision of the "new and additional funding" have not been made clear in the FCCC. Although the preparation of inventories of greenhouse gases does not necessarily incur huge costs and thus will be funded fully, the implementation of remedial measures could require enormous resources. Some means of defining eligibility for limited funds was therefore preceived as necessary.

Partly for this reason, the FCCC states that the developed country parties are obligated only to meet the "agreed full incremental" cost. The term "incremental cost" is interpreted to mean the additional cost of capturing global benefits. However, because the emissions of greenhouse gases by the vast majority of developing countries is currently very low, the potential global benefits to be reaped by such remedial measures are, presumably, also very low. Furthermore, in practice, there are no clear guidelines as to what constitutes the "global benefit" component of a proposed project or programme. Thus, there continues to be confusion regarding the concept of incremental costs and the circumstances under which developing countries can benefit from the financial flow from the developed countries.

The Role of the Global Environment Facility (GEF)

The FCCC designated, on an interim basis, the Global Environment Facility (GEF) as the international entity entrusted with the operations of the financial mechanism to meet the agreed incremental costs of activities under the FCCC. Initiated by France in the meeting of the World Bank-IMF ministerial advisory group in November 1990, the GEF was subsequently formally established as a three-year Pilot Phase (July 1991 to June 1994) for "helping developing countries to contribute toward solving global environmental problems" (World Bank, 1991). The GEF is a body of three UN organisations: UNEP provides scientific and technical guidance; UNDP is responsible for technical assistance activities and, through its worldwide network of offices, helps to identify projects for funding; and The World Bank administers the GEF trust fund and is responsible for investment projects as distinct from those for technical assistance.

Membership of the GEF is not universal, like the UN or the World Bank. As of 1994, there were 69 members in the GEF, including Bangladesh, and 113 projects worth US$727 million. Forty-five of these projects, totalling $288.7 million, received the final approval of the implementing agencies. Some $60.7 million was designated for approved projects that combat climate change (World Bank, 1994). Currently, there are no GEF projects specifically undertaken solely by Bangladesh, although Bangladesh is a participant in a region-wide project of Asian countries —ALGAS (Asia Least-cost Greenhouse gas Abatement Strategy project).

BOX 2.10: The Framework Convention on Climate Change

The Framework Convention on Climate Change (FCCC) was signed by 155 nations, including Bangladesh, at the United Nations Conference on Environment and Development in June, 1992. The ultimate objective of the FCCC is the "...stabilization of greenhouse gas concentrations in the atmosphere at a level that would prevent dangerous anthropogenic interference with the climate system...". Bangladesh ratified the FCCC on 15 April 1994. The commitments under the Convention (Article 4) include developing a national inventory of anthropogenic sources and sinks of greenhouse gases, and formulating programmes containing measures to mitigate climate change and to facilitate adaptation. Bangladesh may propose projects for financing, as through the Global Environment Facility, the international entity which operates the financial mechanism of the FCCC on an interim basis.

2.9 SUMMARY AND RESEARCH DIRECTIONS

In this chapter, we have reviewed the scientific basis for concern about climate change and examined the implications for Bangladesh. It is clear that the scientific uncertainties are very large. However, this should not cloud the fact that the very large majority of the world's climate experts believe that the increasing atmospheric concentrations of greenhouse gases will cause the world to warm. There is little or no doubt with respect to the following:

- that the atmospheric concentrations of greenhouse gases have been increasing;
- that these increases are due largely to emissions resulting from human activities;
- that, as a consequence, the average global temperature should increase;
- that the world has warmed by about 0.3°C to 0.6°C over the last 100 years, and so has Bangladesh (although it is not yet possible to attribute this warming unequivocally to greenhouse gases);

- that greenhouse gas concentrations will continue to increase well into the twenty-first century, even if emissions were stabilised;

- that, in the future, the world will, on average, be warmer than today.

It is sobering to note that no climate model shows global cooling with increasing greenhouse gas concentrations; all models show global warming.

However, there are major uncertainties about the rate and magnitude of the projected warming, and about the specific changes in climate that will occur regionally. It is not possible to say precisely how the climate of Bangladesh will change. It is only possible to present scenarios of what could occur based on the model results currently available. With some degree of confidence, these model results show

- that Bangladesh is expected to warm in all seasons;

- that the monsoon rainfall is expected to increase.

Little can be said currently about future changes in the frequency and magnitude of severe tropical cyclones, floods or droughts — climatic events that are of great significance to Bangladesh.

Nonetheless, whether or not climate change will actually occur in Bangladesh, it is clear that the international concern and emerging agreements will require heightened research and policy activities in fields with direct relevance to global change issues. As Bangladesh has signed the Framework Convention on Climate Change, it has signaled its intention to meet a number of commitments, particularly the

- development of a national inventory of the sources and sinks of greenhouse gases;

- formulation of programmes containing measures to mitigate climate change and to facilitate adaptation.

Bangladesh is formally obligated to meet these commitments. Immediate priority should therefore be given to research that directly supports these commitments. The following research directions are suggested.

First, there is a pressing need *to determine the magnitudes of the natural and anthropogenic sources, sinks and fluxes of greenhouse gases in Bangladesh.* For instance, it is particularly important to resolve the uncertainties regarding methane. It has been claimed by some that Bangladesh is perhaps a relatively large contributor to the global annual emission of methane because of its extensive paddy rice cultivation and animal husbandry. Yet, it is not precisely clear how the emission rates of methane differ amongst the various agricultural activities and land uses in the country. Moreover, it is unclear what the emission rates of methane were from the natural wetlands (globally, one of the largest sources) that agriculture replaced, and, therefore, what the *net change* in methane emissions has been over time. These are significant research areas in their own right, but they also have important policy implications since they pertain directly to the issue of Bangladesh's contribution to global warming and its potential for mitigation.

Second, there is an urgent need *to develop integrated impact models for Bangladesh that link existing climate, environmental and socio-economic models and data.* The broad goal of such research would be to enhance the internal capacity within Bangladesh for analysing the sensitivity of the environment and resources to climate change, for the purposes of education, planning and policy-making. As discussed above (and in other chapters), the obstacle to conducting such sensitivity analyses for Bangladesh is not necessarily the lack of models or data; for some sectors, like climate and agriculture, information is relatively abundant. Rather, a major hindrance has been the general lack of methods for integrating *existing* models and data at different scales — global, regional, national, and local — in a way that allows analyses which are relevant to policy development and strategy implementation. The development of an integrated climate change impacts model would greatly facilitate Bangladesh's task of complying with the requirements of the FCCC.[3]

In general, such tasks increasingly require international cooperation, for two reasons. First, the rapid development of databases, research methods and new knowledge throughout the world means that international collaboration is often the most effective way of achieving progress. It is vital, therefore, for Bangladesh to continue to strengthen its international linkages in the global change field in order to take advantage of expertise from overseas where required, and to contribute directly to international scientific activities.

Second, Bangladesh, having limited financial resources, cannot hope to carry out many of the necessary tasks of its own accord. Bangladesh must be in a position to take best advantage of the international support available to conduct climate-related research and to meet its obligations under the FCCC. Enhanced international linkages could ensure that opportunities are not inadvertently missed.

In summary, climate change is a global problem. It requires global solutions involving cooperative actions by all nations. The main obstruction to the implementation of measures to slow global warming is deeply rooted in the geo-politics of equity between the industrialised and developing world — between the poor and rich. In general, it appears that the developed nations should adopt measures to reduce greenhouse gas emissions immediately, while extending economic and technological support to developing countries so that their future contribution to global warming is curtailed without impeding development. Equity will surely be a key component in any agreements on an emission reduction strategy involving international cooperation and participation of all nations. In the meantime, Bangladesh should consider additional strategies for adaptation to climate variability and change — for both the present and the future.

ENDNOTES

[1] The seasonal definitions for Bangladesh vary. It should be noted that Chapter 4 defines four seasons: pre-monsoon (March to May), monsoon (June to September), post-monsoon (October to November) and winter (December to February). Technically, the monsoon season, when Bangladesh receives about two-thirds of its mean annual rainfall, is the period between early June and August-September. The heavy showers in the transitional pre-monsoon and post-monsoon seasons extend the period when rainfall exceeds evapo-transpiration, so the "wet" (or rainy) season lasts longer than the monsoon season proper.

[2] The monsoon precipitation data were checked for normality using Minitab. The monsoon rainfall data were plotted against normal scores calculated from the data. The points in the plot fall in a straight line and the very high correlation coefficient (r=0.995) is consistent with normality.

[3] The need for such integrative methods is urgent at the international scale as well, particularly with regard to Article 2 of the Framework Convention on Climate Change. Article 2 states that the objective of the Convention is to achieve "...stabilization of greenhouse gas concentrations in the atmosphere at a level that would prevent dangerous anthropogenic interference with the climate system. Such a level should be achieved within a time frame sufficient to allow ecosystems to adapt naturally to climate change, to ensure that food production is not threatened and to enable economic development to proceed in a sustainable manner". The terms "dangerous", "time frame sufficient", "threatened" and "sustainable" are not defined. The need to consider ways of integrating science, impacts and policy in order to address this issue has been recognised as critical by the Convention of the Parties and by the Intergovernmental Panel on Climate Change (IPCC).

REFERENCES

Agarwal, A. and Narain, S. (1991) *Global Warming in an Unequal World: A Case of Environmental Colonialism.* Centre for Science and Environment, New Delhi.

Asaduzzaman, M. (1991) The climatic changes in Bangladesh and their impact. *Proceedings of the Asian-Pacific Seminar on Climate Changes.* Environment Agency, Japan, pp. 102-108.

Barnola, J.M., Raynaud, D., Korotkevitch, Y.S. and Lorius, C. (1987) Vostok ice core: a 160,000 year record of atmospheric CO_2. *Nature*, 329, 408-414.

Bolin, B., (1986) How much CO_2 will remain in the atmosphere? In B. Bolin, B.R. Döös, R.A. Warrick and J. Jaeger, *The Greenhouse Effect, Climate Change and Ecosystems.* SCOPE 29. John Wiley and Sons Inc., Chichester, pp. 93-156.

Bolin, B., Döös, B.R., Jager, J. and Warrick, R.A. (1986) *The Greenhouse Effect, Climatic Change and Ecosystems.* SCOPE 29. John Wiley and Sons, Chichester.

Bretherton, F.P., Bryan, K., and Woods, J.D. (1990) Time-dependent greenhouse-gas-induced climate change. In J.T.Houghton, G.J. Jenkins and J.J. Ephraums (eds.) *Climate Change: The IPCC Scientific Assessment.* Cambridge University Press, Cambridge, pp. 173-194.

Chappellaz, J., Barnola, J.M., Raynaud, D., Korotkevich, Y.S. and Lorius, C. (1990) Ice-core record of atmospheric methane over the past 160,000 years. *Nature*, 345, 127-131.

Charlson, R.J., Schwartz, S.E., Hales, J.M., Cess, R.D., Coakley, J.A. Jr., Hansen, J.E., and Hofmann, D.J. (1992) Climate forcing by anthropogenic aerosols. *Science*, 255, 423-430.

Coastal Zone Management Subgroup (CZMS) (1990) *Strategies for Adaptation to Sea Level Rise.* IPCC Response Strategies Working Group. Ministry of Transport and Public Works, Rijkswaterstaat, The Hague.

Commonwealth Group of Experts (1989) *Climate Change: Meeting the Challenge.* Commonwealth Secretariat, London.

Cubasch, U. and Cess, R.D. (1990) Processes and modelling. In J.T. Houghton, G.J. Jenkins and J.J. Ephraums (eds.) *Climate Change: The IPCC Scientific Assessment.* Cambridge University Press, Cambridge, pp. 69-91.

Cubasch, U., Hasslemann, K., Hock, H., Maier-Reimer, E., Mikolajewicz, U., Santer, B.D. and Sausen, R. (1992) Time-dependent greenhouse warming computations with a coupled ocean-atmosphere model. *Climate Dynamics*, 8, 55-69.

Folland, C.K., Karl, T.R., Nicholls, N. Nyenzi, B.S., Parker, D.E. and Vinnikov, K.Ya. (1992) Observed climate variability and change. In J.T. Houghton, B.A. Callander and S.K. Varney (eds.) *Climate Change 1992: The Supplementary Report to the IPCC Scientific Assessment.* Cambridge University Press, Cambridge, pp. 135-170.

Folland, C.K., Karl, T.R. and Vinnikov, K.Ya. (1990) Observed climate variations and change. In J.T. Houghton, G.J. Jenkins and J.J. Ephraums (eds.) *Climate Change: The IPCC Scientific Assessment*. Cambridge University Press, Cambridge, pp. 195-238.

Gates, W.L., Mitchell, J.F.B., Boer, G.J., Cubasch, U. and Meleshko, V.P. (1992) Climate modelling, climate prediction and model validation. In J.T. Houghton, B.A. Callander and S.K. Varney (eds.) *Climate Change 1992: The Supplementary Report to the IPCC Scientific Assessment*. Cambridge University Press, Cambridge, pp. 97-134.

Government of Bangladesh (GOB) (1973) *The First Five Year Plan 1973-1978*. Planning Commission, Dhaka.

Government of Bangladesh (GOB) (1989) *Study on the Causes and Consequences of Natural Disasters and Protection and Preservation of the Environment*. Bangladesh Country Report. Government of Bangladesh, Dhaka.

Government of Bangladesh (GOB) (1990) *The Fourth Five Year Plan 1990-1995*. Planning Commission, Dhaka.

Grove, J.M. (1988) *The Little Ice Age*. Methuen, London.

Gupta, S. and Pachauri, R.K. (1989) Proceedings of the International Conference on Global Warming and Climate Change - Perspectives from Developing Countries. New Delhi.

Hare, F.K. (1979) Climate variation and variability: empirical evidence from meteorological and other sources. In WMO, *Proceedings of the World Climate Conference*. WMO - No. 537, WMO, Geneva, pp. 51-87.

Houghton, J.T., Callander, B.A. and Varney, S.K. (1992) *Climate change 1992: the Supplementary Report to the IPCC Scientific Assessment*. Cambridge University Press, Cambridge.

Houghton, J.T., Meira Filho, L.G., Bruce, J., Hoesung Lee, Callander, B.A., Haites, E., Harris, N. and Maskell, K. (1995) (eds.) *Climate Change 1994: Radiative Forcing of Climate Change and an Evaluation of the IPCC IS92 Emission Scenarios*. Published for the Intergovernmental Panel on Climate Change. Cambridge University Press, Cambridge.

Houghton, J.T., Jenkins, G.J., and Ephraums, J.J. (eds.) (1990) *Climate Change: The IPCC Scientific Assessment*. Cambridge University Press, Cambridge.

Houghton, R.S. (1991) Tropical deforestation and atmospheric carbon dioxide. *Climate Change* 19, 99-118.

IPCC (1990) Policymakers summary. In J.T. Houghton, G.J. Jenkins and J.J. Ephraums (eds.) *Climate Change: The IPCC Scientific Assessment*. Cambridge University Press, Cambridge.

Isaksen, I.S.A., Ramaswamy, V., Rodhe, H. and Wigley, T.M.L. (1992) Radiative forcing of climate. In J.T. Houghton, B.A. Callander, and S.K. Varney (eds.) *Climate Change 1992: the Supplementary Report to the IPCC Scientific Assessment*. Cambridge University Press, Cambridge, pp. 47-67.

Jaeger, J. (1988) Developing policies for responding to climatic change. WCIP-1, WMO/TD-No.225. WMO Secretariat, Geneva.

Jaeger, J. and Ferguson, H.L. (1991) *Climate Change: Science, IMPACTS and Policy*. Proceedings of the Second World Climate Conference. Cambridge University Press, Cambridge.

Jonas, P.R., Charlson, R.J. and Rodhe, H. (1995) Aerosols. In J.T. Houghton, L.G. Meira Filho, J. Bruce, Hoesung Lee, B.A. Callander, E.Haites, N. Harris and K. Maskell (eds.) *Climate Change 1994: Radiative Forcing of Climate Change and an Evaluation of the IPCC IS92 Emission Scenarios*. Published for the Intergovernmental Panel on Climate Change. Cambridge University Press, Cambridge, pp. 127-162.

Leggett, J., Pepper, W.J. and Swart, R.J. (1992) Emission scenarios for the IPCC: an update. In J.T. Houghton, G.J. Jenkins, and J.J. Ephraums (eds.) (1990) *Climate Change 1992: The IPCC Scientific Assessment*. Cambridge University Press, Cambridge, pp. 69-96.

Leuenberger, M. and Siegenthaler, U. (1992) Ice-age atmospheric concentration of nitrous oxide from an Antartic ice core. *Nature,* 360, 449-451.

Lighthill, J., Holland, G., Gray, W., Landsea, C. Craig, G., Evans, J., Kurihara, Y., and Guard, C. (1994) Global climate change and tropical cyclones. *Bulletin of the American Meteorologoical Society*, 75, 2147-2157.

Marland, G. (1990) Global and national CO_2, emission from fossil fuel burning, cement production, and gas flaring. In T.A. Boden, P. Kanciruk, and M.P. Farrell (eds.) *TRENDS '90: A Compendium of Data on Global Change. ORNL/CDIAC-36*. Carbon Dioxide Information Analysis Centre, Oak Ridge National Laboratory, Oak Ridge, Tennessee, pp. 92-93.

Meehl, G.A. and Washington, W.M. (1993) South Asian summer monsoon variability in a model with doubled atmospheric carbon dioxide concentration. *Science,* 260, 1101-1104.

Mintzer, I.M. (1992) *Confronting Climate Change: Risks, Implications and Responses*. Cambridge University Press, Cambridge.

Mitchell, J.F.B. and Gregory, J.M. (1992) Climate consequences of emissions and a comparison of IS92a and SA90. In J.T. Houghton, B.A. Callander, and S.K. Varney (eds.) *Climate Change 1992: the Supplementary Report to the IPCC Scientific Assessment*. Cambridge University Press, Cambridge, pp. 171-5.

Mitchell, J.F.B., Manabe, S., Tokioka, T. and Meleshko, V. (1990) Equilibrium climate change. In J.T. Houghton, G.J. Jenkins and J.J. Ephraums (eds.) *Climate Change: The IPCC Scientific Assessment*. Cambridge University Press, Cambridge, pp. 131-172.

Mooley, D.A. and Parthasarathy, B. (1984) Fluctuations in all-Indian summer monsoon rainfall during 1871-1978. *Climatic Change*, 6, 287-301.

Murphy, J.M. (1992) A prediction of the transient response of climate. *Meteorological Office Climate Research Technical Note 32*. UK Meteorological Office, Bracknell.

Ramaswamy, V., Schwarzkopf, M.D. and Shine, K.P. (1992) Radiative forcing of climate from halocarbon-induced global stratospheric ozone loss. *Nature*, 355, 810-812.

Raper, S.C.B. (1993) Observational data on the relationships between climatic change and the frequency and magnitude of severe tropical storms. In R.A. Warrick, E.M. Barrow and T.M.L. Wigley (eds) *Climate and Sea Level Change: Observations, Projections and Implications*. Cambridge University Press, Cambridge, pp. 192-214.

Rashid, H.E. (1991) *The Geography of Bangladesh*. University Press Limited, Dhaka.

Ropelewski, C.F. and Halpert, M.S. (1987) Global and regional scale precipitation patterns associated with the El Nino/Southern Oscillation. *Monthly Weather Review*, 115, 1606-1626.

Ropelewski, C.F. and Halpert, M.S. (1989) Precipitation patterns associated with the high index phase of Southern Oscillation. *Journal of Climate*, 2, 268-284.

Schimel, D., Enting, I.G., Heimann, M., Wigley, T.M.L., Raynaud, D., Alves, D. and Siegenthaler, U. (1995) CO_2 and the Carbon Cycle. In J.T. Houghton, L.G. Meira Filho, J. Bruce, Hoesung Lee, B.A. Callander, E. Haites, N. Harris, and K. Maskell (eds.) *Climate Change 1994: Radiative Forcing of Climate Change and an Evaluation of the IPCC IS92 Emission Scenarios*. Published for the Intergovernmental Panel on Climate Change. Cambridge University Press, Cambridge, pp. 35-71.

Shine, K.P., Derwent, R.G., Wuebbles, D.J. and Morcrette, J-J. (1990) Radiative forces of climate. In J.T. Houghton, G.J. Jenkins, and J.J. Ephraums (eds.) *Climate Change: The IPCC Scientific Assessment*. Cambridge University Press, Cambridge, pp. 41-68.

Steele, L.P., Dlugokencky, E.J., Lang, P.M., Tans, P.P., Martin, R.C. and Masarie, K.A. (1992) Slowing down of the global accumulation of atmospheric methane during the 1980s. *Nature*, 358, 313.

Stouffer, R.J., Manabe, S. and Bryan, K. (1989) Interhemispheric asymmetry in climate response to a gradual increase of atmospheric CO_2. *Nature*, 342, 660-62.

Tans, P.P, Fung, I.Y. and Takahashi, T. (1990) Observational constraints on the global atmospheric carbon dioxide budget. *Science*, 247, 1431-1438.

Tegart, W.J.McG., Sheldon, G.W. and Griffiths, D.C. (1990) *Climate Change: the IPCC Impacts Assessment*. Australian Government Publishing Service, Canberra.

UNEP (United Nations Environment Programme) (1987) *Montreal Protocol on Substances that Deplete the Ozone Layer*. UNEP, Nairobi.

Warrick, R.A. (1993) Slowing global warming and sea-level rise: the rough road from Rio. *Transactions of the Institute of British Geographers*, 18, 140-148.

Warrick, R.A., Barrow, E.M. and Wigley, T.M.L. (1990) *The Greenhouse Effect and its Implications for the European Community CEC EUR 12707 EN*. (Available from HMSO).

Warrick, R.A., Barrow, E.M. and Wigley, T.M.L. (1993) *Climate and Sea Level Change: Observations, Projections and Implications*. Cambridge University Press, Cambridge.

Warrick, R.A. and Rahman, A.A. (1992) Future sea level rise environmental and socio-political considerations. In I. Mintzer (ed.) *Confronting Climate Change: Risks, Implications and Responses*. Cambridge University Press, Cambridge, pp. 97-112.

Washington, W.M. and Meehl, G.A. (1989) Climate sensitivity due to increased CO_2: experiments with a coupled atmosphere and ocean general circulation model. *Climate Dynamics*, 4, 1-38.

Watson, R.T., Rodhe, H., Oeschger H. and Siegenthaler U. (1990) Greenhouse gases and aerosols. In J.T. Houghton, G.J. Jenkins and J.J. Ephraums (eds.) *Climate Change: The IPCC Scientific Assessment*. Cambridge University Press, Cambridge, pp. 1-40.

Watson, R.T., Meira Filho, L.G., Sanhueza, E. and Janetos, A. (1992) Greenhouse gases: sources and sinks. In J.T. Houghton, B.A. Callander and S.K. Varney (eds.) *Climate Change 1992: The Supplementary Report to the IPCC Scientific Assessment*. Cambridge University Press, Cambridge, pp. 25-46.

Wigley, T.M.L. (1989) Possible climate change due to SO_2 -derived cloud condensation nuclei. *Nature*, 339, 365-367.

Wigley, T.M.L. and Barnett, T.P. (1990) Detection of the greenhouse effect in the observations. In J.T. Houghton, G.J. Jenkins and J.J. Ephraums (eds.): *Climate Change: The IPCC Scientific Assessment*. Cambridge University Press, Cambridge, pp. 239-256.

Wigley, T.M.L., Pearman, G.I. and Kelly, P.M. (1992) Indices and indicators of climate change: issues of detection, validation and climate sensitivity. In I.M. Mintzer (ed.) *Confronting Climate Change: Risks, Implications and Responses*. Cambridge University Press, Cambridge, pp. 85-96.

Wigley, T.M.L. and Raper, S.C.B. (1991) Detection of the enhanced greenhouse effect on climate. In J. Jaeger, and H. Ferguson (eds.) *Climate Change: Science, Impacts and Policy*. Cambridge University Press, Cambridge, pp. 231-242.

Wigley, T.M.L. and Raper, S.C.B. (1992) Implications for climate and sea level of revised IPCC emission scenarios. *Nature*, 357, 293-300.

Wigley, T.M.L. and Raper, S.C.B. (1993) Future changes in global mean temperature and sea level. In R.A. Warrick, E.M. Barrow and T.M.L. Wigley (eds.), *Climate and Sea Level Change: Observations, Projections and Implications*. Cambridge University Press, Cambridge, pp. 111-113.

WMO (World Meteorological Organization) (1992) *Scientific Assessment of Ozone Depletion*. WMO/UNEP, WMO Global Ozone Research and Monitoring Project, Report No. 25, Geneva.

World Bank (1991) *The World Bank and the Environment*. World Bank, Washington D.C.

World Bank (1994) *Communication to the Administrator*. Global Environment Facility (GEF).

WRI (World Resources Institute) (1990) *World Resources 1990-1991 - A Guide to the Global Environment*. Oxford University Press, Oxford.

SEA-LEVEL CHANGES IN THE BAY OF BENGAL

R.A. Warrick, A.K. Azizul Hoq Bhuiya, W.M. Mitchell, T.S. Murty and K.B.S. Rasheed*

3.1 INTRODUCTION

Global warming from increasing atmospheric concentrations of greenhouse gases could well cause global sea level to rise. As discussed in Chapter 2 of this book, it is the firm consensus of the scientific community that, in the decades to come, the increasing atmospheric concentrations of greenhouse gases should cause the world to warm. Global warming, in turn, should cause sea level to rise by expanding the oceans and increasing the melt rates of ice sheets and glaciers. Global climate change could also affect the frequency and intensity of cyclones and storm surges, and thus coastal flooding, in the Bay of Bengal.

The possibility of future sea-level rise should be a source of grave concern to the many inhabitants of low-lying coastal areas, including those of Bangladesh. A major portion of the world's population lives in coastal areas and is potentially vulnerable to changes in sea level. At local and regional levels, present variations in sea level are reminders that the sea can be highly destructive. Short-term extreme events, like the storm surge that swept over the south-western coast of Bangladesh in April 1991 or Hurricane Andrew that pounded the shores of Florida in the United States of America in August 1992, can bring death, social dislocation and economic damage of crippling magnitude. In some locations, an average long-term trend in sea-level rise can exacerbate the problems of extreme events and slowly erode the foundations of coastal settlements, as in Venice or in Bangkok where the sea is rising at a rate of 10cm per year in places due to land subsidence. Any change in the

* The authors are respectively: Research Director, CEARS, University of Waikato, Hamilton, New Zealand; Professor of Geography, Jahangirnagar University, Bangladesh and Research Scholar at Bangladesh Unnayan Parishad (BUP), Dhaka; Deputy Director, National Tidal Facility, The Flinders University of South Australia, Adelaide, Australia; Director, National Tidal Facility; and Professor of Geography, University of Dhaka, and Visiting Scholar at BUP, Dhaka, Bangladesh.

volume of the global ocean could compound the problem of sea-level variations experienced locally.

In considering the implications of these issues for Bangladesh, this chapter is organised into four main sections, each devoted to addressing a key question of relevance to Bangladesh: What are the causes of sea-level change? Has global sea level been rising and will it rise in future? How are relative sea levels changing in the Bay of Bengal? How do cyclone-induced storm surges affect short-term sea-level changes in Bangladesh? The concluding section highlights the main gaps in knowledge and research needs.

3.2 WHAT CAUSES SEA LEVEL TO CHANGE?

Sea level varies on all time-scales, from seconds to millions of years. Sea level also varies spatially, from one place to another. In order to understand the role of climate in these variations, some basic distinctions between the causes of sea-level fluctuations have to be made.

For occupants of the coastal zone, what matters is *relative* sea level, that is, the level of the sea surface in relation to the land. As shown in Figure 3.1, relative sea level can change either as a result of vertical movement of the land or as a result of changes in the level of the ocean surface itself.

Vertical land movements can be large and sudden. In tectonically-active areas, vertical displacements along fault lines during earthquakes have been known to cause abrupt coastal uplift or submergence of a metre or more (Pugh, 1987). Vertical land movements can also be slow and inexorable. Globally, one of the most important effects has been the continuing isostatic adjustment of the land surface following the disappearance of the large continental glaciers which marked the end of the last ice age about 10,000 years ago. In parts of Sweden and Finland, for instance, the land surface is still rising (hence, relative sea level is falling) at a rate approaching 1m per century due to the isostatic rebound following the release of pressure exerted by the ice mass (Emery and Aubrey, 1985).

In the major deltas of the world, sedimentation processes play a crucial role in relative sea-level change. In the Ganges-Brahmaputra-Meghna delta of Bangladesh, relative sea level depends on the balance between deltaic subsidence due to the massive weight of sediments, and the accretion of land as additional sediments are deposited in the coastal areas (see Chapter 7). Elsewhere, human activities, such as groundwater pumping, can cause land subsidence. The building of dams, embankments, channels and canals in wetlands and deltas can seriously disrupt the flow of the much-needed sediments and cause relative sea level to rise; such has been the case in the Mississippi delta where relative sea level has been rising at the rate of about 1m per century and up to $100km^2$ of wetlands are being lost each year (Gagliano *et al.*, 1981; Day *et al.*, 1993).

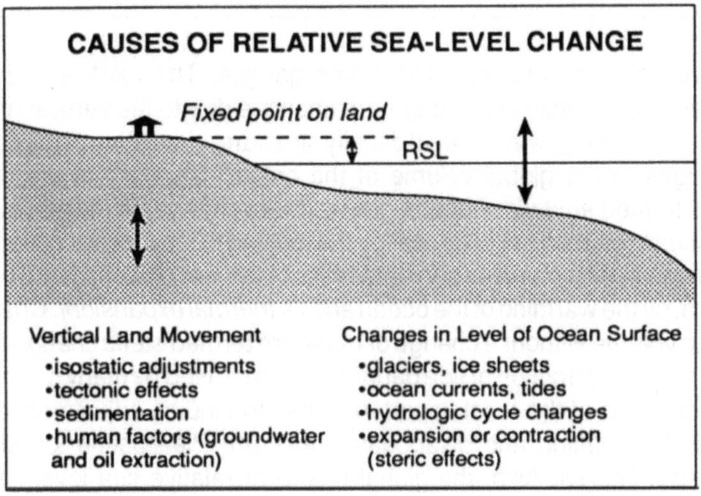

FIGURE 3.1: The causes of relative sea-level changes

Source: Warrick, 1993a

Relative sea level can also be affected by changes in the level of the ocean surface. Over the very long-term, gradual changes in sea level take place through changes in the volume of the ocean basin; for example, by slow spreading of the sea floor (plate tectonics) or deposition of sediment on the ocean floor. Globally, such changes may be causing sea level to rise in the order of 1mm per century — negligible in human terms, but very important on long geological time-scales (millions of years).

In contrast, at the very shortest time-scales (seconds, minutes, hours, days), the surface topography of the sea is highly changeable. Variations in gravitational attraction of the Moon and Sun (tides) or meteorological phenomena, such as atmospheric pressure, wind or precipitation variations, can cause abrupt changes in local sea level (e.g. storm surges). The dynamic influence of the ocean circulation creates an ocean surface that, spatially, is far from uniform. It is not uncommon to find differences of a metre or more in the level of the sea surface across major ocean currents (Levitus, 1982). Such temporal and spatial variations in sea level will continue to occur naturally.

BOX 3.1: Relative sea-level change

> Sea-level changes are recorded at tide gauges. The *relative sea level* (RSL) at a gauge may show long-term changes due to the vertical motion of the gauge, changes in the local density structure, circulation of the ocean, or changes in the global volume of the ocean. Changes in the ocean volume, termed *eustatic changes*, can be due to changes in the mass of the world ocean (caused, for instance, by the melting of ice masses on land) or by changes in the volume at constant mass of the world ocean (caused, for instance, by the warming of the ocean and its *thermal expansion*). Changes in ocean volume without a change of mass are termed *steric changes*. The long-term vertical motion of tide gauges can be caused by many geological processes: one of the most important is the continuing adjustment of the Earth to the melting and disappearance of the Pleistocene ice sheets (*isostatic rebound*). Also, the global record of relative sea level can be affected by *changes in the volume of the ocean basin* due to tectonic or sedimentary processes, but the effects of such processes are small on the time-scales of interest here.

Source: Modified from Polar Research Board, 1985, p. 12.

In the present context, we are concerned with the possibility of a globally-coherent rise in sea level due to global warming on the decade-to-century time-scale. The possible causes of such a rise involve an actual change in the volume of the ocean itself. There are two major factors to consider in this regard: oceanic thermal expansion and changes in the mass of land-based ice (including glaciers and small ice caps, and the polar ice sheets of Antarctica and Greenland). Below, each is examined in more detail.

3.2.1 Oceanic Thermal Expansion

Warming the oceans can cause sea level to rise — a "steric" change in sea level.

Such steric changes are inversely related to changes in the density of water — that is, the higher the density, the smaller the volume (and vice versa). Ocean water density is influenced primarily by three factors: temperature, salinity and pressure. In the context of the enhanced greenhouse effect, the ocean, as well as the land, would be expected to warm. And as the ocean warms, its density would decrease and its volume would increase — oceanic thermal expansion. Changes in salinity, too, may affect the ocean volume, but such changes would be minor on a global scale compared to thermal effects (Wigley and Raper, 1987, 1993).

The rate of thermal expansion (both past and future) due to increasing greenhouse gas concentrations is difficult to ascertain. The uncertainties are large. In general, there are three major sets of uncertainties: the changes in "radiative forcing" due to greenhouse gases; the sensitivity of climate; and the rate of heat uptake by the oceans. Since these have direct implications for global warming and land-ice changes as well as for thermal expansion, each is discussed briefly below (see Chapter 2 for additional detail).

Radiative Forcing Change

The change in radiative forcing refers to the change in the heating of the climate system, either by changing the energy input (e.g. solar insolation) or the energy emitted back to space (e.g. infrared radiation). Greenhouse gases affect the latter. The uncertainties about future changes in radiative forcing are large, for three reasons. First, it is necessary to be able to predict the future emissions of greenhouse gases. The uncertainties are not primarily scientific, but rather socio-economic, involving such factors as population growth, rates of economic development, technological innovations, or the use of alternative energy sources. Prediction of such factors for many decades ahead is fraught with difficulties. Recent estimates of future emissions indicate a wide range of possibilities (Leggett *et al.*, 1992).

Second, the calculation of atmospheric concentration changes from greenhouse gas emissions is subject to error. For example, reliance has to be placed on carbon cycle models for predicting the future carbon dioxide concentration. The problem is that model results do not always agree with observations (within observational uncertainties). Carbon cycle models have tended to over-predict the contemporary carbon dioxide concentration, implying that the mechanisms for carbon dioxide removal from the atmosphere have not all been accounted for (the so-called "missing sink" problem). Recent models (e.g. Wigley and Raper, 1992) have attempted to overcome this problem, but there are still uncertainties (Schimel *et al.*, 1994).

Third, as discussed in Chapter 2, the basis for calculating radiative forcing changes has recently undergone revision to take into account the cooling effect of sulphate aerosols (from sulphur dioxide emissions associated with fossil fuel burning) (Isaksen *et al.*, 1992; Wigley and Raper, 1992; Jonas *et al.*, 1994). The inclusion of sulphate aerosols tends to lower predictions of future global warming and oceanic thermal expansion.

Climate Sensitivity

The second set of uncertainties has to do with the climate sensitivity. As discussed in Chapter 2, conventionally, the climate sensitivity refers to the equilibrium change in average global surface air temperature for an instantaneous doubling of carbon dioxide. The word "equilibrium" is important, for it denotes the eventual (or potential) steady-state change that would result without reference to the time required to reach that condition (in contrast to the "transient", or "time-dependent", changes in temperature that would actually occur due to lags in the climate system caused by the thermal inertia of the oceans — our next point below). There are feedbacks in the climate system — most notably, changes in water vapour, snow/ice extent and clouds — that, on the whole, probably enhance the initial warming caused by changes in greenhouse gas concentrations. The magnitude — and in some cases even the direction — of these feedbacks is uncertain, which gives rise to a rather broad range of estimated values for the climate sensitivity. The range accepted by the International Panel on Climate Change is 1.5°C to 4.5°C (Mitchell *et al.*, 1990; Gates *et al.*, 1992). In terms of estimating past and future thermal expansion effects, the value of the climate sensitivity is crucial, for it determines the amount of additional heat potentially available to be absorbed by the oceans.

Oceanic Heat Uptake

The third set of uncertainties has to do with the rate of oceanic heat uptake. In large part, the geographical differences in the Earth's temperature, and the related density differences in the ocean, drive the ocean circulation (the thermo-haline circulation). In the high latitudes, areas of deep-water formation exist where the heavier, colder waters sink, or "downwell". This water flows equator-ward near the bottom and "upwells" elsewhere in the ocean. Furthermore, the heat from the surface mixed layer (the well-mixed upper layer, about 70 to 110 metres deep on average, which interacts with the atmosphere) slowly penetrates downward to the colder, deeper layers of the ocean. In combination, such processes act to transport heat, both vertically and horizontally, within the ocean. Some of the heat from greenhouse gas forcing changes meets the same fate. However, the uptake of heat by the oceans does not occur instantaneously. Time is required. It is this thermal inertia of the oceans — the ocean's heat capacity and its ability to warm and cool slowly — and the uncertainties about the mechanisms for mixing and transporting heat that have important implications for both global warming and oceanic thermal expansion.

In terms of global warming, the oceanic thermal inertia creates a lag in the climate response to radiative forcing changes. This principle is well-illustrated throughout the year. In the mid and high latitudes, the coldest and warmest months of the year occur a month or two after the solstices (the times of maximum and minimum receipt of solar radiation) as the oceans "catch up" to the temperature changes on land. In the same way, global warming from greenhouse gas forcing would be retarded as the oceans absorb heat.

BOX 3.2: Oceanic thermal expansion

As the ocean warms, it expands and causes a rise in sea level. The future rates of both global warming and oceanic thermal expansion depend critically on three major factors: the future rates of greenhouse gas emissions and hence changes in concentrations and radiative forcing; the sensitivity of the climate system to radiative forcing change; and the rate of heat uptake by the ocean and its distribution. There is some trade-off between global warming and thermal expansion; the higher the rate of oceanic heat absorption and thermal expansion, the slower the rate of global warming. And vice versa.

In terms of thermal expansion, the more heat absorbed by the oceans the greater the expansion effect. Thus, in general, there is some trade-off between global warming and thermal expansion: the faster the penetration of heat to the oceans, the lower the rate of global warming but the higher the rate of expansion-related sea-level rise. With regard to total sea-level rise,the picture is further complicated by the fact that land-ice responses (see below) are strongly dependent on the rate of surface temperature increase. In short, the rates of global warming and sea-level rise are inter-related and the ocean circulation plays a key role in determining the rates of changes of both.

3.2.2 Changes in Glaciers and Ice Sheets

The volume of the oceans can also be affected through changes in the mass of seawater. It is possible, for instance, that changes in groundwater or surface water storage due to human activities have altered sea level noticeably during the present century (Sahagian *et al.*, 1994). It is more likely that changes in the mass of land-based ice — namely, glaciers and small ice caps, and the ice sheets of Antarctica and Greenland — will be more important in changing the volume of the ocean as the world warms.

Glaciers and Small Ice Caps

Mountain (or alpine) glaciers are not confined solely to the higher latitudes of the world. They exist within a wide range of latitudes, including the sub-tropics where mountains are sufficiently high in elevation to allow the accumulation and maintenance of ice. Nonetheless, most of the glacier area of the world is to be found in the higher latitudes of North America (including the small Arctic ice caps) and in Eurasia. Mountain glaciers and small ice caps account for less than one per cent of the land ice of the world (Haeberli *et al.*, 1989).

BOX 3.3: Glaciers and ice sheets

> Whether a glacier or an ice sheet, the volume of ice is controlled by two opposing components of the *mass balance — ice accumulation* (e.g. through snowfall) and *ablation* (e.g. through melting/run-off, evaporation, or calving of icebergs). If the rate of accumulation exceeds that of ablation, the mass balance is positive and the glacier or ice sheet will grow, effectively lowering sea level. If, however, the ablation rate exceeds the accumulation rate, the mass balance is negative and ice volume will shrink, raising sea level. In the present context the issue is, *how will climate change affect the mass balance components of the world's glaciers and ice sheets, and, hence, sea level?*

Because they are located in climates that are considerably warmer than those of polar ice sheets, many glaciers are very sensitive to changes in temperature. With relatively high rates of ablation (e.g. melting/runoff and evaporation), increases in temperature would be expected to expand the ablation zone. If ablation exceeds accumulation the glacier volume will shrink. As will be discussed below, many estimates of past and future sea-level rise attribute a relatively large contribution to the increased melting of glaciers. But, since the total global volume of mountain glaciers is relatively small, the contribution to sea level in a warming world would eventually be a diminishing one. Altogether, glaciers contain the equivalent of somewhere between 0.3 and 0.6 metres of sea-level rise.

Antarctic Ice Sheet

In contrast to the world's glaciers, the Antarctic ice sheet contains an enormous amount of ice (Figure 3.2). If the grounded ice of Antarctica were to be melted and spread evenly over the surface of the global ocean, sea level would rise by about 65 metres. Fortunately, this is not possible on the time-scale of a few centuries. In fact, the most likely effect of global warming in the short-term would be to increase the mass of the Antarctic ice sheet, thus *lowering* sea level. The reason is that accumulation rates (through snowfall) could be expected to increase, while the rate of "loss" of ice to the sea would remain unchanged. To understand this, it is necessary to examine the climate of Antarctica in relation to the ice sheet.

Antarctica is a large, extremely cold desert. On average, precipitation is scanty because the cold air can hold very little moisture. Accumulation rates are therefore low, but they occur over a vast area, so that the total annual accumulation is very large. Because of the prevailing cold temperatures, there is very little melting and runoff to the sea; ice wastage occurs mainly by calving (the breaking off of icebergs). Calving rates are determined by dynamic ice flow. Grounded ice moves downslope in ice streams towards the sea and becomes a floating ice shelf from the point at which it crosses the grounding line. As the shelf ice is "pushed" from behind, large chunks of ice break off and float away. At present, from observational data it is difficult to say whether the Antarctic ice sheet is currently growing or shrinking, since the data are relatively sparse and the potential error is large.

What would be the result of a warmer climate on the Antarctic ice sheet? The immediate effect would most likely be an increase in accumulation rates, since the warmer air would be able to hold more moisture. Even with a warming of several degrees Celsius, however, the temperature would not be sufficiently high to allow substantial melting and runoff to occur. Moreover, on the decadal time-scale, calving rates would probably also remain unaffected, since the dynamic processes that determine calving rates respond only slowly to changes in climate, with a response time of the order of thousands of years. Thus, with an increase in "input" and no change in "output", a positive mass balance would be expected. By itself, this would tend to lower sea level.

One major uncertainty has to do with the West Antarctic ice sheet. It contains a large marine ice sheet (grounded below sea level), where a large portion of covered ground slopes landward (Figure 3.3). Large floating ice shelves serve to

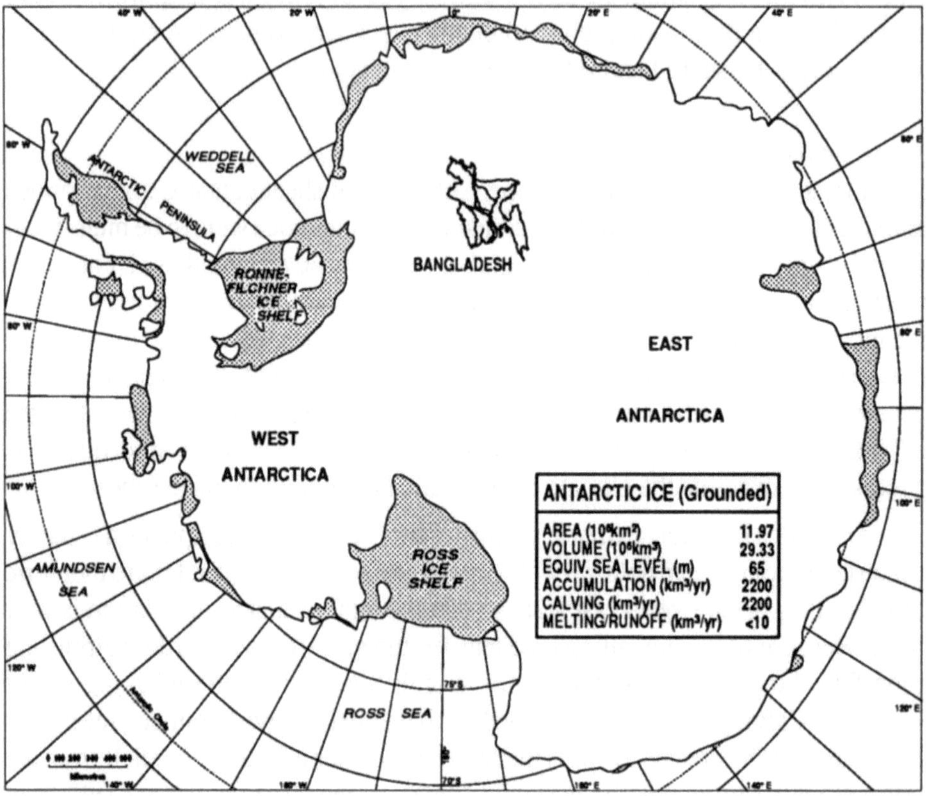

FIGURE 3.2: Antarctic ice sheet. The shaded areas are floating ice shelves, while the white areas denote grounded ice. The inset map of Bangladesh gives an indication of the vast size of Antarctica. The ice in Antarctica contains the equivalent of about 65 metres of sea-level rise.

Source: Inset data - Warrick and Oerlemans, 1990.

buttress the ice sheet, applying "back pressure" to slow the ice flow. For this reason, some experts believe that the West Antarctic ice sheet is inherently unstable and that any thinning of the ice shelves from global warming could ease the back pressure and cause the ice sheet to "surge". Some claim that this is what accounted for the 5 to 6 metre rise in sea level that occurred during the last warm interglacial around 120,000 years ago (Thomas *et al.*, 1979), although this claim has recently been disputed (Andrews, 1991). However, even if true, it should be borne in mind that such a dynamic response probably occurred over hundreds, if not thousands of years. It is unlikely that the West Antarctic ice sheet will cause a rapid rise in global sea level over the next century as a result of greenhouse gas-induced warming, although the uncertainties are large.

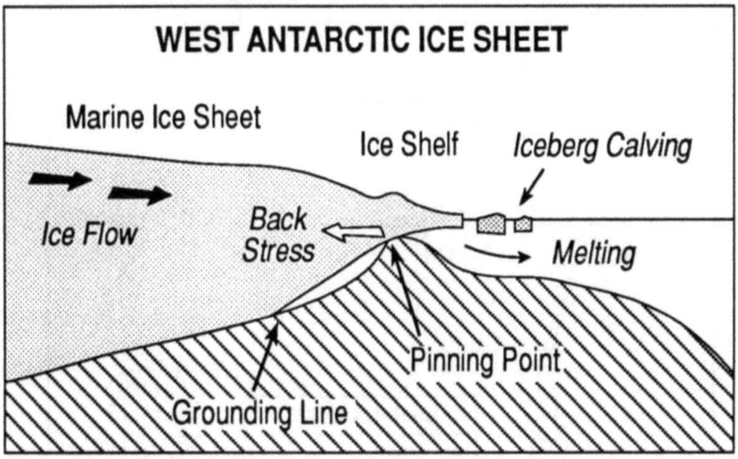

FIGURE 3.3: Processes affecting the stability of the West Antarctic ice sheet

Source: Warrick, 1993a.

Greenland Ice Sheet

The mass of the Greenland ice sheet is about one-tenth that of Antarctica (Figure 3.4). If entirely melted, the Greenland ice sheet would raise sea level by about seven metres. In comparison to Antarctica, the climate of Greenland is much warmer. As a consequence, it is believed that ice wastage occurs from both melting/runoff and calving, in nearly equal proportions, although the data are very uncertain. In the event of an increase in temperature, melting rates would increase and most likely dominate over any increase in accumulation rates. This would decrease the ice mass and raise sea level.

There is some evidence to suggest that during the recent past the Greenland ice sheet has shrunk. First, the average summer (June, July, August — the main ablation period) temperature has increased over the last 100 years, reaching a peak of nearly +2°C during the 1930s (in comparison to the turn of the century). Second, outlet glaciers, mainly in the central and southern portions of the west coast, have

Figure 3.4: Greenland ice sheet. The dashed line shows the extent of the ice sheet. The inset map of Bangladesh gives an indication of the relative size of Greenland. The ice in Greenland contains the equivalent of 7 metres of sea level rise.

Source: Inset data - Warrick and Oerlemans, 1990.

retreated strongly over the last 100 years. However, such evidence is only circumstantial. Confirmation requires observational data on accumulation, ablation and calving rates, which are meagre. Estimates of the mass budget of the Greenland ice sheet are therefore subject to large error.

3.3 HAS GLOBAL SEA LEVEL BEEN RISING AND WILL IT RISE IN THE FUTURE?

3.3.1 The Geological Past

On geological time-scales, large changes in sea level are evident. The clues to past changes in sea level can be found, for example, in the form of ancient wave-cut terraces or other features like coral reefs; in the unearthing and dating of life forms characteristic of marine, estuarine or terrestrial habitats; or, in more recent times, in human settlements, artefacts or historical accounts. From such evidence, it is clear that there have been large, globally-coherent changes in climate and sea level in the geological past.

For example, it is indicated from palaeo-climatic information that the Earth's climate has experienced changes in the past on time-scales ranging from several million years to a few years. In the relatively recent history of the Earth, the most notable phase of climate change began about 2 million years before the present (BP), generally known as the Pleistocene, characterised by glacial/interglacial cycles. These cycles occurred on time-scales of about 100,000 years when the climate was generally cooler than at present. During the Pleistocene, global surface temperatures have typically varied by 4°C to 6°C, and even up to 10°C to 15°C in some middle and high latitude regions of the northern hemisphere (see Chapter 2). These cool and warm phases of the glacial cycles were invariably associated with large accumulations of ice volume in the cool phase and sea-level rise in the warm phase (from glacial melting).

During the last glacial epoch, which reached a maximum around 18,000 years BP and ended around 10,000 years BP, large sections of northern North America and northern Europe were covered with continental glaciers. At the time of glacial maximum, the global temperature was 4°C to 5°C colder than today and global sea level was more than100 metres lower. Between 15,000 and 7,000 years BP, sea level rose quickly as the global temperature increased and the continental glaciers receded. From 7,000 years BP, sea level continued to rise due to isostatic crustal adjustments and tectonic effects, but at a slower, diminishing rate. Over much of the most recent millenium (excluding the present century), the rate has probably been around 0.0mm/yr to 0.2mm/yr on average.

On such geological time-scales, the connections between climate change and sea level are clearly apparent. Difficulties arise, however, in attempting to measure global sea level and to establish the climate-sea level connections on shorter time-scales and at finer temporal resolutions. In order to say how sea level might change in the decades ahead, it is necessary to ask how sea level has changed in the past decades and whether there is any evidence to link such changes with climate. What do the observational data tell us?

3.3.2 The Recent Past

For the last 100 years, sea-level records for many parts of the world are available from tide gauges. Tide gauges measure relative sea-level change with respect to a fixed benchmark on the land. As noted above, this means that any vertical movements in the land are included, and most tide gauges are located at the continental margins where such effects can be pronounced. In seeking to determine the eustatic component of sea-level change, these vertical movements "contaminate" the record and preferably should be removed from the data. Often, attempts are made to account for the long-term geological effects by examining Holocene (last 10,000 years) sea-level indicators (or using a geophysical Earth model) and substracting the long-term trend from the tide-gauge records. Furthermore, attempts to measure global mean-annual sea level involve averaging the tide-gauge measurements over the year, as well as over space, in order to minimize the effects of tidal, tectonic or other residual local and regional variations. Even so, the resultant estimates of global mean sea level are very "noisy", with substantial variations from year to year and from region to region.

Bearing in mind the data problems, many attempts have been made to estimate the trend in global sea level from tide-gauge records. Practically all studies agree that global sea level has been on the rise over the last 100 years. The best estimate is that over this period sea level has probably risen by 10cm to 20 cm. The implied rate of rise is probably higher than that of the preceding centuries (Woodworth, 1990). If, in fact, the rate of rise has increased in the present century, could it be that climate change is responsible? There is some evidence to suggest that this is the case.

During the last 100 years, the global temperature has increased by about 0.3°C to 0.6°C (although the warming has been far from uniform around the world). Mountain glaciers have, on the whole, also retreated significantly over that time period, and various analyses have indicated that the global glacier mass has also decreased as a consequence of warming (Meier, 1984; Grove, 1988). Although the observational data are scanty, various estimates suggest that melting of mountain glaciers could have added 1cm to 7cm to sea level over the last 100 years, although recent analyses support values at the lower end of the range (e.g. Oerlemans and Fortuin, 1992). In terms of thermal expansion, direct observations are too few to allow meaningful statements about global effects. Instead, reliance has to be placed on computer simulation models, which suggest that thermal expansion may have accounted for about 2cm to 6cm of the observed sea-level rise, given the history of greenhouse gas forcing changes over the last 100 years (Wigley and Raper, 1987, 1993). As mentioned above, the temperature and observations of the Greenland ice sheet tentatively suggest that Greenland has also been contributing positively to sea level, perhaps as much as 1cm to 4cm over the last 100 years (Warrick and Oerlemans, 1990).

BOX 3.4: **Past changes in global sea level**

On geological time-scales, large changes in sea level have resulted from climate change, especially as a consequence of glacial-interglacial cycles. At the height of the last ice age some 18,000 years ago, for example, sea level was more than 100 metres lower than today. Over the past century, sea level has risen by 10cm to 20cm, most likely in response to a global warming of about 0.3°C to 0.6°C over the same period. Thus, based on the past, there is an apparent connection between changes in global climate and sea level which should not be ignored in projecting the future.

Altogether, the evidence, though sparse, suggests a connection between climate and sea-level changes in the recent past. This provides some basis for believing that future global climate change will accelerate the rate of sea-level rise in the future.

3.3.3 The Near Future

In making future projections of global temperature and sea-level changes, it is preferable to use a single climate model (or set of linked models) in order to ensure internal consistency. Moreover, future rates of greenhouse gas emissions are extremely uncertain, so the results should be presented as scenarios, rather than actual "predictions". Let us look at some recent "best-estimate" scenarios of sea-level rise for "business-as-usual" emission scenarios (i.e. if no explicit attempts are made to reduce greenhouse gas emissions).

In 1990, the IPCC estimated that sea level could rise at an average rate of about 6mm/yr under business-as-usual conditions, reaching 66cm by the year 2100 (within a large range of uncertainty: 31cm to 110cm) (Warrick and Oerlemans, 1990). The 1990 IPCC projections are shown in Figure 3.5 and Table 3.1. (As the most recently published international scientific assessment, the IPCC projections of future sea-level rise are appropriate for use as scenarios for other chapters in this book and are consistent with the temperature projections provided in Chapter 2). The various contributions to the IPCC projections of sea-level rise are illustrated in Figure 3.6. The largest contributions are derived from oceanic thermal expansion, followed by mountain glaciers and small ice caps. The contributions from the

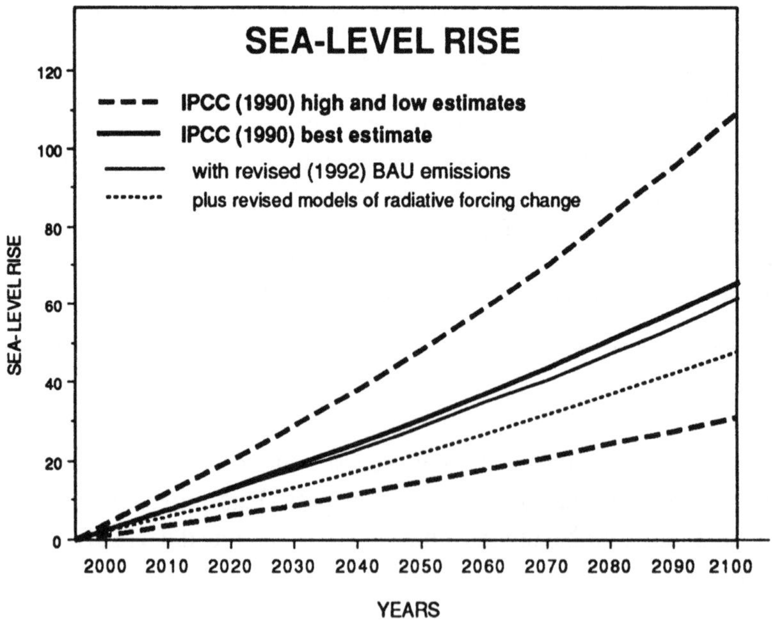

FIGURE 3.5: Projections of future sea-level rise

Source: IPCC projections are from Warrick and Oerlemans, 1990; revised curves from Wigley and Raper, 1992.

polar ice sheets are projected to be minor — negative in the case of Antarctica. But, it should be noted that the uncertainties regarding the ice sheets are the largest, given the massive annual turnover of ice involved and the lack of knowledge concerning their response to climate change.

TABLE 3.1: The 1990 IPCC projections of future global mean sea-level rise for the business-as-usual emissions scenario

Model Assumptions	Sea-level rise scenario by year (cm)			
	2010	2030	2050	2100
Low	4	8	15	31
Best-estimate	8	18	30	66
High	13	29	48	110

Source: Warrick and Oerlemans, 1990.

Research developments since the 1990 IPCC assessment have tended to lower projections of sea-level rise for several reasons (Warrick, 1993b). First, the IPCC, in its supplemental report (Houghton *et al.*, 1992) revised its emission scenarios, inferring a lower business-as-usual case (Leggett *et al.*, 1992). By itself, this implies a slower rate of global warming and, consequently, sea-level rise (as indicated by the thin solid line in Figure 3.5, from Wigley and Raper, 1992). Second, there has been progress in quantifying the effects of three factors which reduce somewhat the projections of future radiative forcing changes, namely: the terrestrial biospheric sink for carbon dioxide, the cooling effect of sulphate aerosols, and the ozone-depletion feedback — all of which tend to partially off-set greenhouse warming (see Section 3.2.1 above and Chapter 2). When these factors are taken into account as well, the projected sea-level rise is about 27 per cent lower than the IPCC 1990 estimate (as described by Wigley and Raper [1992, 1993] and as shown by the dotted line in Figure 3.5).

However, it should be emphasised that the uncertainties are very large. This applies particularly to the Antarctic ice sheet where the possibilities of ice-flow instabilities and relatively sudden "surges" of ice to the sea — although unlikely — cannot be lightly dismissed.

Although the range of estimates of future sea-level rise for business-as-usual (and other) scenarios is large and subject to revision, there are two general points that should not be overlooked. First, despite the large uncertainties, it is generally agreed that global warming should cause global sea level to rise. Few would argue that global warming should cause global sea level to fall. Second, even the implied lower projections would still represent rates of sea-level rise that are higher than those experienced over the last 100 years and that should, therefore, be of major concern to Bangladesh. It is in this global context that the specific implications for relative sea-level rise in the Bay of Bengal and its effects on Bangladesh must be considered.

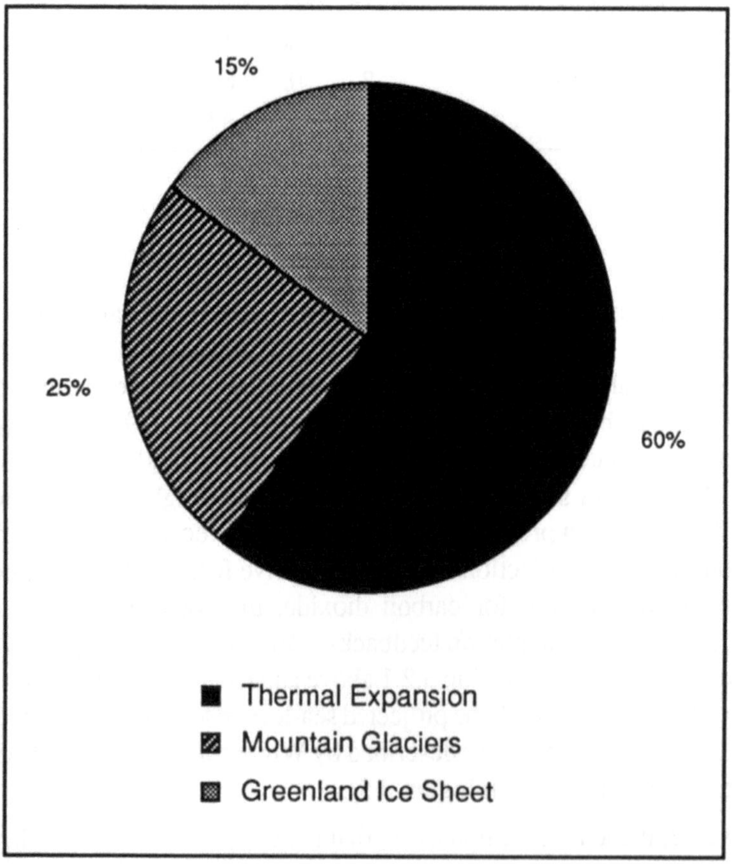

FIGURE 3.6: The relative contributions to the IPCC (1990) best-estimate projection of the 1990-2100 rise in sea-level (excludes Antarctica)

BOX 3.5: Projections of future sea-level rise

> The current scientific consensus is that global warming should cause global sea level to rise. The rate of rise, however, is very uncertain. In 1990, the Intergovernmental Panel on Climate Change (IPCC) estimated that with a "business-as-usual" scenario of greenhouse gas emissions, global sea level would be 66cm higher (and the world 3.3°C warmer) by the end of the next century, with a range of uncertainty of 31cm to 110cm. Subsequent projections suggest somewhat lower rates of rise. Nonetheless, these estimates suggest a future rate of sea-level rise that is 2 to 6 times higher than that experienced over the last 100 years.

3.4 HOW ARE RELATIVE SEA LEVELS CHANGING IN THE BAY OF BENGAL?

Future scenarios of global sea-level rise are not sufficient for estimating changes in relative sea levels in the Bay of Bengal. For the coast of Bangladesh, relative sea level is affected by three main factors: the level of the sea surface itself; the rates of sediment supply to the coast and its distribution and deposition; and the long-term rates of deltaic subsidence (or uplift) from tectonic and isostatic (from the weight of sediments) forces. The first two factors can be influenced by climate change on the decadal time-scale, whereas the latter cannot.

As discussed in Chapter 7, the coast of Bangladesh appears, in general, to have been in a state dynamic equilibrium over the last 100 years. In other words, the sediment supply to the coast has, on average, been sufficient to balance deltaic subsidence. In future, the direction and rate of relative sea-level change will depend not only on global sea-level rise, but also on the extent to which sediment supply — which itself is affected by climate and sea-level change — can keep up with subsidence (Resource Analysis, 1992; see Chapter 7 for discussion of sediment supply).

Thus, knowledge of the rates of vertical land movement is crucial to understanding the relative importance of possible changes in global sea level for Bangladesh. This requires an examination of the geological structure of the region, including the tectonic forces and sediment loadings that affect such land movements.

3.4.1 Geological Setting

In broad terms, the geology of Bangladesh is controlled by the north-eastwards movement of the Indian plate and its collision with the Eurasian plate (Figure 3.7). The result of the collision of the Indian and Eurasian plates is the formation and continued uplift of the Himalayas.

Bangladesh occupies most of the Bengal Basin, a major geotectonic element of the Assam-Himalayan region (Figure 3.8). The Basin is bounded in the west by the Indian Platform, in the north by the Precambrian Shillong Massif, and in the east by the Arakan Yoma-Naga folded system (part of the Indo-Burma orogenic belt). The Basin's southern boundary is unclear, although geophysical evidence suggests that the Basin continues south for at least 200km from the Bangladesh coast. The Basin can be divided into three units: shelf, hinge and foredeep. The shelf and hinge zones are in the north-east of Bangladesh and well inland from the coastal zone. The foredeep, or the active part of the Basin, occupies the present central coastal zone and the adjacent part of the west coastal zone.

3.4.2 Tectonic Influences

Because of its location between one of the world's major subduction faults in the north and a major transform fault in the east, the Bengal Basin and its adjacent area is an active tectonic region. Large areas within Bangladesh have been uplifted in recent times and some areas are still sinking. The tectonic activity may be due to the presence of a major fault at depth or a subsiding trough along the axis of the main rivers. This subsiding structural zone may be the foredeep of the transform fault in the east. The regional tectonic setting and major tectonic elements of the Bengal Foredeep include the following: the Sylhet Trough; the Tangail Tripura Uplift; the continental part of Patuakhali Depression; the offshore part of Patuakhali Depression; and the central part of Folded Flank.

The geology of Bangladesh is dominated by sedimentary formations resulting from deposits accumulated over the last two million years of the Pleistocene. Four distinct regions of Bangladesh are apparent, two of which flank the east and west, and two of which lie within the Basin. The two major Pleistocene units in the central part of the Bengal Basin, the Barind and Madhupur Tracts, exhibit evidence of active faulting. The Barind Pleistocene unit has apparently experienced domal uplift. The western part is higher toward its centre and slopes both to the north and south. The eastern part is lower and consists of a non-uniform surface tilted towards the southwest. The Madhupur Tract, bounded on the west by a series of en-echelon

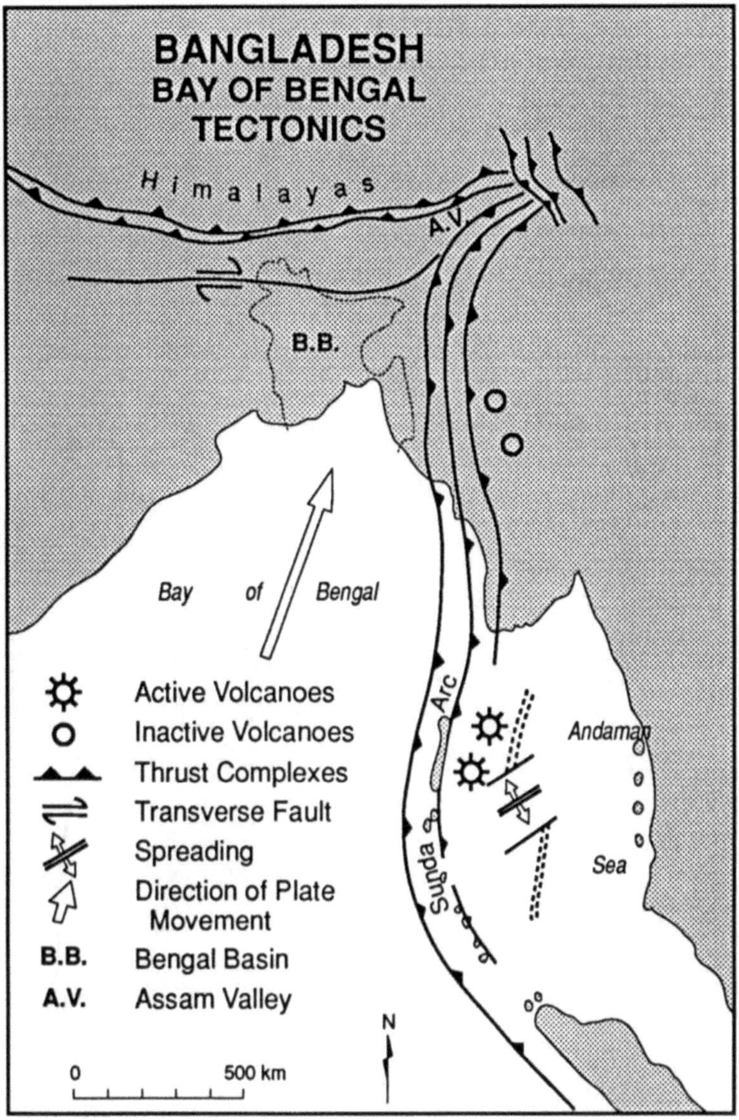

FIGURE 3.7: Diagrammatic sketch of the present tectonics of the head of the Bay of Bengal, the Bengal Basin (B.B.), the Assam Valley (A.V.) and the Andaman Sea

Source: Anwar, 1989.

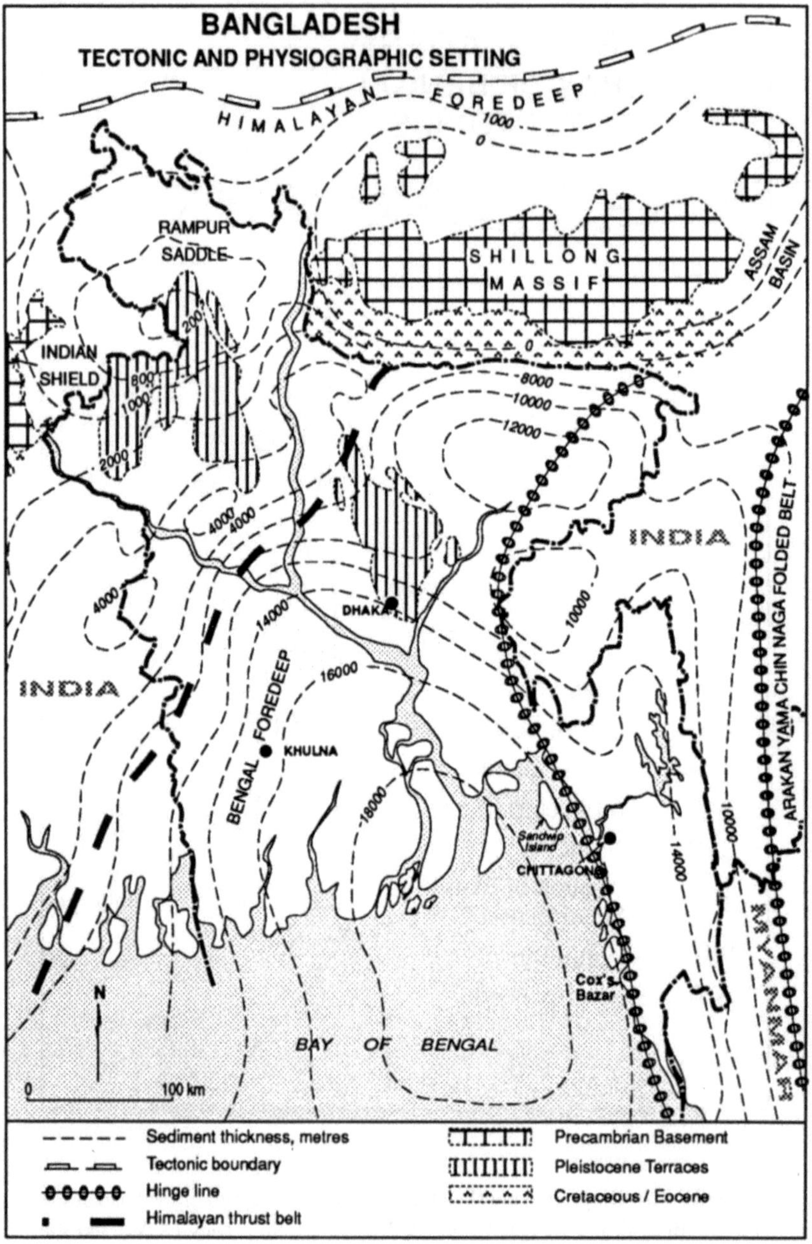

FIGURE 3.8: Regional tectonic setting and major elements of the Bengal Foredeep including sediment depth.

Source: Anwar, 1989.

faults, has been uplifted and tilted towards the northeast. The uplift has caused some of the old Brahmaputra distributaries to become antecedent which probably helped effect the diversion of the river to its present course along the Jamuna.

Areas adjacent to the Madhupur unit have also been affected by structural activity. The Sylhet Basin is sinking; the areas of rapid subsidence are thought to be related to movement along the major fault flanking the uplifted Shillong Plateau. A large area of recent floodplain south and east of the Madhupur Tract has been slightly uplifted, resulting in the establishment of a distinct and mappable portion of the floodplain which has been named the Tippera surface. Uplift of the Barind Pleistocene unit and the Tippera surface coupled with subsidence of the Sylhet Basin and the deltaic plain have resulted in torsion of the crust in Bangladesh. It is believed that surface faulting of the Madhupur has resulted either from the torsion of the region, the effect of shear along a major fault at depth, or a subsidence following approximately the axis of the Jamuna-Ganges-Meghna river System.

The Chittagong coastal plain of the east coastal zone (see Chapter 7) has a geology that is distinct from the rest of the coastal zone. This area is part of the Arakan Yoma-Naga folded system. As a result, it may be expected to have characteristic subsidence/uplift rates. There has been no analysis of geological subsidence/uplift for this zone. Consequently, the recent tectonic movements of this zone can only be inferred from its geomorphology. Evidence from raised shorelines suggests that this zone has been relatively stable, or possibly uplifting, during the recent geological past. However, detailed studies are required to verify this observation.

3.4.3 Sediment Loading

As they are uplifted, the Himalayas are continuously reduced by erosion. Rocks and sediment eroded from the mountains by ice, snow and rain are transported towards the sea by rivers. The Ganges and Brahmaputra rivers, originating in the Himalayas, have provided a huge supply of sediment to the Bengal Basin in the south.

In general, the oldest deposits in the Bengal Basin are the sandstones, claystones and shales of the Cretaceous period (MPO, 1986). Much later, the Pleistocene sediments were deposited as a floodplain of the former Ganges-Brahmaputra river system. The Pleistocene sediments are highly oxidised and are typically reddish-brown, forming the Madhupur clays noted above. Because of their antiquity and long periods of exposure, they are characteristically more compacted and weathered than the subsequent overlay of more recent alluvium. The deltaic region is constantly subsiding due to the compaction of recent sediments as well as to structural down-warping.

The maximum loading, and thus the maximum crustal subsidence, can be expected to occur where the thickness of sediment is greatest. This occurs around the Hatia-Barisal-Patuakhali area where the sediment is up to 18km thick (MPO, 1986). The thickness of sediment in the Bengal Basin is shown in Figure 3.8. In general, crustal subsidence may be expected to be less outside the zone of maximum sediment thickness. However, there are likely to be significant local variations in subsidence rates that are determined by tectonic activity and deep crustal features associated with the collision of the Indian and Eurasian plates.

Gravity surveys have resolved a gravity high which runs south-west to north-east across the central coastal zone (Khan, 1991). Gravity highs within relatively homogeneous deltaic sequences can mean raised deep crustal rock. Such modification to the basement of the Bengal Basin could have reduced the deposition of sediment there in the past, thereby reducing present-day subsidence rates, although the extent of this influence is uncertain.

3.4.4 Estimated Rates of Subsidence and Uplift

The rates of subsidence and uplift are not known very well for Bangladesh. Estimates have been derived from early historical and anecdotal evidence as well as from modern methods.

For example, Fergusson (1863) made an elaborate study of the ancient capital cities of the Eastern Gangetic Plain, and found that 3000 years ago no city existed on the bank of the Ganges as far down the river as Allahabad. He also found that 5000 years ago the sea (not the tide) extended up to Rajmahal (about 320km north of Calcutta), and that Bengal was a vast lagoon or shallow sea. Using this information, Col. Burrard (1933) estimated the rise of the plain in the neighbourhood of Calcutta as 0.33 metres in 100 years. The Geodetic Survey of India (1929 to 1930) reported that the alluvial plains of northern Bengal and Bihar had risen by at least 0.3 metres during the previous 70 years. Some 40 years of tidal records at Kidderpore (1881 to 1920) showed that Calcutta was not sinking.

The difficulty with such estimates is that they are inferred from observations of relative sea-level change. Because relative sea level can also be affected by sedimentation and the actual level of the sea surface, as well as by vertical land movement, isolating the causal factors by such methods is fraught with difficulty. More reliable estimates require modern scientific methods.

Recent scientific evidence of subsidence rates in the Bay of Bengal is extremely limited. In general it is known that major areas of subsidence occur in Sylhet Basin, Faridpur Trough, Chalan Beel, Dhaka Depression, and Khulna-Sundarbans. However, the only systematic study to determine the rates of subsidence was carried out by Morgan and McIntire (1959) who found evidence of both subsidence and uplift. In the deltaic coastal zone they found only subsidence, estimated to be 0.6mm/yr to 5.5mm/yr, and possibly up to 20mm/yr in the central coast (also see MPO, 1986). However, it must be stressed that these estimates were derived from hand-made borings in coastal sediments, and from archaeological evidence. Nevertheless, their estimated subsidence rates are consistent with those inferred from the bore hole sediments and radiocarbon dates obtained by Umitsu (1993), notwithstanding the difficulties of distinguishing between sediment processes and structural downwarping. More studies using modern geological and geochronology techniques, including carbon-dating, need to be carried out.

Recent scientific evidence of uplift is difficult to find in Bangladesh. Geodetic levelling over the Shillong Plateau shows a rise of 25mm over the period of 1910 to 1977, for which the average annual rate of rise is 0.37mm (MPO, 1986).

3.4.5 Measurements of Relative Sea-Level Change from Tide Gauges

Over shorter time-scales, tide gauges, which measure relative sea level, can provide important information for estimating the rates of vertical land movement and changes in the level of the sea surface. The Bangladesh Water Development Board, the Bangladesh Inland Water Transport Authority, and the Chittagong Port Authority have been operating an extensive hydrological network for several decades. A subset of the Bangladesh Water Development Board tide-gauge stations that could potentially be useful for these purposes in the Bay of Bengal is shown in Figure 3.9.

The existing tide-gauge data have shortcomings that presently limit their usefulness for interpreting trends in relative sea level. First, the time span of most of these datasets would be considered too short for reliable determination of trends in relative sea level on the decade-to-century time-scale. Many of the station records are little more than a decade in length, whereas ideally much longer periods are required in order to account for the effects of seasonal, interannual and decadal variability in the data (Shennan and Woodworth, 1992).

Second, a preliminary trend analysis of selected records from Bangladesh reveals some very large rates of change (a range of about +60 mm/yr to -50 mm/yr) which imply relative sea-level changes of the order of 5 to 6 metres per century. In view of the fact that the coast of Bangladesh has been relatively stable over the last 100 years, it is unlikely that such large rates could be attributed to a combination of absolute sea-level change and long-term vertical land movements. Otherwise, vast tracts of Bangladesh would have appeared or disappeared during this century! Rather, it is likely that the data are "contaminated" by other factors, such as: movement in the jetty or other structure on which the tide gauge is attached in relation to local survey benchmarks; the effects of sedimentation on tidal regimes; changes in channel and off-shore morphology; and changes in river flow which could affect water density and dynamic water heights.

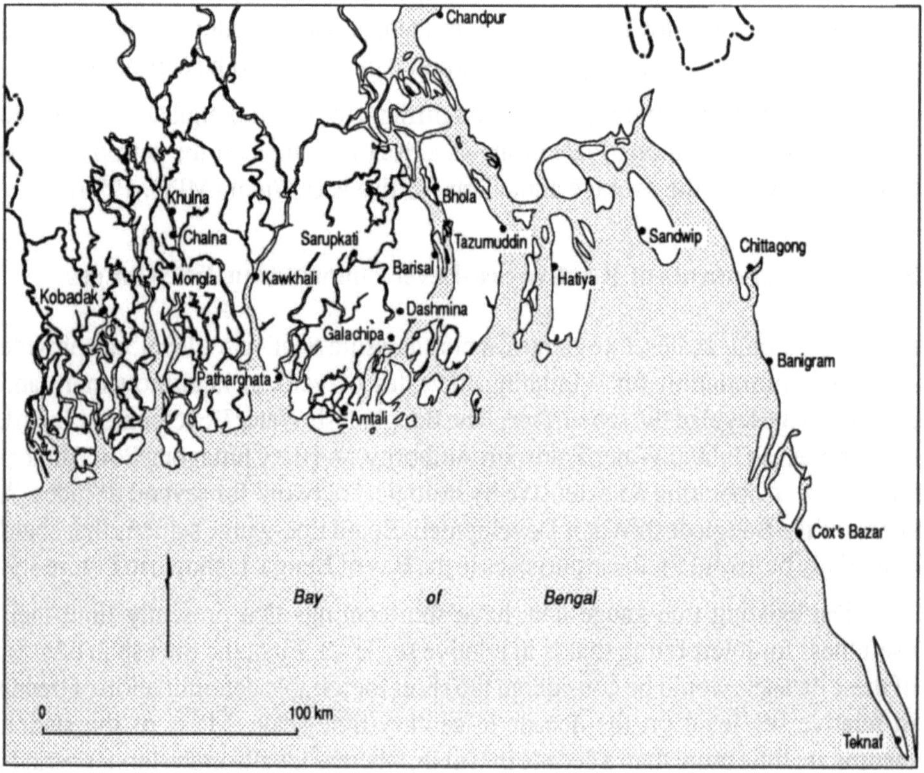

FIGURE 3.9: The location of selected water level gauges in the Bay of Bengal that are potentially useful for examining trends in sea level

Source: Sites extracted from the Bangladesh Water Development Board hydrological network.

For these reasons, a careful assessment of data quality needs to be carried out on existing tide-gauge records in order to evaluate their utility for estimating trends in relative sea level for Bangladesh. For instance, it is vital to determine whether each station has been sinking or rising with respect to local survey benchmarks. This is necessary in order to extract local land or gauge movements from the data. A reference point at Cox's Bazar was established in 1908 following geodetic and triangulation surveys carried out for the subcontinent including Burma (Myanmar), and no attempt or survey has so far been made for a correction or recalibration of this reference point. These are potentially important deficiencies of exising tide-gauge data which need to be rectified for purposes of identifying trends and their contributing factors.

In short, the tide-gauge data do not yet allow an unambiguous estimate of the regional and local trends in relative sea-level change and their causes.

3.5 HOW DO CYCLONE-INDUCED STORM SURGES AFFECT SEA LEVEL?

Coastal flooding associated with storm surges — the changes in water level generated by storms passing over the sea — is one of the world's foremost natural hazards in terms of loss of life and property damage. Nowhere are these losses more serious than in the Bay of Bengal. Globally, about 60 per cent of the fatalities due to storm surges have occurred in the low-lying arable coastal areas of the countries bordering the Bay and the adjoining Andaman Sea, with Bangladesh alone accounting for about 40 per cent (Murty, 1984). The steady growth in population continuously increases the potential for disaster.

BOX 3.6: Relative sea-level change in the Bay of Bengal

> For Bangladesh, relative sea-level change depends only partially on changes in the volume of the ocean. Geological factors — including tectonic movements, rates of subsidence/uplift and sedimentation rates — also play an important role, as they affect the level of the land in relation to the sea. However, locally and regionally, the rates of vertical land movements and their causes are not very well known. Increased knowledge of the geological components of sea-level change is crucial for making future projections of sea-level change for Bangladesh.

3.5.1 Factors Related to Surge Generation

Storm surges are atmospherically-forced oscillations of the water level in a coastal or inland water body, lasting for periods of a few minutes to a few days. By this definition, storm surges are distinct from wind waves and swells, which have periods in the order of a few seconds. Storm surges belong to the same class of wave as tides and tsunamis, that is, long gravity waves.

The surges are caused by tropical cyclones that develop in the southeast part of the Bay or in the Andaman Sea during the pre-monsoon and post-monsoon seasons. The cyclones move along tracks that frequently cross the continental shelves bordering the Bay. A storm surge is generated partly by the resulting variations in atmospheric pressure, but the main contribution comes directly from the winds, often of exceptional strength, acting over shallow water (Murty *et al.*, 1986).

The results of studies of cyclone-surge generation show that the wind stress is less effective in generating surges in deep water than in shallow water. In shallow water, and in particular on continental shelves and near to coasts, wind stress forcing dominates. An accurate knowledge of the wind field is therefore essential in the prediction of storm-surge height since, for example, a 10 per cent error in wind speed translates into a 20 per cent error in the wind stress that generates the surge.

In deep water surges are produced mainly by changes in atmospheric pressure. The effect of atmospheric pressure change on surge height is independent of water depth. In simple terms, a 1mb ($=10^2$Pa) decrease in pressure produces a 1cm increase in surge height. For a hypothetical cyclone with central pressure 100mb below that at the periphery, the maximum pressure-induced surge would be 100cm.

The interactions of storm surges and tides can be critical. In particular, where the tides are large and the water shallow, interaction occurs between the tide and surge making the two components inter-dependent. Thus, the surge produced by given meteorological forces at a certain state of the tide may differ significantly from the surge resulting from identical forcing at a different stage of the tide or under the assumption that there is no tide. This means that the surge and tide cannot be treated independently without causing substantial errors in the predicted total change in water level. To solve the complete problem, including these important effects, numerical techniques are required. (Details of these techniques, boundary conditions, and related topics are beyond the scope of this review but can be found in Murty, 1984.)

3.5.2 Surge Generation in the Bay of Bengal

The coasts of Bangladesh and Myanmar (Burma) and the Bay of Bengal coast of India have been subjected to frequent severe cyclones and storm surges (Figure 3.10 and Annex 3.1). Studies of the meteorology of the region provide a good knowledge of the origins, frequency of occurrence and behaviour of the cyclones that generate the surges. The Bay of Bengal is a breeding ground for tropical storms and depressions, the yearly average from 1890 to 1969 being 13 (Raghavendra, 1973). A warm pool of air usually forms at the 200mb level, 3 to 4 days before the advent of cyclogenesis. The associated low pressure centres over the Bay intensify into depressions, some of which intensify further into cyclonic storms. A few of these in turn develop into severe cyclonic storms of hurricane intensity. Low pressure areas mainly form in the Andaman Sea, on the eastern side of the Bay of Bengal. Occasionally, lows which are remnants of cyclones from the China Sea move into the Bay from the east.

Tropical cyclones capable of generating surges usually occur during the monsoon transition periods, but rarely during the monsoon season itself. For the pre-monsoon storms (April to May) the strongest winds are usually in the south-east or eastern sector, whereas for the post-monsoon storms (September to December) the region of maximum winds is to the north of the storm centre.

The Bay of Bengal is essentially a northward extension of the Indian Ocean, with depths decreasing from 4000m in the south to 2000m in the north. The Andaman and Nicobar Islands form an almost continuous barrier which nearly makes the eastern part of the Bay (referred to as the Andaman Sea) into a separate water body, at least from the point of view of tides. The width of the shelf is fairly small, typically 35km for both eastern and western margins of the Bay. However, at the head of the Bay, the Ganges delta forms a much wider shelf, typically 175km wide. This shelving effect, combined with the funnel-like configuration of the coastline, amplifies the tides and surges at the head of the Bay, especially on the Bangladesh coast. The only significant morphological feature of the shelf at the head of the Bay is a canyon-like structure, referred to as the "Swatch of No Ground". This canyon (about 150km long, 22km wide, over 1000m deep at its southern end and some 270m deep at its northern end) extends in a north-eastward direction to a point south of the Pussur River estuary and must obviously influence tide and surge propagation into the area, although its exact role is uncertain.

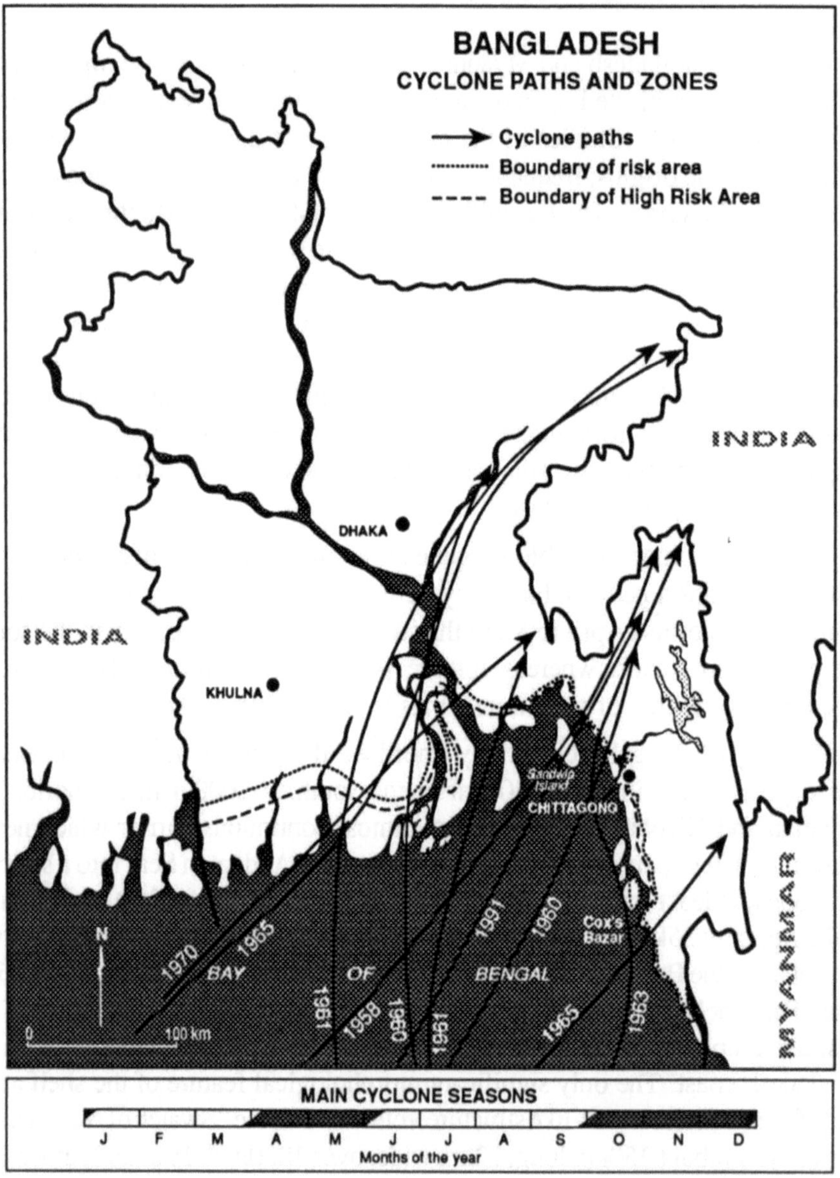

FIGURE 3.10: High risk cyclone hazard areas of Bangladesh and cyclone paths for the period 1958-1991

Source: FAP7, 1992; MCSP, 1993.

3.5.3 Cyclone Movement in the Bay of Bengal

The movement of a cyclone in the Bay of Bengal follows a sequence of stages as it approaches the coast. In view of the large water depth, atmospheric pressure effects will dominate over the deeper part of the Bay. The surge elevation reflects the variation in surface pressure in producing a mound of water which is centred under and moving with the cyclone. Very strong wind-generated currents may also occur and can influence sea level where they impinge on the continental shelf.

Closer to shore, the main surge effect is generated by the wind acting over the continental shelves. The maximum onshore winds, and hence also the maximum surge levels, tend to occur on the right of the storm track as it crosses the coast. A negative surge associated with offshore winds on the left of the storm centre may also appear. The surge maximum occurs near to the time at which the storm centre crosses the coast. The speed and angle of approach of the storm, relative to the coastline, influences the magnitude of surge, and convergence of coastlines in a bay or estuary can produce substantial amplification of the surge. Other dynamical factors may also modify the response. In particular, the resonant coupling mechanism (which occurs when the speed of a travelling atmospheric disturbance matches that of free gravity waves) could be significant in the shallow coastal waters of the Bay of Bengal.

The extreme shallowness of the water in some near-shore areas of the Bay of Bengal coast, the focussing effect of inlets and estuaries, and the absence of shore defences to protect the low-lying coastal lands may lead to the formation of a 'storm wave'. This is frequently described by eye-witnesses as a 'wall of water', which can travel several kilometres inland, flooding affected areas to a depth of as much as 13m. Some accounts suggest that the accumulating water associated with the surge may be held back by the river flow and ebbing tide in estuaries, eventually advancing upstream as a bore, overtopping the river banks and inundating low-lying land as a storm wave (e.g. the Calcutta cyclone of October, 1864). Clearly, non-linear interactions with tides play a vital part in this stage in the development of the surge.

In the northern reaches of the Bay of Bengal and in the Andaman Sea, the tidal range is comparable with probable surge amplitudes. The relative timing of high tide and peak surge then becomes very important. For instance, in the disastrous inundation of Bangladesh in November 1970, the surge peak occurred at high tide and killed more than 220,000 people (see Chapter 5). In contrast, a comparable surge on October 31, 1960, occurred at low tide and caused far fewer casualties. The apparent importance of non-linear effects suggests that dynamical tide-surge interaction must also play a part here (Flather and Khandker, 1993). The tidal range decreases to the south so that along much of the east coast of India the tides are small.

With respect to the future, a key issue is whether or not the frequency and intensity of the severe cyclones that generate storm surges will increase as a consequence of global warming. Although it is well known that a minimum sea-surface temperature of about 27°C is required for their development and intensification, there is little observational evidence to support the notion that further warming of the sea surface will increase either the frequency (Raper, 1993) or intensity (Evans, 1993) of tropical cyclones, especially in the Northern Indian Ocean. The latest international reviews of the subject indicate that there is as yet no sound basis for predicting an increase in severe tropical cyclones due to global warming (Mitchell *et al.*, 1990; Gates *et al.*, 1992; Lighthill *et al.*, 1994; also refer to Chapter 2).

3.5.4 Increasing the Accuracy of Storm-Surge Prediction

So far as surge prediction methods are concerned, it seems highly probable that numerical models offer the best solution in the near and medium term. Far too few past surges have been recorded adequately enough to permit statistical or empirical surge forecasting, particularly in cases where the state of the tide is significant. At present, in fact, there are too few records to allow verification of numerical models or other prediction techniques. Of the few useful surge records that exist, most are from Bangladesh, a few are from Myanmar and virtually none are from India.

In this regard, progress in storm surge data collection could be made rapidly at reasonable cost. In a report (Murty and Henry, 1983) to the World Meteorological Organisation-Economic and Social Commission for Asia and the Pacific Panel on Tropical Cyclones, it was proposed that an international effort be organised for the purpose of collecting and publishing complete dossiers of relevant data on future major surges in the Bay of Bengal. On average, about four large surges occur per year, but since each affects only a small portion of the vulnerable 4000km of coastline, this data-gathering scheme would have to be pursued for one or two decades (at least) to accumulate an adequate body of verification data. For a moderate annual cost, a programme of this type, begun in good time, could provide a better basis for improved surge-forecasting capability than any short-term effort, no matter how heavily funded.

Collecting data on a storm surge requires a fair degree of prior organisation, since most local services are disrupted or destroyed in the inundated area. Complete meteorological records are required for several days before and sometimes after the time of cyclone landfall. Aerial or satellite photographs that show the extent of inundation, together with confirmatory ground observations, are also needed. Water level records during surge episodes pose one of the most serious difficulties, due to the relatively sparse network of permanent tide gauges around the Bay and the fact that the types of gauge in use often fail to record water levels much beyond their normal working range. Accurate observation of flood marks can nevertheless give highly useful information. For instance, flood marks inside large buildings should give a good estimate of water levels which are relatively unaffected by the influence of wind waves.

BOX 3.7: Cyclones, storm surges and sea level

> On short time-scales (minutes to days), changes in sea level from cyclone-induced storm surges have devastating consequences for Bangladesh. The last major storm surge in Bangladesh, in April 1991, left more than 138,000 dead in its wake. Such surges are caused by the change in atmospheric pressure and the wind stress (especially in the shallower water near the coast) associated with severe tropical cyclones. At present, there is no firm evidence to conclude that global warming will increase either the frequency or intensity of cyclones in the Bay of Bengal.

When all available relevant data for a surge have been collected, it is important that it should be assembled to permit convenient distribution to interested researchers and to ensure safe preservation of the data. It is not extravagant to think that records collected carefully in our time could be immensely useful to surge forecasters a century or more from now.

3.6 SUMMARY, KNOWLEDGE GAPS AND RESEARCH NEEDS

For Bangladesh, the issue of sea-level rise is of vital concern. The health and livelihood systems of virtually millions of people who inhabit the vast tracts of low-lying coastal land are under constant threat from changes in sea level. The effects of sea-level rise can be slow and insidious, as through the erosion and salinisation of land. Or they can be abrupt and catastrophic, as through the sudden destructive inundation from storm surges caused by severe tropical cyclones. For Bangladesh, the key issue is whether global warming will cause sea level to rise and worsen the problems currently being encountered at these time-scales.

Globally, a warmer climate is likely to cause global sea level to rise. There are two major contributing factors to such a rise: expansion of the oceans as they warm, and increased melting of land-based ice masses. Projections of future sea-level rise vary widely and the uncertainties are large. The current best estimates are that global sea level will rise by 22cm to 30cm by the middle of the next century, at a rate of about 4mm/yr to 5mm/yr. This compares to a rate of rise of 1mm/yr to 2mm/yr over the past 100 years.

Regionally, however, relative sea level depends on the vertical movement of the land as well as the ocean surface. In the Bay of Bengal, such land movements can be high. Because of the enormous weight of sediments deposited over millennia by the Ganges, Brahmaputra and Meghna river system, the coastal region is subsiding. If it were not for the continual accumulation of sediments that build up the land, vast areas of Bangladesh would have long ago disappeared under the sea. Knowledge of the dynamic balance between subsidence and sedimentation is thus crucial to estimating possible future changes in relative sea level in Bangladesh. At present, such knowledge is meagre. *It is therefore imperative that research be carried out to determine the rates of subsidence within the coastal zone of Bangladesh, as well as the rates of sediment supply and its distribution within the coast zone.*

Locally, attempts to measure the trends in relative sea level depend on records from tide gauges. Tide gauges measure the level of the sea in relation to a fixed benchmark on the land. A number of tide-gauge records are available, although the lengths of record are usually too short to establish the trends with confidence. These limited data indicate that some very large changes (positive and negative) in relative sea level are occurring locally (an order of magnitude larger than global sea-level rise). However, the data quality is questionable. *Investigations are required to establish the veracity of the tide-gauge data and to rectify any problems in instrumentation, recording and coverage.*

As we look to the future, monitoring of sea level should be a crucial element in Bangladesh's management strategies for coastal areas. In this regard, *there needs to be an assessment of the feasibility of implementing the latest technologies for precision measurement of sea-level variations, with the aim of identifying the various contributions from the sea and land components of relative sea-level changes.*

A key issue for consideration in Bangladesh is how global warming and sea level rise could affect storm surges. Nowhere in the world do storm surges occur with such destructive force and loss of life as in the Bay of Bengal. The storm surges are caused by the low pressures and, especially, the high winds associated with severe tropical cyclones in the Bay of Bengal. In the long term, a crucial issue is thus whether the frequency and magnitude of cyclones would increase in a warmer world. Unfortunately, climate modellers are not yet able to provide clear answers. In the short term, however, the forecasting of individual cyclone-induced surges can be improved through development of numerical models. Improvements in the prediction of surge heights and locations have the potential for saving lives and property. Currently, the lack of observational data is a hinderance to model development. Thus, *it is crucial that systematic collection of data on individual storms and surges be undertaken for purposes of storm surge model development and validation.*

Overall, sea-level rise on various time-scales (from hours to decades) is a major problem confronting Bangladesh. Whether global warming will exacerbate the problems presently encountered is not clear. It should not be assumed that global warming will necessarily lead to a permanent flooding of the coastal zone or to an increase in storm surge events. The processes involved are complex and not yet well understood. There is an urgent need to take the precautionary steps of improving the data and knowledge necessary to develop a clearer understanding of the likely implications for the future of Bangladesh.

ANNEX 3.1

Storm Surges in the Bay of Bengal Affecting Bangladesh During the Period 1584 to 1991

Date	Location	Damage
1584	Near Backergunge, Bangladesh	200,000 deaths
6 June, 1822	Barisal/Backergunge, Bangladesh	100,000 deaths 100,000 cattle lost
16 May, 1869	Khulna	Surge entered through the Pussur and Haringhata Rivers and inundated a vast area causing damage to life and property. 250 people died in Morelganj.
October, 1872	Cox's Bazar, Bangladesh	Many deaths, numerous cattle lost
31 October, 1876	Backergunge, Bangladesh	This was the infamous Backergunge cyclone. Peak water levels in Bangladesh varied from 3m to 15m, including tide and wind waves. 100,000 deaths in Bangladesh (then part of India). The India Meteorological Department was expanded following this surge. Gill (1975) and Khalil (1992) put the death toll at 400,000, but this could be an over estimate.
27 October, 1893	Noakhali	Lot of destruction Inundation subsided in 2 hours
October, 1895	Sundarban/Bagerhat Subdivision, Bangladesh	Damage estimates not available
24 October, 1897	Kutubdia Island, Chittagong, Bangladesh	175,000 deaths, thousands died in the epidemic which followed the calamity
May, 1898	Teknaf, Bangladesh	Damage estimate not available
November, 1901	Western Sundarbans, Bangladesh	Damage estimate not available
October, 1909	Chittagong, Bangladesh	Damage estimate not available

Date	Location	Damage
December, 1909	Cox's Bazar, Bangladesh	Damage estimate not available
April, 1911	Teknaf, Bangladesh	Damage estimate not available
1912		40,000 deaths in Bangladesh
May, 1917	Sundarbans, Bangladesh	Damage estimate not available
20-25 September, 1919	Barisal, Bangladesh	40,000 deaths in Bangaldesh
May, 1926	Cox's Bazar, Bangladesh	Damage estimate not available
26 May, 1941	Between Barisal and Noakhali, Bangladesh	4m surge, 7,500 deaths
17-19 May, 1948	Between Chittagong and Noakhali, Bangladesh	Damage estimate not available
15-20 November, 1950	Patuakhali, Bangladesh	Damage estimate not available
16-19 May, 1958	East Meghna Estuary, Bangladesh	Damage estimate not available
21-24 October, 1958	Noakhali and West Meghna Estuary, Bangladesh	Up to 2m surges in Bangladesh and West Bengal State of India
29 May, 1960	Sundarbans, Bangladesh	Peak surge of 3.2m
9 October, 1960	Noakhali, Meghna Estuary, Bangladesh	1.5m tide and 5.1m surge 3,000 deaths, 200 kph winds
30 October, 1960	Noakhali, Chittagong, Bangladesh	6.6m surge, 8,149 deaths
9 May, 1961	Noakhali, West Meghna Estuary, Bangladesh	3m surges, 11,468 deaths
30 May, 1961	Chittagong, Cox's Bazar, Noakhali, offshore islands, Bangladesh	6m to 8m surge, peak water levels up to 9m. 10,466 deaths
26-30 October, 1962	Feni, Bangladesh	Damage estimate not available

Date	Location	Damage
28 May, 1963	Noakhali, Chittagong, Bangladesh	Peak water levels up to 8.1m, 11,520 deaths
5-8 June, 1963	Jessore, Bangladesh	Peak surge 3.1m
25-29 October, 1963	Teknaf, Bangladesh	Damage estimate not available
11 April, 1964	Bangladesh	196 deaths
11 May, 1965	Barisal and Noakhali, Bangladesh	6m water levels, 19,270 deaths
31 May, 1965	Chittagong, Bangladesh	Up to 7.6m surges, 12,000 deaths
5 November, 1965	Bangladesh	Up to 3.7m surges
11-15 December, 1965	Cox's Bazar, Bangladesh	Up to 8.8m water levels 873 deaths
1 October, 1966	Chittagong, Bangladesh	Peak water levels up to 10m 850 deaths
1 November, 1966	Bangladesh	Peak water levels up to 10m
12 December, 1966	Cox's Bazar, Bangladesh	Damage estimate not available
10 October, 1967	Noakhali, Bangladesh	Peak water levels up to 9.3m
23-24 October, 1967	Cox's Bazar, Bangladesh	Up to 7.6m surges
17 April, 1969	Bangladesh	75 deaths
10 October, 1969	Khulna, Bangladesh	Peak water levels up to 8m
7 May, 1970	Cox's Bazar, Bangladesh	Up to 5m surges
22-23 October, 1970	Chandpur, Bangladesh	4.7m surge, 300 deaths
12 November, 1970	Khulna to Chittagong and offshore islands, Bangladesh	Up to 9m surges, 275,000 deaths The often quoted 500,000 deaths could be an over-estimate

Date	Location	Damage
8 May, 1971	Meghna Estuary, Bangladesh	Up to 5m water levels
30 September, 1971	Chandpur, Bangladesh	Surge plus tide of 5m
5-6 November, 1971	Chittagong, Bangladesh	Water levels up to 5.5m
28-30 November, 1971	Sundarban coast, Khulna, Bangladesh	1m surge
16-18 November, 1973	Barisal, Bangladesh	1.0m surge. Peak water level including wind waves 3.8m
9 December, 1973	Patuakhali coast and offshore islands, Bangladesh	Peak water level 6.2m Peak surge 4.5m, 183 deaths
13-15 August, 1972	Khulna, Bangladesh	6.7m surge
28 November, 1974	Chittagong, Bangladesh	4.9m surge 20 deaths 50 injured, 280 persons missing, 1,000 cattle killed, 2,300 houses destroyed
9-12 May, 1975	Bhola, Cox's Bazar, Khulna, Bangladesh	5 deaths, 4 injured, 36 fishermen missing
5-7 June, 1975	Chittagong, Bangladesh	Peak surge 4.0m
24-28 June, 1975	Bangladesh	Tide and surge 4.8m
8-12 November, 1975	Barisal, Noakhali, Bangladesh	Maximum surge 3.1m
20 October, 1976	Meghna Estuary, Bangladesh	Tide plus surge 5.0m at Companiganj
20 November, 1976	Chittagong, Bangladesh	1.0m surge, 2.1m tide
12-13 May, 1977	Sundarban, Bangladesh	0.6m surge, 0.7m tide
30 September to 3 October, 1978	Khulna and Sundarban coast, Bangladesh	Damage estimate not available

Date	Location	Damage
10 December, 1981	Bangladesh	2m surge, 15 deaths
15 October, 1983	Chittagong coast near the Feni River, Bangladesh	43 deaths, substantial damage
9 November, 1983	Chittagong, Cox's Bazar, Bangladesh	300 fishermen missing, substantial damage
24 May, 1985	Cox's Bazar, Chittagong, Sandwip, Bangladesh	4.3m surge, 11,069 deaths 94,000 houses destroyed
9 November, 1986	Chittagong, Bangladesh	14 deaths, substantial damage
29 November, 1988	Khulna coast near Raimangal, Bangladesh	4.4m surge, 5,708 deaths 6,000 missing, 65,000 cattle killed, 15,000 deer killed, 9 royal Bengal tigers killed, substantial crop damage
7-8 October, 1990	Barisal, Bangladesh	150 fishermen missing, substantial damage
29 April, 1991	Chittagong, Cox's Bazar, Bangladesh	4m to 8m surges, 150,000 deaths, 70,000 cattle killed, great damage
2 June, 1991	Bangladesh	1.2m surge

Sources: Jayantih and Sarma, 1986; Murty *et al.*, 1986; Katsura *et al.*, 1992; Khalil, 1992; and Talukder *et al.*, 1992.

REFERENCES

Andrews, J.T. (1991) Association of ice sheets and sea level with global warming: a geological perspective on aspects of global change. In R.S. Bradley (ed.) *Global Changes of the Past.* UCAR Office of Interdisciplinary Earth Studies, Boulder, Colorado, pp. 321-339.

Anwar, J. (1989) Geology of coastal area of Bangladesh and recommendations for resource development and management. In H.J. Moudud, H.E. Rashid and A.A. Rahman (eds.) *National Workshop on Coastal Area Resource Development and Management, Part II.* Coastal Area Resource Development and Management Association (CARDMA), Dhaka, pp. 36-66.

Bolin, B., Döös, B.R., Warrick, R.A. and Jaeger, J. (1986) *The Greenhouse Effect, Climatic Change and Ecosystems* (SCOPE 29). John Wiley and Sons Inc., Chichester.

Chowdhury, M.I. (1959) Morphological analysis of Bengal Basin. Unpublished Masters Thesis, Cambridge University, Cambridge.

Cubasch, U. and Cess, R.D. (1990) Processes and modelling. In J.T. Houghton, G.J. Jenkins and J.J. Ephraums (eds.) *Climate Change: The IPCC Scientific Assessment.* Cambridge University Press, Cambridge, pp. 69-91.

Cubasch, U., Santer, B.D., Hellback, A., Hegerl, G., Hock, K.H. Maier-Reimer, E., Mikolajewicz, U., Stossel, A. and Voss, R. (1992) *Monte-Carlo Climate Change Forecasts with a Global Coupled Ocean-atmosphere Model.* Max-Planck-Institut fur Meteorologie, Report no. 97, MPI, Hamburg.

Day, J.W., Conner, W.H., Costanza, R., Kemp, G.P. and Mendelssohn, I.A. (1993) Impacts of sea level rise on coastal systems with special emphasis on the Mississippi River deltaic plain. In R.A. Warrick, E.M. Barrow and T.M.L. Wigley (eds.) *Climate and Sea-Level Change: Observations, Projections and Implications.* Cambridge University Press, Cambridge, pp. 276-296.

Emery, K.O. and Aubrey, D.G. (1985) Glacial rebound and relative sea levels in Europe from tide-gauge records. *Tectonophys.*, 120, 239-255.

Emery, K.O. and Aubrey, D.G. (1991) *Sea Levels, Land Levels, and Tide Gauges.* Springer-Verlag, New York.

Evans, J.L. (1993) Sensitivity of tropical cyclone intensity to sea surface temperature. *J. Climate*, 6, 1133-1140.

Fergusson, J. (1863) On recent changes in the delta of the Ganges. *Q. J. Geol. Soc.*, London 19, 321-354.

Flather, R.A. and Khandker, H. (1983) The storm surge problem and possible effects of sea level changes on coastal flooding in the Bay of Bengal. In R.A. Warrick, E.M. Barrow and T.M.L. Wigley (eds.) *Climate and Sea Level Change: Observations, Projections and Implications*. Cambridge University Press, Cambridge, pp. 229-245.

Flood Action Plan (FAP) 7 (1992) *Cyclone Protection Project II: Feasibility and Design Studies* (Main Report). Government of Bangladesh, Ministry of Irrigation, Water Development and Flood Centre, Dhaka.

Gagliano, S.M., Meyer-Arendt, K.J. and Wicker, K.M. (1981) Land loss in the Mississippi River deltaic plain. *Trans. Gulf Coast Assoc. Geol. Soc.*, 31, 295-300.

Gates, W.L., Mitchell, J.F.B., Boer, G.J., Cubasch, U. and Meleshko, V.P. (1992) Climate modelling, climate prediction and model validation. In J.T. Houghton, B.A. Callander and S.K. Varney (eds.) *Climate Change 1992: The Supplementary Report to the IPCC Scientific Assessment*. Cambridge University Press, Cambridge, pp 97-134.

Gregory, J.M. (1993) Sea-level changes under increasing atmospheric CO2 in a transient coupled ocean-atmosphere GCM experiment. *J. Climate*, 6, 2247-2263.

Grove, J.M. (1988) *The Little Ice Age*. Methuen, London.

Haeberli, W., Bosch, H., Scherer, K., Ostrem, G. and Wallen, C.C. (1989) *World Glacier Inventory. Status 1988*. IAHS (ICSI)-UNEP-UNESCO.

Houghton, J.T., Callander, B.A. and Varney, S.K. (1992) *Climate Change 1992: The Supplementary Report to the IPCC Scientific Assessment*. Cambridge University Press, Cambridge.

Houghton, J.T., Jenkins, G.J. and Ephraums, J.J. (1990) *Climate Change: The IPCC Scientific Assessment*. Cambridge University Press, Cambridge.

Isaksen, I.S.A., Ramaswamy, V., Rodhe, H. and Wigley, T.M.L (1992) Radiative forcing of climate. In J.T. Houghton, B.A. Callander and S.K. Varney (eds.) *Climate Change 1992: The Supplementary Report to the IPCC Scientific Assessment*. Cambridge University Press, Cambridge, pp. 47-67.

Jayanthi, N. and Sarma, A.K.S. (1986) *A Catalogue of Storm Surges in India, Rept. 86-1*. Regional Meteorological Center, Madra, India.

Jonas, P.R., Charlson, R.J. and Rodhe, H. (1995) Aerosols. In J.T. Houghton, L.G. Meira Filho, J. Bruce, Hoesung Lee, B.A. Callander, E. Haites, N. Harris and K. Maskell (eds.) *Climate Change 1994: Radiative Forcing of Climate Change and an Evaluation of the IPCC IS92 Emission Scenarios*. Published for the Intergovernmental Panel on Climate Change. Cambridge University Press, Cambridge, pp. 127-162.

Katsura, J., Hyashi, T., Nishimura, H., Isobe, M., Yamashita, T., Kawata, Y., Yasuda, T., and Nakagawa, H. (1992) *Storm Surges and Severe Wind Disasters Caused by the 1991 Cyclone in Bangladesh.* (Rept. No. B-2, Research Rept. on Natural Disasters). Supported by the Japanese Ministry of Education, Science and Culture, Grant No. 03306020.

Khalil, G.M. (1992) Cyclones and storm surges in Bangladesh: some mitigative measures, *Natural Hazards,* 6(1), 11-24.

Khan, F. H. (1991) *The Geology of Bangladesh.* Dhaka University Press, Dhaka.

Khan, S. R. and Mridha, A.S. (1992) *Geological Evidence of Late Holocene Sea Level Fluctuations in the Coast of Bangladesh.* (IOC-UNEP Workshop on Impacts of Sea Level Rise Due to Global Warming, Ed. Dhaka). National Oceanographic and Maritime Institute (NOAMI), Bangladesh, pp. 1-6.

Leggett, J., Pepper, W.J. and Swart, R.J. (1992) Emission scenarios for the IPCC: an update. In Houghton, J.T., Callander, B.A. and Varney, S.K. (eds.) *Climate Change 1992: The Supplementary Report to the IPCC Scientific Assessment.* Cambridge Unversity Press, Cambridge, pp. 69-95.

Levitus, S. (1982) *Climatological Atlas of the World's Oceans.* NOAA Professional Paper 13, US Government Printing Office, Washington D.C.

Lighthill, J., Holland, G., Gray, W., Landsea, C., Craig, G., Evans, J., Kurihara, Y. and Guard, C. (1994) *Global Climate Change and Tropical Cyclones.* Findings of the ICSU/WMO Symposium on Tropical Cyclones and Climate Change, 22 November - 1 December 1993, Mexico City. (Submitted for publication to the Bulletin of the American Meteorological Society.)

Master Plan Organisation (MPO) (1986) *National Master Plan, Final Report,* Volume 1. MPO, Dhaka.

Meier, M.F (1984) Contribution of small glaciers to global sea level. *Science,* 226, 1418-1421.

Meier, M.F. (1990) Reduced rise in sea level. *Nature,* 343, 115-116.

Mitchell, J.F.B., Manabe, S., Tokioka, T. and Meleshko, V. (1990) Equilibrium climate change and its implications for the future. In J.T. Houghton, G.J. Jenkins and J.J. Ephraums (eds.) *Climate Change: The IPCC Scientific Assessment.* Cambridge University Press, Cambridge, pp. 131-172.

Morgan, J. P. and McIntire, W.G. (1959) Quaternary geology of the Bengal Basin, East Pakistan and India. *Bulletin of the Geological Society of America,* 70, 319-342.

Multi-purpose Cyclone Shelter Project (MCSP) (1993) *Summary Report.* Bangladesh University of Engineering and Technology and the Bangladesh Institute of Development Studies, Dhaka.

Murty, T.S., (1984) Storm surges - meteorological ocean tides. *Canadian Journal of Fisheries and Aquatic Sciences* (Bulletin 12), 897p.

Murty, T.S., Flather, R.A. and Henry, R.F. (1986) The storm surge problem in the Bay of Bengal, *Progress in Oceanography*, 16, 195-233.

Murty, T.S. and Henry, R.F., (1983) *Recommendations for Improving Storm Surge Predictions in the Bay of Bengal Area.* Report to WMO-ESCAP Panel on Tropical Cyclones.

National Research Council (NRC) (1990) *Sea-level Change.* Studies in Geophysics. National Academy of Sciences, Washington, D.C.

Oerlemans, J. and Fortuin, J.P.F. (1992) Sensitivity of glaciers and small ice caps to greenhouse warming. *Science*, 258, 115-117.

Polar Research Board (PRB) (1985) *Glaciers, Ice Sheets and Sea Level: Effect of a CO_2-induced Climatic Change.* Report of a workshop held in Seattle, Washington, September 13-15, 1984. US DOE/ER/60235-1. U.S. Department of Energy, Washington, D.C.

Pugh, D.T. (1987) *Tides, Surges and Mean Sea Level.* John Wiley and Sons Inc., Chichester.

Raghavendra, V.K. (1973) A statistical analysis of the number of tropical storms and depressions in the Bay of Bengal during 1890 to 1969. *Indian Journal of Meteorology and Geophysics*, 24, 125-130.

Raper, S.C.B. (1993) Observational data on the relationships between climatic change and the frequency and magnitude of severe tropical storms. In R.A. Warrick, E.M. Barrow and T.M.L. Wigley (eds.) *Climate and Sea Level Change: Observations, Projections and Implications.* Cambridge University Press, Cambridge, pp. 192-212.

Resource Analysis (1992) *Assessment of the Vulnerability of Coastal Areas to Sea-level Rise: Case Study Bangladesh.* Report on preparatory analyses in Bangladesh and the Netherlands. Resource Analysis, Delft.

Sahagian, D.L., Schwartz, F.W. and Jacobs, D.K. (1994) Direct anthropogenic contributions to sea level rise in the twentieth century. *Nature*, 367, 54-56.

Schimel, D., Enting, I.G., Heimann, M., Wigley, T.M.L., Raynaud, D., Alves, D. and Siegenthaler, U. (1995) CO_2 and the carbon cycle. In J.T. Houghton, L.G. Meira Filho, J. Bruce, Hoesung Lee, B.A. Callander, E. Haites, N. Harris, and K. Maskell (eds.) *Climate Change 1994: Radiative Forcing of Climate Change and an Evaluation of the IPCC IS92 Emission Scenarios.* Published for the Intergovernmental Panel on Climate Change. Cambridge University Press, Cambridge, pp. 35-71.

Shennan, I. and Woodworth, P.L. (1992) A comparison of late Holocene and twentieth century sea-level trends from the UK and North Sea region. *Geophysical Journal International*, 109, 96-105.

Shine, K.P., Derwent, R.G., Wuebbles, D.J. and Morcrette, J-J. (1990) Radiative forcing of climate. In J.T. Houghton, G.T. Jenkins and J.J. Ephraums (eds.) *Climate Change: The IPCC Scientific Assessment*. Cambridge University Press, Cambridge, pp. 41-68.

Talukder, J., Roy, G. and Ahmad, M. (1992) *Living with Cyclones: Study on Storm Surge Prediction and Disaster Preparedness*. Community Development Library, Dhaka.

Tans, P.P., Fung, I.Y. and Takahashi, T. (1990) Observational constraints on the global atmospheric carbon dioxide budget. *Science*, 247, 1431-1438.

Thomas, R.H., Sanderson, T.J.D. and Rose, K.E. (1979) Effects of a climatic warming on the West Antarctic ice sheet. *Nature*, 227, 355-358.

Umitsu, M. (1985) Natural levees and landform evolutions in the Bengal Lowland. *Geographical Review of Japan*, 58, Ser. B., (2), 149-164.

Umitsu, M. (1987) Late Quaternary sedimentary environment and landform evolution in the Bengal Lowland, *Geographical Review of Japan*, 60, Ser. B., (2), 164-178.

Umitsu, M. (1993) Late Quaternary sedimentary environment and landforms in the Ganges Delta, *Sedimentary Geology*, 83, 177-186.

Warrick, R.A. (1993a) Sea-level rise. *Earthquest* (Spring vol.), Science Capsule. Office of Interdisciplinary Earth Studies, Boulder.

Warrick, R.A. (1993b) Projections of future sea level rise: an update. In R. McLean and N. Mimura, (eds.) *Vulnerability Assessment to Sea-Level Rise and Coastal Zone Management*. Proceedings of the IPCC Eastern Hemisphere Workshop, Tsukuba, Japan, 3-6 August 1993. Secretariat of the Eastern Hemisphere Workshop, Tokyo.

Warrick, R.A., Barrow, E.M. and Wigley, T.M.L. (1993) *Climate and Sea Level Change: Observations, Projections and Implications*. Cambridge University Press, Cambridge.

Warrick, R.A. and Oerlemans, J. (1990) Sea level rise. In J.T. Houghton, G.J. Jenkins and J.J. Ephraums (eds.) *Climate Change: The IPCC Scientific Assessment*. Cambridge University Press, Cambridge, pp. 257-281.

Watson, R.T., Meira Filho, L.G., Sanhueza, E. and Janetos, A. (1992) Greenhouse gases: sources and sinks. In J.T. Houghton, B.A. Callander and S.K. Varney (eds.) *Climate Change 1992: The Supplementary Report to the IPCC Scientific Assessment*. Cambridge University Press, Cambridge, pp. 25-46.

Wigley, T.M.L. and Raper, S.C.B. (1987) Thermal expansion of sea water associated with global warming. *Nature*, 330, 127-131.

Wigley, T.M.L. and Raper, S.C.B. (1992) Implications for climate and sea level of revised IPCC emission scenarios. *Nature*, 357, 293-300.

Wigley, T.M.L. and Raper, S.C.B. (1993) Future changes in global mean temperature and sea level. In R.A. Warrick, E.M. Barrow and T.M.L. Wigley (eds.) *Climate and Sea Level Change: Observations, Projections and Implications*. Cambridge University Press, Cambridge, pp. 111-133.

Woodworth, P.L. (1990) A search for accelerations in records of European mean sea level, *Int. J. Clim.*, 10, 129-143.

EFFECTS OF CLIMATE AND SEA-LEVEL CHANGES ON THE NATURAL RESOURCES OF BANGLADESH

H. Brammer, M. Asaduzzaman and P. Sultana*

4.1 INTRODUCTION

The 110 million people who live in the low-lying deltaic country of Bangladesh are highly dependent upon an agricultural system that is finely attuned to a tropical monsoon climate and associated water regime. Any significant change in the climate and water regimes of Bangladesh is likely to have profound effects on biological resources. This in turn could have serious effects on the socio-economic systems of the country (see Chapter 5). Here, the effects of climate and sea-level changes on the hydrological and biological resources of Bangladesh are explored through four main questions: What is the current resource setting of Bangladesh? How might the water resources be affected by climate warming? How might crop agriculture be affected? How might other bio-resources be affected by changing climate and water regimes? The final section of this chapter draws conclusions from this review and identifies follow-up actions that could be taken in response to the question: What needs to be done?

4.2 WHAT IS THE CURRENT RESOURCE SETTING OF BANGLADESH?

At the outset, this chapter reviews the existing situation, as well as the probable future trends in resource uses in Bangladesh over the next 30 to 50 years during which the first impacts of global warming might begin to appear. First, the physical and socio-economic setting of the country's land use is outlined as an introduction to later sections in which the impacts of climate change on agriculture, forestry, fisheries and livestock production are described in more detail.

* The authors are respectively: Formerly FAO Agricultural Development Adviser, Bangladesh; Senior Fellow, Bangladesh Institute of Development Studies (BIDS), Dhaka, Bangladesh and Member, BUP, Dhaka; and former Consultant, BUP, Dhaka.

4.2.1 The Physical Setting

Bangladesh's physical environment is both diverse and complex, and both traditional and modern systems of land use are very closely adapted to the heterogeneous conditions. This heterogeneity has important implications for assessing the possible impacts of future climate change and sea-level rise. The impacts will vary from area to area, even within villages, according to specific local conditions. This makes generalization difficult.

Additionally, neither the physical environment nor technology are static. Dynamic changes are taking place in the soil and hydrological conditions which influence land use. This is due, in part, to natural changes taking place in river channels, river flow and sedimentation within the country's floodplains; partly it is due to human interventions. Coupled with population pressure and technological changes in, for example, agricultural production, this means that land use is subject to changes with time. This makes it difficult to forecast environmental impacts due to climatic change that may be taking place several decades in the future.

Climate

The climate is tropical monsoon. Seasonal changes in temperature and moisture supply are primary determinants of biological activity. They particularly influence farmers' choice of crops and planting and harvesting periods. Four main seasons are recognized.

- Pre-monsoon (March to May), when temperatures are highest (5 to 10 days with maximum >40°C in the west; <1 day in the centre and east), evaporation rates are highest, erratic heavy rainshowers and line-squalls occur, and coastal areas are exposed to the risk of tropical cyclones and storm surges.
- Monsoon[1] (June to September), when about two-thirds of the mean annual rainfall occurs, cloud cover and humidity are highest, and maximum temperatures generally are lower than in the pre-monsoon period.
- Post-monsoon (October to November), a transitional period when rain fall amounts, minimum temperatures and cloudiness decrease, and coastal areas are again exposed to the risk of tropical cyclones and storm surges.
- Winter (December to February), when rainfall, temperatures and cloudiness are lowest; minimum temperatures can fall below 5° C in the north

(though frost is extremely rare).

Significant variations in climate occur across the country (Food and Agriculture Organization [FAO], 1988). Mean annual rainfall ranges from about 1250mm in the west to 2500 to 5000mm in the north-east, and the rainy season from 4 to 5 months in the west to 6 to 7 months in the north-east. The cool winter period is longer in the north (>100 days) than near the coast (30 to 40 days). Temperatures and evapo-transpiration rates in the pre-monsoon season are higher in the west than in the centre and east; the risk of drought in the pre-monsoon, monsoon and post-monsoon seasons is also higher in the west than in the east. These variations influence crop suitability and the crop rotations practised in different parts of the country, as described in the third section of this chapter.

Monthly and annual rainfall totals vary from year to year, leading periodically to floods in the monsoon season and to drought in the pre-monsoon and post-monsoon seasons. Drought is rare in the monsoon season proper, though there is evidence in some years of a bimodal rainfall pattern in the north-west and centre which provides insufficient excess water for transplanting *aman* paddy on time in July to August (FAO, 1988). Representative climatic data for Dhaka, in the centre of the country, are given in Table 4.1. Note the large standard deviation (St D) values for monthly and annual rainfall[2]. More details of Bangladesh's climate are given in Chapter 2.

Physiography and relief

Floodplains occupy about 80 per cent of the country, hills about 12 per cent and uplifted fault blocks (sometimes referred to as terraces) about 8 per cent. A total of 34 physiographic units and sub-units are recognized (FAO, 1988). These have been grouped into seven main units in Figure 4.1.

Floodplains of the country's main rivers generally have smooth relief comprising broad and narrow ridges (former river levees) and depressions. Differences in elevation between adjoining ridge tops and depression centres range from less than 1m on tidal floodplains to 1m to 3m on the main river and estuarine floodplains, and up to 5m to 6m in the Sylhet Basin in the north-east. Such elevation differences can occur within the boundaries of individual villages. They are important in determining local differences in the depth and duration of seasonal flooding. Only in the extreme north-west do land elevations exceed 30m above mean sea level (MSL) (Figure 4.2).

The uplifted blocks (Madhupur and Barind Tracts) are formed in an unconsolidated clay, possibly of Pleistocene age. These tracts generally stand 1m to 5m above adjoining floodplains, but the western edge of the Barind Tract is up to 25m higher than the adjoining Mahananda floodplain. Most of the Madhupur

Table 4.1: Climatic data for Dhaka

Parameter	Month												
	J	F	M	A	M	J	J	A	S	O	N	D	Annual
Temperature (°C)													
Mean	18.4	21.5	24.9	27.1	29.3	28.5	28.4	28.4	28.5	27.1	24.0	19.7	25.5
Mean max.	25.0	28.2	29.9	31.1	34.3	31.3	30.9	30.8	31.3	30.7	28.8	25.8	29.8
Mean min.	11.8	14.7	19.9	23.1	24.2	25.7	25.9	26.0	25.7	23.4	19.2	13.6	21.1
Rainfall (mm)													
Mean	8	22	56	130	285	378	409	340	267	165	26	7	2038
Standard Deviation	13	25	51	82	152	154	176	153	106	124	35	17	362
Penman PET (mm)	67	91	140	160	178	137	143	140	127	114	82	62	1440
Sunshine (%)	82	80	74	70	63	36	39	45	50	65	78	84	64

Source: FAO [1988]; rainfall from Bangladesh Agricultural Research Council [BARC] Agroecological Zones [AEZ] Database.

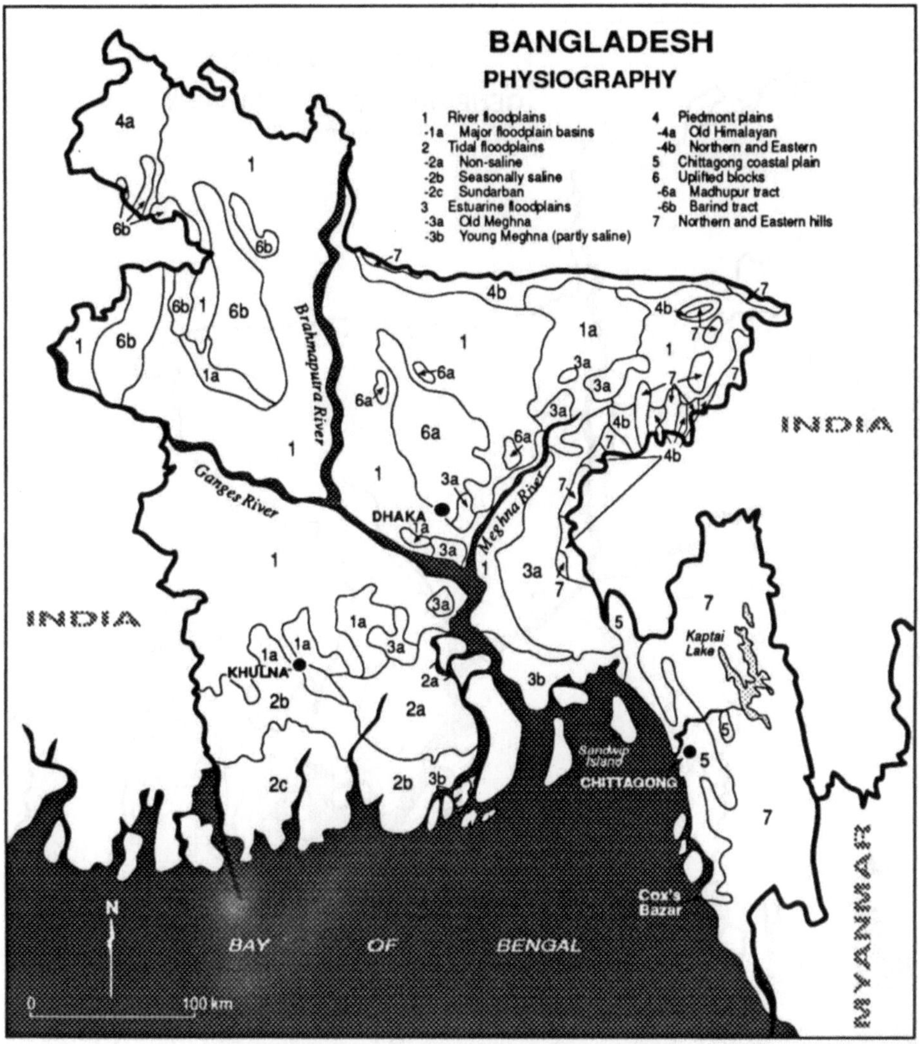

FIGURE 4.1: Bangladesh physiography

Source: Adapted from FAO, 1988.

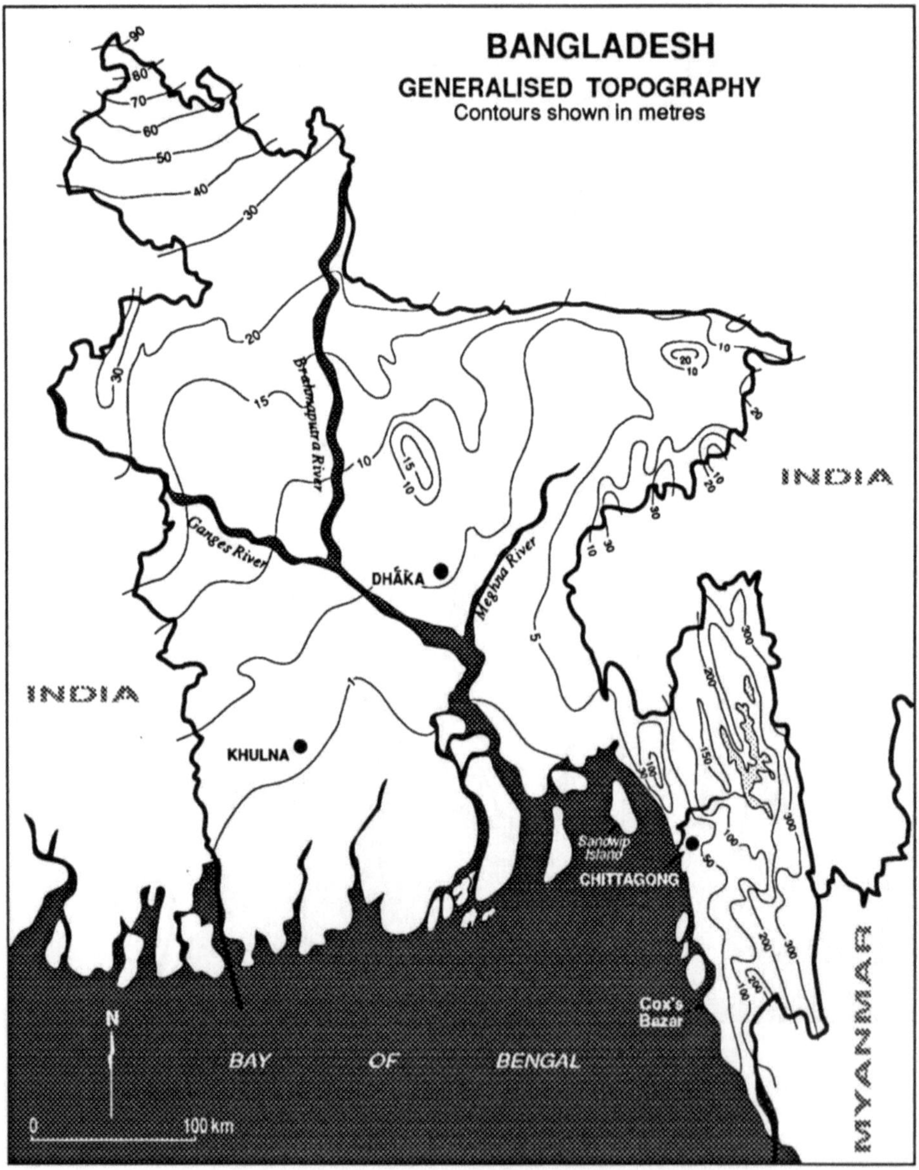

FIGURE 4.2: Bangladesh topography

Source: Adapted from the Master Plan Organisation [MPO], 1985.

Tract is closely dissected by valleys (which occupy 30 per cent of the total area), whereas most of the Barind Tract (except in the west) is almost level.

Hills along the northern and eastern borders are formed mainly in unconsolidated Tertiary sands and shales. These have been uplifted, folded, faulted and dissected to form linear ranges variously reaching 10m to1000m above MSL. Slopes generally are very steep, but there are some areas with moderate or gentle slopes.

BOX 4.1: Dynamic environment

Greenhouse warming will impinge on a diverse, complex and dynamic environment. Significant differences in climate occur across Bangladesh. Year-to-year variability causes periodic floods and droughts. Floodplains occupy about 80 per cent of the country, hills 12 per cent and uplifted blocks 8 per cent. Micro-relief within floodplains produces differences in depth and duration of seasonal flooding even within villages. Soils also vary with micro-relief. Cropping patterns are closely adapted to these heterogeneous conditions, as well as to the widespread availability of water for irrigation. Dynamic changes in land use are taking place with the spread of irrigation and the provision of flood protection works.

Soils

Soil patterns often are complex, but generally less so on tidal and estuarine floodplains than on river floodplains, piedmont plains, uplifted blocks and hills. More than 500 soil series grouped into 21 General Soil Types have been recognized (FAO, 1988). Soil conditions strongly influence land use, and need to be considered in assessing possible impacts of climate change and sea-level rise in different parts of the country.

On floodplains, there is a general pattern of grey or brown, permeable, loamy (sometimes sandy) soils on the higher parts of ridges, grading into grey or dark grey, less permeable, silty clay loams or clays in depressions. The relative proportions of sandy, silty and clay soils vary between physiographic units. Proportions also vary sub-regionally or locally within some units. The depth and duration of seasonal flooding also vary between ridges and basins, as well as between physiographic units.

Floodplain soils vary from calcareous to strongly acid in different regions. As is indicated on Figure 4.1, some tidal and estuarine floodplain soils near the coast become saline to varying degrees in the dry season. Most soils are low in organic matter, but those in depressions are better provided and some are peaty. Although generally considered 'fertile', all soils need the addition of fertilizers or manures in order to sustain the high annual crop yields needed to support the country's large and growing population. Most floodplain soils have been altered by human interventions, such as paddy cultivation, intensive cropping, expansion of settlements, flood protection, irrigation and drainage.

The different soil and flooding conditions which occur from place to place strongly influence crop suitability and crop rotations, both regionally and locally. The pattern of lighter-textured, shallowly or non-flooded soils on ridges and heavier, more-deeply flooded, depression soils means that several different crops and crop rotations commonly are found within the bounds of floodplain villages. Also, because farmers' fragmented holdings generally occur scattered over the different soil and land types, farmers typically grow several crops in several different rotations[3]. This has the effect of diversifying their market opportunities, spreading their labour and cash inputs over the year, and reducing the risks of total loss of crops and investments due to natural hazards.

The Madhupur and Barind Tracts have a wide range of soils, differing in colour, texture, depth and permeability. They range from deep, well drained, permeable red clays — most prevalent on the Madhupur Tract and best suited for dryland crops or forest — to shallow, poorly drained, grey silty soils which are most extensive on the Barind Tract and best suited for paddy cultivation. Seasonally-flooded valleys have grey silty and dark grey clay soils, both best suited for paddy cultivation. Some valley soils contain buried peat layers up to about 6,000 years old. Natural fertility and moisture-holding capacity are generally low.

Hill soils also are very varied, sometimes within short distances. Most are deep, red to yellow, rapidly permeable, sandy loams to clay loams, but shallower, paler, silty soils occur over some shales. Most soils occur on very steep slopes, are strongly acid and low in natural fertility. Organic matter contents are moderate or high under closed forest or tree crops, but become low under shifting cultivation, especially where soils have been eroded. Most hill soils are best suited for forest or tree crops; they are poorly suited for terrace cultivation of arable crops.

Hydrology

Both surface water and groundwater are important determinants of agricultural land use and potential in Bangladesh. Differences in depth and duration of seasonal flooding on different soil and land types strongly influence the kinds of crops and crop rotations which farmers can use; and surface water and groundwater are widely used for irrigation in the dry season. Accordingly, any changes in these water resources caused by climate change and sea-level rise could have impacts on agriculture; they could also affect forestry, fisheries and livestock, as well as domestic and industrial water supplies and water communications.

As is described in more detail in the next major section, seasonal flooding is caused by a combination of high local rainfall and incoming floodwater in the Ganges, Brahmaputra, Meghna and smaller rivers entering the country from surrounding territories[4]. The extent of flooding varies considerably from year to year, from about 25 per cent of the land area in a 'normal' year to as much as 60 per cent in an exceptionally high flood year, such as in 1988.

Increasingly in recent years, the extent of seasonal flooding has been reduced by flood control schemes, and this trend must be expected to continue. Most rivers already are embanked for part or all of their length; and some areas are drained by pumps or tidal sluices. Embankments in coastal areas protect considerable areas from salt-water flooding at high tide; they are not designed to protect land from high storm surges. However, not all river and coastal embankments are fully effective, because of breaches caused by river or coastal erosion, 'public cuts'[5], storm surges or poor maintenance (Hunting Technical Services, 1992).

At present, relatively little water is diverted or pumped from the main rivers for irrigation purposes. However, a substantial part of the dry-season flow of the Ganges is diverted at the Farakka barrage in India to maintain an adequate depth of water in the Hooghly River at the Calcutta port. This has reduced flow through the western part of the Ganges floodplain within Bangladesh, allowing salt water to penetrate further inland than previously in the south-west. This has aggravated a natural trend taking place for several centuries in this 'moribund' part of the Ganges delta. Proposals to divert water within Bangladesh down the Gorai River channel to restore freshwater supplies to the south-west are presently being studied under the Flood Action Plan[6].

The salinity balance in the Meghna estuary and in the south-west would be affected by any major interventions made for flood control or irrigation anywhere in the Ganges-Brahmaputra-Meghna catchment during future decades. The salinity balance could also be affected by changes in river flow or sea-level rise resulting from global warming.

Most floodplain areas, and most of the Madhupur and Barind Tracts, are underlain by fairly good aquifers at shallow or moderate depths, and monsoon rainfall combined with seasonal flooding normally is sufficient to recharge them annually. Aquifers are less satisfactory in a few floodplain areas (mainly in the north-east), in the northern part of the Madhupur Tract and in the west of the Barind Tract. Saline groundwater occurs near the coast and locally inland in the south-east, but sweet water occurs below about 300m in at least some coastal areas and is tapped for industrial and domestic supplies.

4.2.2 The Socio-Economic Setting

Bangladesh's socio-economic environment is characterized by high population density, widespread poverty and predominantly rural settlement with a majority of households dependent mainly on agriculture, fisheries and other forms of primary production. These aspects are described in greater detail in Chapter 5. Here, brief attention is focused on those aspects that particularly affect land use, especially agriculture and fisheries.

Population

The 1991 population census gave a total population of 110 million (adjusted figure), growing at an annual rate of 2.02 per cent (Bangladesh Bureau of Statistics (BBS), 1992). The urban population was 9.32 per cent of the total. However, the official classification of 'urban' includes only those townships with a municipal administration. For this reason, and because of definitional changes over time, the actual size and growth-rate of the country's urban population are uncertain. The present size of the urban population is thought to be 16 per cent of the total (Chapter 5).

Projections of the census figures indicate that the country's population could more than double by around 2030, by which time the urban population might be more than one-third of the total. The present population density of 741 per km^2 might then be more than 1500 per km^2, and the present major cities (Dhaka, Chittagong, Khulna) and many rural towns may have grown considerably in size. By 2030, therefore, climate change and sea-level rise could be impacting on a very different society from today's.

At present, 80 per cent of the population are considered rural. Some 74 per cent of those employed are mainly dependent on agriculture, either as land-owners, cultivating tenants or labourers. However, many household members engage in other means of subsistence through fishing, trading, service or either temporary or quasi-permanent employment outside their home area (from which they remit funds to the household). Unemployment in rural Bangladesh is 36 per cent or more in terms of available labour time (Chapter 5).

BOX 4.2: Social Complexity

Bangladesh's socio-economic environment is no less diverse, complex and dynamic than its physical environment. Presently, 80 per cent of the country's 110 million people are rural, engaged largely in farming, fishing and many artisanal pursuits (in which women often play important roles, unseen and unpaid). Land ownership is highly skewed, and about 56 per cent of rural households are functionally landless. Unemployment, poverty, and malnutrition are widely prevalent. Rural population densities are high (>1000/km^2 in some parts) and rural to urban migration rates are accelerating. With population projected to double by 2030, the urban population could by then reach one-third of the total. Greenhouse warming could then be impacting on a different society from today's.

Land Tenure

Land-ownership is highly skewed. In 1983-1984 (the latest year for which national data are available) 56 per cent of the rural households were functionally landless[7]. Among the farm households, 4.9 per cent with greater than 3ha owned 26 per cent of the land; 24.7 per cent with 1 to 3ha owned 45 per cent; while 70.3 per cent with less than 1ha owned 29 per cent of the total (BBS, 1992). The 1983-1984 Agricultural Census showed that there was a tendency for the number of the largest land holdings to decrease and for medium and small holdings to increase.

The relative proportions of different ownership classes vary across the country. In general, there are relatively higher proportions of small and medium farmers in the most densely settled central and south-eastern floodplain areas, and relatively higher proportions of larger farmers (often absentee owners) in cyclone-prone coastal areas, deeply-flooded basin areas, and drought-prone western areas. Considerable numbers of migrant labourers are employed in seasonal rice planting and harvesting in the latter areas.

Land holdings typically are fragmented into a number of scattered fields averaging about 0.06ha in size. In part, this is due to the Muslim laws of inheritance; but it is also attributable to farmers purchasing land of different types so as to diversify their cropping opportunities. There is also a considerable amount of renting in and renting out of land for the latter purpose; and for this reason, many farmers are both land-owners and tenants.

Traditionally, the main form of land tenancy was share-cropping, with the tenant paying half (or more) of each crop to the land owner. In recent years, particularly with the use of HYV seed, fertilizers and irrigation, there has been an increase in other forms of short-term lease tenancy, paid for in cash or a fixed quantity of production. However, share-cropping is still widely prevalent, especially in areas where absentee land-ownership is common.

Fishing

As with farming, fishing is both a subsistence and a commercial activity. The majority of rural households engage in casual fishing for subsistence during the period of seasonal flooding. Additionally, the relatively larger households usually catch fish from naturally or artificially stocked ponds adjoining their houses. Professional fishermen (who, despite their recognized skills, usually are a depressed social class) catch fish for sale in rivers, beels and the ponds of big land-owners. Marine fisheries are operated by both small artisanal groups and bigger commercial interests.

In recent years, shrimp farming has expanded rapidly in some brackish-water coastal areas. Usually, operations are controlled by big land-owners or commercial interests employing seasonal labour (either local or migrant). Often, these operations are in conflict with neighbouring farmers whose cropping opportunities are affected to varying degrees by the breaching of embankments to allow entry of brackish water onto the land.

Other Characteristics

Three other socio-economic characteristics deserve attention in this brief introduction. One is the role of women, who perform important tasks within the household (usually unpaid) in tending grain and seed stocks, homestead gardens and livestock, while increasing numbers of poor women are becoming involved in field operations as well as in rural construction activities.

The second is that, faced with the increasing difficulty of feeding cattle because of grazing and fodder shortages, increasing numbers of farmers are moving to the use of power tillers for cultivation and haulage, a trend which may be expected to continue and which has implications for the economic size of farm holdings and for future employment.

A third important point to note is that it is estimated (United Nations Development Programme, 1989) that agriculture may not be able to absorb more than about 30 per cent of the future incremental growth in the labour force. This implies that the greater part of an expanding future labour supply will need to find (and be found) employment in industry, commerce, services or abroad.

These socio-economic pressures exist, with or without global warming impacts. It needs to be assessed to what extent global warming might exacerbate or ameliorate on-going socio-economic trends.

4.2.3 Land Use

Land use in Bangladesh is determined mainly by the monsoon climate and the seasonal flooding which affects the greater part of the country. These physical determinants are reinforced by high population pressure and, increasingly, by alterations to the natural environment through flood protection, drainage and irrigation interventions.

Agriculture dominates both land use and the national economy. The areas occupied by major forms of land use are indicated in Table 4.2. Further details are given in subsequent parts of this section.

TABLE 4.2: Land use and cropping intensity 1990-1991 (000 ha)

Cropped area				Other land uses			
Single	Double	Triple	Sub-Total	Waste+ fallow	Forest	NAC[1]	Total
3294	3899	981	8174	1546	1899	3221	14840

[1]/ NAC = not available for cultivation. This includes settlements, industrial land, roads and water. Soil surveys in the 1960s and 1970s showed that, at that time, settlements occupied 1.205 million ha and major rivers and *beels* 0.973 million ha (FAO, 1988).

Source: BBS, 1992, Table 5.08.

Settlements and water occupy about 22 per cent of the land area. In rural areas, settlements mainly comprise individual homesteads on raised mounds adjoining a small tank (pond) used for domestic water supply and often for fish production. Buildings are surrounded by multipurpose trees, bamboos and a small vegetable garden. Floodplain settlements typically are concentrated on the highest land available, on river banks or ridges. They are surrounded by agricultural land extending down to the lowest land which is most deeply flooded in the monsoon season and where one or more *beels* may retain water through most or all of the dry season. This seasonally-flooded agricultural land is also important for fish production.

Despite the huge increase in population in recent decades, rural settlements have increased very little in size (Serno, 1981). Instead, houses have crowded closer together, resulting in a reduction in the number of trees grown around homesteads. It seems probable, however, that the rapidly growing urban population (including that in rural towns) is encoaching more significantly on former agricultural land. This seems likely to provide an increasing problem with the accerlerating rate of urban population growth expected to occur over the next few decades.

4.3 HOW MIGHT THE WATER RESOURCES BE AFFECTED BY CLIMATE WARMING?

Bangladesh's surface-water and groundwater resources play a vital role in agricultural, forestry, fisheries and livestock production. They play equally vital roles in settlement, domestic water supply and communications, and indirectly in sanitation and health. This section describes the country's water resources in more detail than was given in the first section as a basis for examining how they might be affected by climate warming which is discussed in the final section.

4.3.1 Surface Water

This sub-section summarises the essential elements of the surface water hydrology of Bangladesh (major rivers; seasonal flooding; and flood protection) as a basis for considering how they may be affected by climate change.

Major Rivers

Bangladesh's major rivers have their headwaters outside the country. Only 9.1 per cent of their total catchment area of about 1.41 million km^2 lies within the country (Ahmad *et al.*, 1994)[8]. Thus, Bangladesh is exposed to the consequences of rainfall, land degradation and hydrological interventions occurring outside its boundaries. Unless a greater degree of regional cooperation is achieved than in the past, this limitation on Bangladesh's influence over external hydrological events and interventions could further aggravate adverse impacts of future climate change.

The Brahmaputra and Meghna rivers begin to rise in March-April as a result of snow-melt in the Himalayas and pre-monsoon rainfall in Assam and the north-eastern part of Bangladesh. The Ganges starts to rise later, in May. All three rivers rise rapidly in June-July in response to heavy monsoon rainfall over their catchment areas. The Brahmaputra-Jamuna normally peaks in July-August, the Ganges in August-September; occasionally, the two peaks coincide, as in 1988. The Meghna rises and peaks like the Brahmaputra, but high levels often extend into September due to backing up of water above the confluence with the Padma. Discharges in all the rivers fluctuate considerably during the flood season, so that several 'peaks' may occur. All the rivers generally fall rapidly from September through November, then more slowly in the dry season.

Discharge data for the three major rivers are given in Table 4.3. They show the predominant influence of the Brahmaputra, especially in the dry season. Rivers near the coast are tidal throughout the year. Tidal influence extends far inland up the Meghna and Padma rivers in the dry season, though river flow is sufficient to hold back salt water to the Meghna estuary. Only in the south-west does salinity penetrate significantly inland during this season.

TABLE 4.3: Discharge data for the Brahmaputra, Ganges and Meghna rivers (cumecs)

Parameter	Ganges (Hardinge Bridge)	Brahmaputra (Bahadurabad)	Meghna (Bhairab Bazar)
Mean low flow	790[1]	7,572	1,332
Mean peak flow	51,625	65,491	14,047
Highest peak flow	76,000	98,600	19,800
Mean annual flow	8,544	19,359	6,748

[1]/ Mean for 1975-1982 (since operation of the Farakka barrage). Mean low flow for 1934-74 was 1,806 cumec.

Source: Master Plan Organisation (MPO), 1985.

Seasonal Flooding

Monsoon rainfall and river water both contribute to seasonal flooding of floodplain land; tidal floodplains are subject to flooding at high tides, sometimes only in the monsoon season,where embankments do not prevent this. The extent of rain-water and river-water flooding varies from year to year, depending on local rainfall and river levels. In this respect, it is useful to distinguish between 'normal' flooding (Bengali - *barsha*), to which land use and housing are adapted, and 'floods' (*bonna*) which damage crops and property.

BOX 4.3: *Barsha* and *bonna*

> Some 25 per cent of Bangladesh goes under water in a normal year. When high floods occur, more than 60 per cent can be submerged. Bengalis differentiate between normal flooding *(barsha),* to which their land use, settlements and economic activities are well adapted, and abnormal floods *(bonna)* which damage crops and property, and disrupt normal economic activities. A two-pronged Government strategy is providing greater security against flood losses: the spread of irrigation, mainly from groundwater, enables farmers to increase rice production in the safer dry season; and provision of flood protection works, which aim to exclude unwanted floodwater in the monsoon season.

In 'normal' years, most floodplains are flooded by rainwater which is ponded on the land when high river levels impede external drainage. Flooding with silty river water normally occurs only close to river banks, on north-eastern and eastern piedmont plains which are subject to flash floods, and on parts of tidal and young estuarine floodplains subject to tidal flooding (Figure 4.3). River water extends more widely in high flood years, such as in 1988, when river banks are overtopped and embankments may be breached. This pattern of rainwater and river-water flooding has important implications for cropping, soil fertility and the impact of flood control and drainage interventions, as described later.

Because of the ridge and basin floodplain relief, the depth and duration of seasonal flooding vary on a local scale, as described in the first section. They vary also between flood-protected and neighbouring unprotected land. Flooding depth and duration vary also between years, according to local rainfall and river water levels.

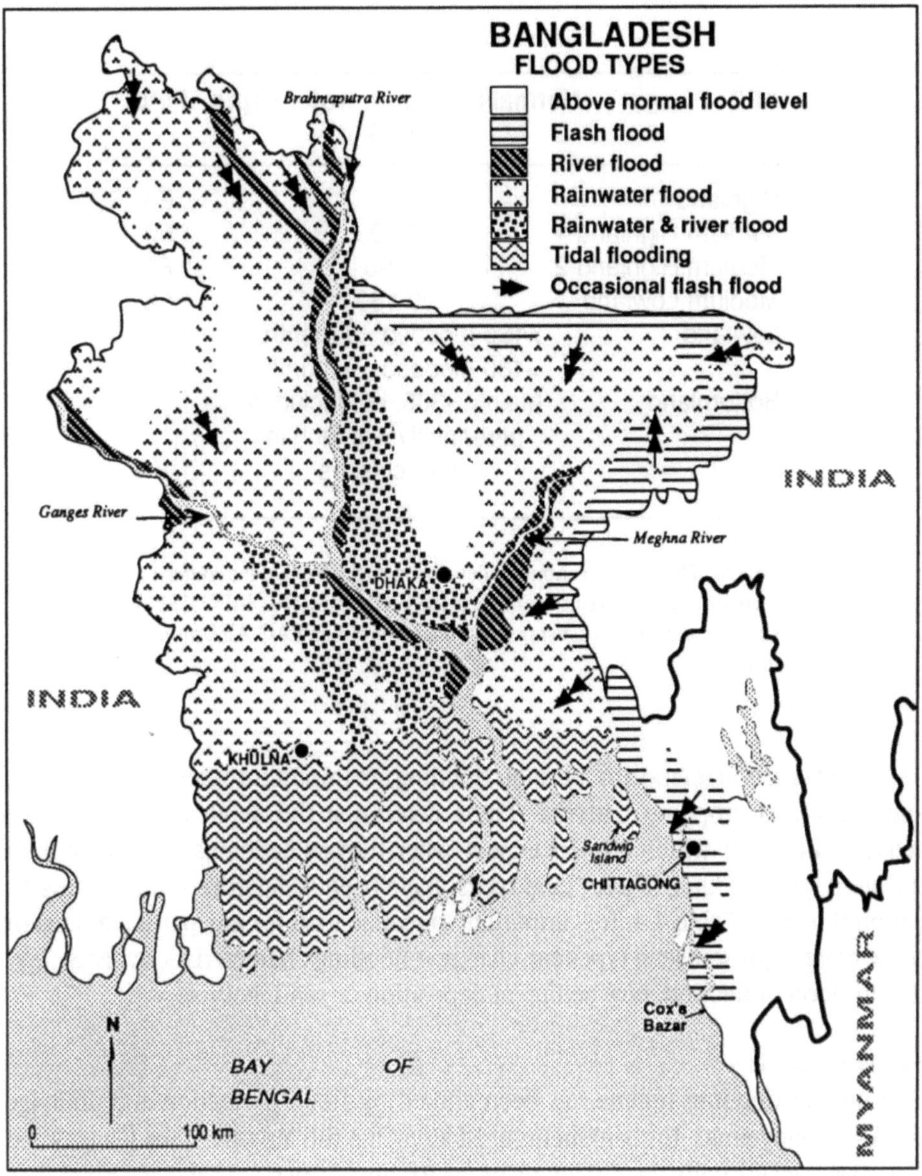

FIGURE 4.3: Flood types

Source: Brammer and Khan, 1991.

TABLE 4.4: Depth-of-flooding land types

AEZ	Normal maximum water depth (cm)	MPO
Highland	0	F0
Medium Highland-1	0-30	F0
Medium Highland-2	30-90	F1
Medium Lowland	90-180	F2
Lowland	>180 (flooded <9 months)	F3
	>180 (flooded >9 months)	F4
Bottomland	Mainly >300, but includes perennially wet land in other classes	F4

Source: This classification was developed on the basis of the country's reconnaissance soil surveys made in the 1960s and 1970s. MPO (1986) simplified the classes, using data obtained from an FAO AEZ project in 1984.

Land levels in relation to 'normal' flooding are important determinants of crop suitability and cropping patterns. These determine the depth and duration of seasonal flooding which farmers expect when they select crops to grow on their different types of land. These 'land types' have been standardized in the classification shown in Table 4.4. It is important to realize that, for any particular area or piece of land, present land types can change if flooding characteristics change: e.g. due to flood protection, new sediment deposition or sea-level rise.

Flood Protection

The natural flooding regime has been altered by flood protection and drainage works in many areas. Embankments exist along the full lengths of the Ganges and Teesta rivers; virtually the full length of the Brahmaputra right bank and more than half of the left bank; long stretches of both banks of the Padma and Meghna rivers; as well as along most tidal rivers in the south-west and along many small rivers inland. Submersible embankments have been built along many rivers in and around the Sylhet Basin to protect the *boro* paddy crop against pre-monsoon flash floods; the embankments are over-topped by monsoon floods. Additionally, some embanked areas (e.g. the Chandpur Irrigation Project near the coast) are artificially drained by pumps and/or through tidal sluices to reduce monsoon-season water levels so that transplanted *aman* paddy can be grown. (*Boro* and *aman* paddy types are defined later.)

Flood protection projects have not been successful everywhere, and they remain a somewhat controversial subject. The Brahmaputra right embankment is frequently breached by the erosion of river banks; and polder embankments in the Chalan Beel in the north-west are frequently cut by people who feel their lives or crops are threatened either by higher, confined, river levels or by water ponded inside embankments when high external river levels prevent drainage through sluices. Poor maintenance also leads to breaching of embankments, including submersible embankments; and embankments along tidal rivers are regularly cut by shrimp farmers to let brackish water onto their land, which salinizes adjoining paddy farmers' land.

These problems are being addressed in several ways under the Flood Action Plan by: works to strengthen the Brahmaputra right embankment at critical points; a pilot river training project on the Brahmaputra-Jamuna; introducing the concept of 'controlled flooding' (i.e. by means of regulators, allowing 'normal' flooding to occur, but preventing or reducing unwanted high or untimely floods) instead of aiming at full protection; protecting important urban areas from floods and river-bank erosion; for rural areas, 'compartmentalization' of land behind embankments, to regulate water flow across protected areas; setting back embankments (the 'green river' concept) so as to reduce confinement effects on rivers and riverine populations; foreshore afforestation to protect exposed coastal embankments; and, particularly, by strengthening local participation in project planning, design, operation and maintenance (Flood Plan Coordination Organisation, 1992a).

4.3.2 Groundwater

Excess monsoon rainfall over Bangladesh is everywhere sufficient to recharge groundwater annually where aquifer conditions are suitable, except possibly during one-in-ten year droughts in the extreme west. Monsoon rainfall is, of course, considerably supplemented by seasonal flooding of floodplain land and by recharge through river beds.

Aquifer conditions are widely suitable for suction-mode pumps to be used: the No. 6 hand pump for domestic water; shallow tube-wells (STW) for irrigation. Force-mode pumps need to be used in some areas: the deep-set Tara pump for domestic supplies; and deep tube-wells (DTW) or deep-set 'shallow' tube-wells (DSSTW) for irrigation. By 1991, official figures for the numbers of pumps installed were: 108,345 No. 6 hand pumps; 109,009 STW and 22,917 DTW (BBS, 1992).

Groundwater availability is erratically distributed in the north-east, and aquifers generally are poor in the north of the Madhupur Tract and the high western part of the Barind Tract. Saline groundwater occurs near the coast, especially in the south-west, and in some inland parts of greater Comilla District in the south-east. In coastal areas, domestic hand pumps use near-surface groundwater which may become brackish during the dry season, but sweet-water aquifers exist below about 300m (in some parts at least) and are tapped by DTWs, mainly (because of the expense) for urban use.

4.3.3 Impact of Climate Change

As indicated in Chapter 2, the best-estimate scenarios for precipitation over Bangladesh with greenhouse warming are for an increase in average monsoon rainfall of 10 to 15 per cent by 2030 and 15 to 23 per cent by 2050. The rationale behind these scenarios — an expected increase in the temperature gradient between the Indian Ocean and the Asian continent — might also be inferred to imply an increase in average rainfall in both the pre-monsoon and post-monsoon periods. Predictions for the winter season are more uncertain. Projections of General Circulation Models range between 5 per cent drier and 19 per cent wetter, with a best estimate of plus 5 to 10 per cent by 2030 and plus 8 to 15 per cent by 2050.

BOX 4.4: Climate change: floods or droughts

> Climate models generally agree that mean annual rainfall in the Bangladesh region should increase with global warming. By 2030, the best-estimate projection is for monsoon rainfall to increase by 10 to 15 per cent and winter rainfall by 5 to10 per cent. However, the range of uncertainty is high. The projected increase would have both positive and negative impacts on Bangladesh. Increased rainfall could reduce drought frequency and severity, increase groundwater recharge and decrease irrigation demand. On the other hand, increased rainfall could increase flood frequency and severity, but increased river discharges could decrease salinity penetration in coastal areas, thus helping to counteract salinity increases associated with a rising sea level.

These scenarios imply both potential benefits and potential losses for Bangladesh. On the plus side, both surface and groundwater availability would be increased by the projected increase in precipitation. This could be reinforced by a reduction in irrigation demand, because the projected increases in precipitation in the pre-monsoon season generally are greater than probable increases in evapotranspiration losses due to the projected temperature increase of 0.5°C to 1.7°C by 2030 (see Table 4.5).

TABLE 4.5: Possible changes in average monthly rainfall, evapo-transpiration (PET) and moisture balance (in mm) by 2030 with global warming

Station	Parameter	Month												
		J	F	M	A	M	J	J	A	S	O	N	D	Year
Dinajpur	Rainfall	1	1	2	5	15	31	36	30	27	12	1	0	161
	PET	2	3	4	4	7	5	6	5	4	3	2	2	47
	Balance	-1	-2	-2	1	8	26	30	25	23	9	-1	-2	114
Rajshahi	Rainfall	1	1	2	5	16	33	39	32	29	12	1	0	171
	PET	1	3	4	5	6	6	6	5	5	4	2	2	49
	Balance	0	-2	-2	0	10	27	33	27	24	8	-1	-2	122
Dhaka	Rainfall	1	2	5	10	23	38	36	29	23	14	2	0	183
	PET	1	2	3	3	3	3	3	3	2	2	2	1	28
	Balance	0	0	2	7	20	35	33	26	21	12	0	-1	155
Sylhet	Rainfall	2	4	7	23	45	86	62	58	41	18	1	0	347
	PET	2	2	2	4	3	3	3	3	2	2	1	1	28
	Balance	0	2	5	19	42	83	59	55	39	16	0	-1	319
Chittagong	Rainfall	1	2	5	10	23	50	59	50	28	23	4	1	256
	PET	2	2	3	4	3	2	3	3	3	2	2	2	31
	Balance	-1	0	2	6	20	48	56	47	25	21	2	-1	225

Notes: 1. This table assumes increases of monthly rainfall by 9 per cent and of monthly temperatures by 1.1°C throughout the year by 2020. All figures are positive, unless otherwise indicated

2. Dinajpur is in the north-west (longest winter); Rajshahi in the west (driest; hotter summer); Dhaka in the centre; Sylhet in the north-east (wettest); Chittagong in the south-east (coastal).

Source: Data kindly provided by Research Unit, Agriculture Department, FAO, Rome.

On the negative side, Bangladesh could suffer more severe floods as a result of the projected increase in monsoon rainfall over the Ganges-Brahmaputra-Meghna catchment area together with increased snow-melt in the Himalayas consequent upon greater winter snowfall and higher summer temperatures.

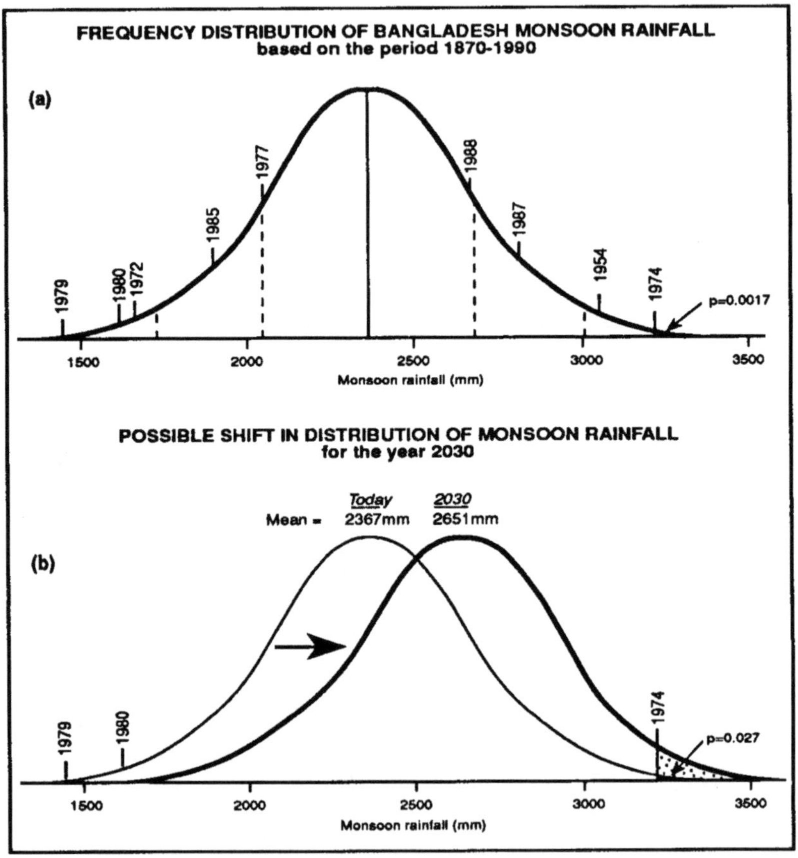

FIGURE 4.4: Frequency distribution of Bangladesh monsoon rainfall(June to Oct). (a) Present distribution based on the period 1870-1990, with selected extreme wet and dry seasons noted. The shaded area denotes the probability of the extremely wet monsoon of 1974 being equalled (or exceeded) in any given year. (b) The shift in distribution for a 12 per cent increase in mean rainfall projected for the year 2030, assuming no change in variability. The stippled area under the curve indicates the increase in probability of occurrence of an extremely wet monsoon like 1974.

Source: Chapter 2. Data courtesy of Climatic Research Unit [CRU], University of East Anglia.

Increased monsoon rainfall over Bangladesh itself could aggravate flood effects. The graphs in Figure 4.4, reproduced from Chapter 2, indicate how a 12 per cent increase in monsoon rainfall could increase the frequency of an extremely wet year (and high-flood year), such as 1974, from approximately a once in 600 years event to once in less than 50 years. Correspondingly, in this simplistic scenario, the frequency of drought years, such as 1979, would become negligible.

One further scenario that needs to be considered is the possibility that, with higher summer temperatures on the Indian subcontinent, the inter-tropical convergence zone might move further north than it does at present. This could bring the southern part of Bangladesh a bimodal rainfall pattern (in some years, at least) with a dry spell occurring in the middle of the monsoon rains. This would have implications for agriculture, which are discussed in the next main section.

4.3.4 Implications for Planning

It must be stressed that the scenarios described above are conjectural. Existing climate models are relatively crude, especially for regional application. It may be several years yet before refinement of the models, coupled with close monitoring and analysis of actual climatic trends, will provide greater certainty on which to base predictions that can be used for planning specific mitigatory measures. The following discussion is based on the best-estimate scenarios discussed above.

The potential increase in flood frequency and severity has four important implications for flood mitigation planning, such as is currently being undertaken under the Flood Action Plan (FAP): embankment design; channel stability; drainage; and salinity. These topics are discussed in turn below, followed by a discussion of the implications of embankments for alluvial sedimentation on adjoining river and tidal floodplains, taking into account also the implications of a rising sea level.

Embankment Design

Flood protection projects now in the planning stage, which might be implemented by around the year 2000, need to take into consideration the possible increase in flood frequency and height during their life-time. This is because major river embankments are presently designed to withstand a one-in-one hundred year flood and, for economic assessment, major water control projects generally are assumed to have a 30-year life-span. However, in practice, because of the very flat discharge-elevation curves for rivers crossing the Bangladesh delta, embankments built to withstand a one-in-one hundred year flood by 2030 might need to be raised by only a few centimetres above present levels [9].

Channel Stability

A more serious consequence of the increased severity of river floods might be greater channel instability and increased sediment loads. Such changes would aggravate the existing problems of selecting and maintaining 'safe' alignments for river embankments and of providing river training works to protect riverside towns and structures. Additionally, the increased flood frequency and severity, together with increased channel instability, would further increase the insecurity of lives and livelihoods of present and future *char* dwellers living in and alongside major river channels, whether embanked or not.

Drainage

Increased monsoon rainfall coupled with higher river levels could increase the depth and duration of seasonal flooding on floodplain land, especially in the southern half of the delta where higher sea levels could further impede drainage from the interior. This would exacerbate existing drainage problems on low-lying land, which could only be relieved by expensive pump-drainage.

Salinity

The projected increase in average flood flows through the Bangladesh delta would tend to push back the salinity front in estuarine and tidal floodplains. That might apply both in the monsoon season and in the dry season.

For the monsoon season, one can be reasonably confident that increased river discharges resulting from increased monsoon precipitation over upstream catchments would push back the existing salinity front. Greater uncertainty exists regarding dry-season conditions. However, it seems reasonable to expect that dry-season discharges would also be higher than at present because of increased residual flows in the rivers following heavier monsoon precipitation and because of the conjectured reduction in the length of the dry season.

The pushing back of seasonal salinity limits, if it occurred, would be beneficial for agriculture in some presently saline coastal areas. Benefits to forestry could also accrue from the stabilization of forest composition and growth in the Khulna Sundarban. However, these benefits are contingent upon the conjectured increase in dry-season flows not being off-set by increased rates of water abstraction for irrigation, domestic and industrial use upstream, which seems to be a highly probable consequence of the projected doubling of population by 2030.

Whether the increased flood flows in the monsoon season and in the dry season might off-set the salinity incursion resulting from the projected 14cm to 24cm rise in sea level by 2030 can only be guessed at. That is because of the dynamic state of the Bangladesh delta and uncertainties regarding future dry-season flows and human interventions in the natural systems.

Alluvial Accretion

The fact that, as is shown in Figure 4.3, most floodplain land is flooded mainly by rainwater has two important implications for existing and possible future flood protection works: the possible impacts of river embankments on soil fertility and on land levels. Since such areas do not receive regular deposits of river alluvium, river embankments do not and will not deprive such land of actual or supposed benefits of continuing sedimentation.

In the case of soil fertility, it is now believed that the benefits of seasonal flooding are derived mainly from biological activity (especially blue-green algae) in the floodwater and not from the deposition of mineral sediments (FAO, 1988; Catling, 1992). Therefore, flood embankments do not necessarily deprive protected land of a traditional source of soil fertility, especially where controlled flooding is practised. Ponding of rainwater and irrigation water on paddy fields also helps to sustain such biological benefits.

Land subject to rainwater flooding may subside due to the compaction of underlying sediments or tectonic movements, whereas sedimentation alongside rivers and in unembanked tidal areas can off-set such subsidence. Over time, these opposing processes can increase differences in land levels between interior floodplain areas and neighbouring rivers, causing a deterioration of drainage from basin land, such as exists at present in the Gopalganj-Khulna *beels* to the west of the Lower Meghna and in the Begumganj depression east of the Meghna. Sedimentation on tidal floodplains and on river levees keeping pace with a rising sea level could increase such elevation differences, thus aggravating drainage problems on interior floodplain land, whether embanked or not.

Embankments around coastal polders, by preventing tidal flooding of the enclosed land, could create similar differences in elevation between protected (and possibly subsiding) land and adjoining tidal river levees (outside the embankments) which would increase in height with a rising sea level. This would impede drainage from polder interiors, which might thus become increasingly subject to waterlogging by rainwater flooding with a rise in sea level. This change would aggravate the existing problem caused by the silting up of tidal channels between embankments.

In the case of coastal polders, it might be feasible to arrange ingress of tidal water periodically, when this would least interfere with cropping or shrimp farming, thus enabling silt deposition within polders to keep pace with a slowly rising sea level. This process (called warping or colmetage, and already practised by some Bangladeshi farmers) might be more difficult to arrange and manage on non-tidal interior floodplains, but the possibilities deserve investigation.

Recommendations for studies needed to improve knowledge about possible impacts of global warming on Bangladesh's water resources are given in the final section of this chapter.

4.4 HOW MIGHT CROP AGRICULTURE BE AFFECTED BY CHANGING CLIMATE AND WATER REGIMES?

Agriculture dominates both land use and the national economy. Arable land occupies about 9 million ha of the country's 14.48 million ha. Agriculture generated about 39 per cent of the GDP over the second half of the 1980s, of which nearly 73 per cent was due to crop agriculture (Chowdhury and Shahabuddin, 1992). Much land is intensively cropped, and there have been dramatic changes in technology and production over the past two decades. These changes are expected to continue under the pressure of continuing population growth.

This sub-section first sets the scene for assessing possible future impacts of greenhouse warming by giving an overview of some of the main characteristics of Bangladesh's present agriculture and envisaged future trends. It then describes what is known about the potential impacts of the factors involved in climate change on crop growth and yields.

4.4.1 The Production Context

Farming is intensive in many parts of Bangladesh, especially in floodplain areas. Much land is double and triple cropped (see Table 4.2), and the average cropping intensity (the number of crops grown on a piece of land in one year) is high (172 per cent). Rice occupies about 80 per cent of the total cropped area.

Cereals production (including wheat and minor cereals) has doubled over the past 30 years, more-or-less keeping pace with the rate of population growth. None the less, annual foodgrain imports have averaged nearly 2 million tons over the past decade, in part to compensate for production shortfalls caused by recurrent natural disasters. For the country's current population of 110 million, the foodgrain demand is 22 million tons (allowing 10 per cent for seed, feed and waste). The official production figure for rice plus wheat in 1992 was 19.25 million tons. In addition, about 1.5 million tons of tuber crops were produced.

Present development efforts aim to increase food production through the expansion of irrigation and flood protection so that, with the support of modern agricultural technology (especially HYV seed and fertilizers) national self-sufficiency in foodgrains production can be achieved and sustained as the population grows. The expected doubling of population by around 2030 would require an annual cereals production of about 44 million tons to provide self-sufficiency[10].

The increased crop production needed to meet the growing population demands will require a considerable intensification of land use, with concommitant intensification of fertilizer use and irrigation, and protection against damaging floods and cyclones. This intensification of land use, accompanied with an expansion of urban settlements to accommodate the growing population, will put increasing pressure on an already strained environment. The ecological implications will need close monitoring and remedial attention to the extent feasible. It is against this dynamic background that the implications of climate change and sea-level rise on agricultural potential need to be assessed.

4.4.2 A Distinctive Agriculture

Bangladesh's climate, with a warm wet summer and a cool dry winter, provides conditions suitable for growing a wide range of annual and perennial crops: tropical (e.g. rice, jute) in summer; temperate (e.g. wheat, potato) in winter; and subtropical (e.g. sugarcane, banana) throughout the year. The high rainfall and seasonal flooding make conditions particularly suitable for paddy[11] cultivation, and rice is the staple cereal. None the less, cropping is highly diversified. Bangladeshi farmers have more than 100 different crops in their collective repertoire. Additionally, it is estimated that there are over 10,000 rice varieties, adapted to specific micro-environments. The great majority of these varieties belong to the *aman* group (see below).

Conventionally, there are two main cropping seasons (*kharif* and *rabi*) and three rice-growing seasons (*aus, aman, boro*). The *kharif* season coincides with the rainy season, the *rabi* season with the dry season. Both seasons can be divided in two: *kharif-1* includes the pre-monsoon season and most of the monsoon season; *kharif-2* includes the later part of the monsoon season and the post-monsoon season. The *rabi* season includes the cool winter period when dryland crops (such as wheat) are grown, and a second part, which includes the hot pre-monsoon season, during which *boro* paddy is grown. These seasons are indicated in Figure 4.5. The *aman* rice group is photoperiod-sensitive: i.e. plants flower in response to changing day-length (in this case, shorter days). The *aus* and *boro* groups are time-bound, maturing in a fixed number of days according to variety and temperature.

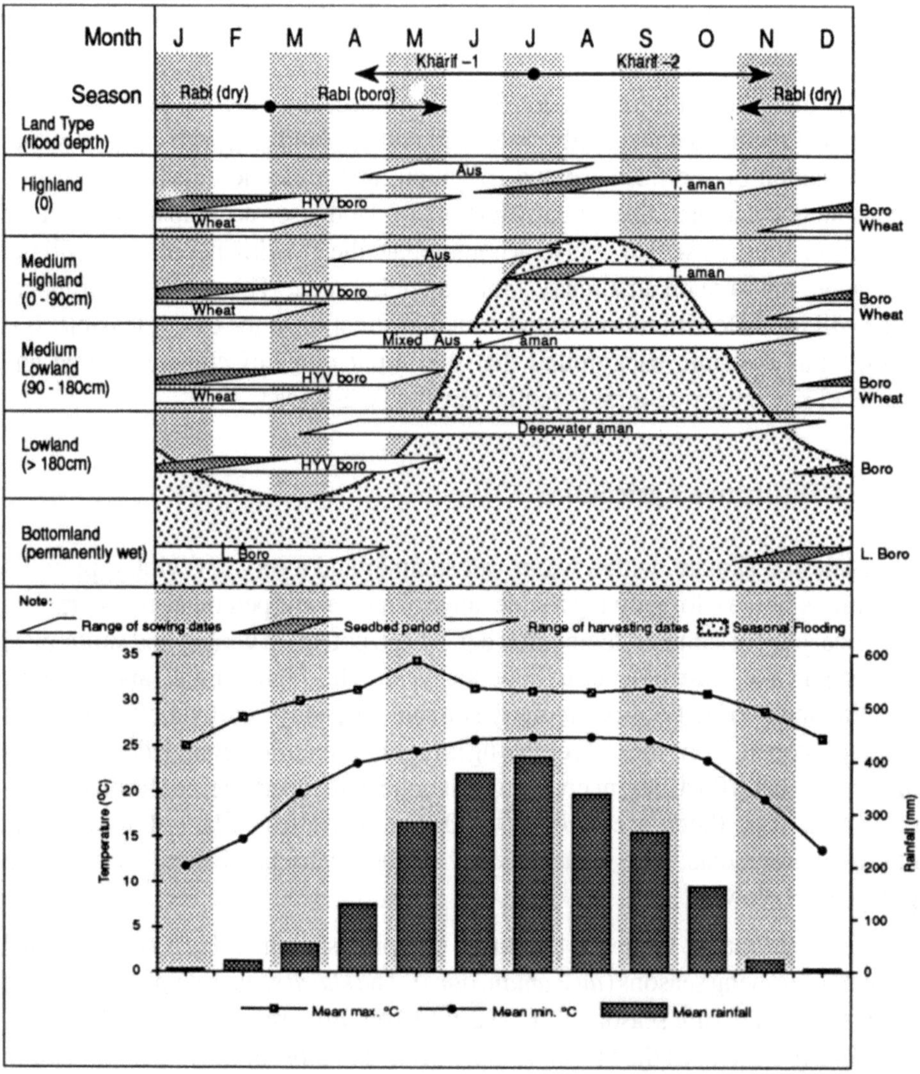

FIGURE 4.5: Rice and wheat crop calendar in relation to seasonal flooding, rainfall and temperature

Aman paddy is divided into two groups. Deepwater *aman* is sown pre-monsoon, grows with the rising floodwater, and is harvested post-monsoon, after floodwater recedes. Different varieties are adapted to different depths of flooding ranging between <1m and >4m. Transplanted *aman* is sown in seed-beds in June-July, transplanted to fields mainly in August-September, and harvested November-December (extending into January in coastal areas). In some areas, farmers transplant deepwater *aman* seedlings in June, usually after harvesting a previous irrigated *boro* paddy crop. Modern, high-yielding varieties (HYV) of rice are mainly photoperiod-insensitive, but sensitivity has been bred into some recent HYVs released by the Bangladesh Rice Research Institute (BRRI).

As is shown in Figure 4.5, the three rice-growing seasons and the dryland *rabi* season overlap with each other. They also overlap with the four climatic seasons (pre-monsoon, monsoon, post-monsoon, winter) described in the first main section of this chapter. Part of this overlap is due to differences in the dates of onset and ending of the rainy season between eastern and western parts of the country, and in the onset and end dates of the cool winter period between the north and the south. Part is due to local differences in soils and in the depths and duration of seasonal flooding between ridge and depression sites, as described in the first section.

4.4.3 Dynamic Agricultural Development

The rapid spread of dry-season irrigation in the last three decades (reaching about 3 million ha in 1991, or nearly 33 per cent of the cultivable area) has considerably modified the natural physical conditions. This has resulted in higher-yielding *boro* paddy displacing considerable areas of dryland *rabi* crops (especially pulses) which formerly were grown, as well as some *aus*, deepwater *aman* and jute. Official area and production figures for the four main rice crops, wheat, jute and some other important crops are given in Table 4.6.

The average cropping intensity is 172 per cent. Several factors determine whether a particular piece of land is single, double or triple (or more)-cropped. Amongst the physical factors are: length of growing season when sufficient moisture is available for plant growth, depending in turn on soil conditions, length of rainy season and whether irrigation is available or not; flooding characteristics, including hazards of too-deep or too-rapid flooding on some land, which may prevent one or more crops from being grown in the rainy season; and time-overlap between crops. For example, cultivation of *boro* paddy precludes jute from being grown; and local transplanted *aman* often prevents dryland *rabi* crops from being grown, because of overlap of seasons and the puddled topsoil conditions. The main cropping patterns on depth-of-flooding land types are shown in Table 4.7.

There is much intercropping, both in traditional and in modern farming systems. Examples are: mixed *aus* and deepwater *aman* on Medium Lowland; dryland *rabi* crops intercropped with sugarcane; and multiple mixtures of *rabi* vegetables and spices. Undersowing one crop before another is harvested can enable four (or more) crops to be grown in a year where soils hold moisture throughout the year and seasonal flooding is not too deep.

Somewhat unexpectedly, the introduction of irrigation can sometimes reduce cropping intensity. Farmers generally provide irrigation specifically in order to grow HYV *boro* paddy (even on permeable soils not well suited for transplanted paddy cultivation), which may prevent dryland *rabi* crops and *aus*, jute or deepwater *aman* from being grown on Medium Lowland and Lowland. This illustrates the fact that farmers appear to be more concerned with total crop production than with cropping intensity. This example also illustrates the fact that farmers are concerned with security of production. In many areas, *boro* paddy, grown with irrigation in the dry season, is less susceptible to loss by natural disasters than are *aus* and *aman* paddy crops grown in the rainy season.

BOX 4.5: The changing agricultural scene

Farming in Bangladesh is diverse, intensive and dynamic, and farmers are flexible, adaptable and sensitive to market prices. Although rice occupies 80 per cent of the cropped area, more than 100 crops are grown as well as many thousands of rice varieties adapted to local micro-environments. Much land grows two or three crops a year, but average yields are low because of insecurity of production. The recent spread of irrigation has greatly increased rice production in the more secure dry season, but at the expense of pulses, oilseeds and grazing land. Flood protection works aim to increase crop security in the monsoon season. The projected doubling of food demand by 2030 will require further intensification and security of production, but will have to be obtained from a shrinking cropped area as urban land expands.

The cultural preference for rice over other cereals explains the priority which farmers generally give to growing rice rather than wheat or maize, even on land where the latter crops might be better suited agronomically than rice. Market prices (up to now, at least) are such that rice cultivation can be more profitable than that of other cereals, except on very permeable soils. *Boro* paddy cultivation is also more profitable than the cultivation of many dryland *rabi* crops that are widely grown on non-irrigated land. For instance, it was estimated in 1989 that prices for *rabi* pulses and oilseeds would need to be two or three times higher for them to compete successfully with *boro* paddy at prevailing prices (UNDP, 1989).

Farmers are highly sensitive to market prices. This has been demonstrated over many decades in swings between *aus* paddy and jute cultivation in different years according to farmers' expectations about relative prices. It must be expected that crop areas and production will continue to vary between years and over longer periods according to relative prices as determined by market demand (which, in turn, can be influenced by production shortfalls and reduced incomes following natural disasters, and by large-scale food imports, often at aid-concessionary prices).

TABLE 4.6: Area and production of major crops in 1990-1991

Crop	Area (million ha)	Production (million t.)	Yield (t/ha)
Aus	2.11	2.32	1.10
Transplanted *aman*	4.84	8.16	1.69
Deepwater *aman*	0.94	0.99	1.05
Boro	2.55	6.36	2.49
Total rice	10.44	17.83	
Wheat	0.61	1.01	1.66
Other cereals	0.11	0.08	0.73
Total cereals	11.16	18.92	
Jute	0.58	0.95	1.64
Pulses	0.72	0.52	0.72
Oilseeds	0.56	0.45	0.80
Potato		1.24	
Sugarcane	0.19	7.69	40.47

Source: Bangladesh Bureau of Statistics (BBS), 1992, and BBS (unpublished data).

TABLE 4.7: Main cropping patterns on floodplain land types

Land type[1]	**Main cropping patterns[2]**	
	Without irrigation	With irrigation
Highland (FO) (Not flooded)	Tree crops, sugarcane; *kharif*, vegeatbles, spices, fruits. B.*aus*/jute-*rabi* crops. HYV/LV t.*aman* or HYV/LV t.*aus*-HYV/LV t.*aman* on impermeable soils.	Similar to non-irrigated, but more cash crops (vegetables, spices, tobacco, cotton); also wheat. HYV *boro*-HYV t.*aman* on impermeable soils.
Medium Highland - I (FO) (0-30cm)	B.*aus*/jute-*rabi* crops or B.*aus*-HYV/LV t.*aman*. On impermeable soils: HYV/LVt.*aman* or HYV /LV t.*aus*-HYV/LVt.*aman*.	Similar to non-irrigated, but more wheat and *rabi* cash crops on permeable soils.
Medium Highland - II (F1) (30-90 cm)	B.*aus*/jute-*rabi* crops. HYV/LV t.*aman* or HYV/LV t.*aus*-LV t.*aman* on impermeable soils.	Similar to non-irrigated, but more wheat and *rabi* cash crops on permeable soils.
Medium Lowland (F2) (90-180cm)	Mixed b.*aus*+b.*aman*-*rabi* crops; jute-*rabi* crops.	HYV *boro*, locally preceded by non-irrigated mustard and/ or followed by transplanted deepwater *aman*.
Lowland (F3) (>180cm)	B.*aman*-*rabi* crops/ fallow	HYV *boro*
Bottomland (F4) (Permanently wet; max. flood depth usually >300cm)	LV *boro*. Grassland; reed swamp.	*boro* (mainly LV)

1. See Table 4.4 for definitions of land types.
2. Abbreviations:
   HYV = high yielding varieties
   LV = local varieties
   B/b = broadcast
   T/t = transplanted

4.4.4 Irrigation and Flood Protection

Variability is an important characteristic of Bangladesh's climate, as described in an earlier section. Appreciable differences occur between years in the start and end dates of the rainy season and of the cool winter season; in the amounts of rainfall received in the *kharif* and in the *rabi* seasons; in the continuity of rainfall within the *kharif* season; and in the incidence of extreme events, such as excessive rainfall, strong winds and hail (FAO, 1988).

Such differences may occur locally, regionally or nationally. In addition, there are differences between years in the extent and severity of floods, drought and tropical cyclones, and in the localities affected by such events. Farmers' cropping patterns and management practices reflect their experience in coping with this uncertainty and exposure to natural hazards. Far from being conservative, they display remarkable flexibility and adaptability in coping with variable conditions and adapting to technological change (see Endnote 3).

It has been a major objective of the Government's development policy in recent decades to make the farmers' cultivation environment more secure by promoting the spread of irrigation and by investing in flood protection and drainage projects. With reduced risks of crop damage or loss, farmers are encouraged to make greater investments in production inputs, such as HYV seed and fertilizers, as well as for labour, thereby increasing their crop yields and production, and also generating additional employment.

By 1991, irrigation had spread to about 3 million ha, mainly using small tube-wells and low-lift pumps, and over 6,000km of embankments had been built along the country's major and minor rivers inland, and in the coastal zone. Nominally, these embankments protect some 1.6 million ha against river floods and a further 950,000ha against salt-water flooding. However, as described earlier, under water resources, not all such embankments are fully functional, and the areas actually protected are substantially less than the figures quoted.

4.4.5 Future Trends

The rapid expansion of irrigation has been the main factor contributing to increased agricultural production in the past two decades. This increase has been achieved predominantly through the expansion of HYV *boro* paddy production in the dry season. There is still considerable potential to expand irrigated *boro* paddy cultivation, but water resources are not infinite.

The present irrigated area is about 3 million ha. It has been estimated that another 2 million ha can be economically developed using only shallow tube-wells (STW) and deep tube-wells (DTW), and an additional 1.2 million ha through surface water development without major river diversions, using low-lift pumps (LLP). Thus, the total irrigated area could be increased to 6.2 million ha, though other estimates range from 5.0 to 7.6 million ha (Task Force, 1991a)[12]. The total arable area in the country is about 9 million ha.

Assuming that, as at present, 95 per cent of this potential expansion of irrigation by 3.8 million ha would be for *boro* paddy production, that could increase its production by 8 million tons (at present yield levels). In turn, that would increase total cereals production to 27.5 million tons (assuming no changes in the production of *aus* and *aman* paddy and in other cereals)[13]. Net of seed, feed and waste, that would support about 135 million people, a population which could be reached before 2005.

Thereafter, where might other potential sources of increased production be sought to feed the growing population? There is very little scope to increase the arable area. Indeed, it must be expected that the cultivable area will shrink as the population increases, though not proportionately. Greater losses may be expected from urban expansion than from the expansion of rural settlements which experience shows tend to become more crowded as population increases rather than expand onto agricultural land (Serno, 1981).

There is still scope to increase yields per hectare above present low average levels, especially through more efficient use of fertilizers and irrigation water. Present average crop yields, as shown in Table 4.6, are considerably below potential yields for the present climate, and average annual rates of fertilizer use are also relatively low at 114kg of fertilizer nutrients per cultivated hectare in 1991 (calculated from BBS, 1992).

However, experience shows that incremental increases in yields through the adoption of better management practices are slow to appear unless greater security of production is assured (as well as credit and secure prices). In floodplain areas, the main cause of insecurity is the risk of flood damage to *kharif* crops and, to some extent, to *boro* paddy also in the pre-monsoon season.

Inevitably, therefore, a continued expansion of flood control and drainage works will be needed to provide a secure environment for investment in intensive, year-round, crop production. Regional planning studies under the Flood Action Plan are currently examining options for improving flood management so as to provide this greater security.

To the extent that the provision of flood protection and drainage might be considered technically, economically or environmentally infeasible (locally, regionally or nationally) then alternative means will need to be sought to generate wealth and employment through which additional food requirements might be purchased outside the area or the country. These are the production scenarios against which the potential impacts of climate and sea-level change on agriculture need to be considered.

4.4.6 Impacts of Global Warming

Four factors linked to the atmosphere or climate strongly influence plant growth. They are carbon dioxide (CO_2), rainfall, light intensity and temperature. Changes in all these factors may occur with global warming. This section examines how changes in carbon dioxide concentration, light intensity, soil moisture and temperature could affect crop growth and yields. A wide range of references to research on these topics is given in Downing, 1992; Parry and Jiachen, 1991; and Rosenzweig *et al.*, 1993.

Carbon-dioxide Concentration

The amount of carbon dioxide in the atmosphere affects plant growth in several ways (Parry and Jiachen, 1991). The increased carbon dioxide levels which contribute to global warming could increase the rates at which plants absorb carbon dioxide from the atmosphere and also increase the efficiency of water use by plants. Therefore, a doubling of carbon dioxide in the atmosphere could significantly increase plant growth and crop yields, if other factors are not limiting.

Photosynthesis is the process by which plants use light to convert atmospheric carbon dioxide into plant tissue. Carbon makes up about 40 per cent of the dry matter weight of higher plants. Therefore, if higher carbon dioxide levels increase carbon uptake, plant productivity should increase. This factor is made use of in temperate countries in the production of tomatoes and vegetables in greenhouses artificially enriched with carbon dioxide (at concentrations several times higher than is envisaged to occur with a doubling of atmospheric carbon dioxide).

Plants differ in their photosynthesis mechanisms. Some, called C_3 plants, benefit more than so-called C_4 plants. C_3 plants include rice, wheat, jute and many other crops grown in Bangladesh. C_4 plants include sugarcane and maize. C_3 crops could also benefit from reduced weed competition, because 14 of the world's 17 most troublesome weeds are C_4 plants.

How much of the increased plant growth is transformed into usable yield remains problematic. In controlled environment studies, doubling of carbon dioxide concentration has increased rice, wheat, barley and sunflower yields by an average of 36 per cent; and there are indications that benefits increase with higher temperatures. Little is yet known about yield quality, but a change in the ratio of carbon to nitrogen in plants could reduce protein levels, and hence reduce their nutritional value for humans and livestock.

BOX 4.6: How higher CO_2 levels increase plant growth

> Laboratory experiments show that increased carbon dioxide (CO_2) concentrations in the atmosphere increase photosynthesis rates by which plants absorb carbon. They decrease transpiration rates, resulting in greater efficiency of water-use by plants. Thus, if other factors are not limiting, expected increases in atmospheric CO_2 levels may significantly increase plant growth rates and yields. Crops such as rice, wheat, jute and pulses would benefit more than maize and sugarcane. However, limiting factors, such as reduced light intensity due to increased cloudiness, might offset some of the benefits of CO_2 increases.

Water use efficiency is improved by increasing carbon dioxide levels. This results from a reduction in transpiration rates caused by a decrease in the size of the openings in plant leaves (stomata) through which plants breathe. A doubling of carbon dioxide reduces the size of these openings by about 40 per cent in both C_3 and C_4 plants. This could reduce transpiration rates by 23 to 46 per cent, giving plants better tolerance against drought and reducing irrigation demand.

Yield benefits from increased photosynthesis and water use efficiency with a doubling of carbon dioxide levels could be 10 to 50 per cent in C_3 crops and 0 to10 per cent in C_4 crops, where other factors are not limiting (Parry and Jiachen, 1991). Rosenzweig *et al.* (1993) report using increments of 22 per cent for wheat, 19 per cent for rice and 7 per cent for coarse grains (mainly maize) in crop modelling studies. However, it needs to be understood that these findings are based on experiments under artificial conditions with controlled environments. They remain to be tested under field conditions. It needs to be kept in view that increased carbon dioxide, temperature and humidity levels associated with global warming could also favour insect, bacterial and fungal pests and diseases of crops which might reduce or negate potential increases in crop yields.

Light Intensity

A warmer ocean would evaporate more moisture. Hence, the number of cloudy days may increase with global warming. Cloud intercepts incoming radiation from the Sun, which could reduce plant growth and yields. This effect is seen in Bangladesh in the difference in rice yields between crops grown in the dry season and those grown in the monsoon season. An example for rice is given in Table 4.8 (although it should be noted that differences in temperature and length of vegetative growing period also influence yields). Therefore, increased cloud cover associated with global warming could off-set some of the benefits derived from increased carbon dioxide concentrations, but it is impossible on present evidence to predict whether the net effect on yields would be positive or negative.

TABLE 4.8: Yields of IR-36 in relation to seasonal differences in temperature and radiation

Factor	Cold dry season	Hot humid season
Minimum temperature (°C), transplanting to flowering	13.5	25.2
Radiation, transplanting to flowering (mwh/cm²/day)	434	427
Transplanting to flowering period (days)	108	72
Minimum temperature(°C), flowering to harvest	23.3	24.7
Radiation, flowering to harvest (mwh/cm²/day)	561	434
Flowering to harvest period (days)	23	23
Panicles/hill	14	19
Yield (t/ha)	5.5	2.6

Source: Bangladesh Rice Research Institute (BRRI), 1983. Annual Report for 1982, BRRI, Joydebpur.

Soil Moisture

The conjectured increase in monsoon and winter rainfall with global warming would increase soil moisture supplies. This is most likely to benefit dryland crops in winter and in the pre-monsoon and post-monsoon seasons. These benefits could be enhanced by the reduction in crop transpiration rates described above, and by increased irrigation supplies.

However, the statements in the previous paragraph refer to average conditions. As is indicated by the standard deviation figures given in Table 4.1, rainfall in Bangladesh is highly variable from year to year, and it must be expected to remain so with global warming. Thus, although droughts might become less frequent with global warming (see Figure 4.4), they could become more intense when they do occur because of higher temperatures and evaporation rates during rainless periods, more especially in the winter, pre-monsoon and post-monsoon seasons.

It needs also to be recognized that excessive amounts of soil moisture can damage crops. Such damage occurs periodically under present climatic conditions: e.g. *rabi* crops such as wheat and pulses can be destroyed if fields become waterlogged following heavy post-monsoon or winter rainfall; and summer vegetables and jute (even newly-sown *aus* and deepwater *aman* paddy) can be drowned when fields are flooded by heavy pre-monsoon rainfall. If the conjectured increases in rainfall take place, farmers will need to pay increased attention to improving field drainage in order to grow dryland crops securely. On the other hand, conditions would become more suitable for transplanting rice.

An increased frequency of years with a bi-modal monsoon rainfall pattern, which it is conjectured could occur, would have implications for *aman* paddy in particular. Either transplanting dates would need to be delayed, or supplementary irrigation used to allow seedlings to be planted on time. However, delayed planting is likely to be less critical than under present conditions because of the higher temperatures expected to prevail: see below. The occurrence of a dry spell in the middle of the monsoon season would enable *aus* paddy to ripen and be harvested under better conditions than generally prevail at present.

In principle, higher temperatures could exacerbate existing drought risks in the pre-monsoon and post-monsoon seasons by increasing evapo-transpiration rates. However, such an outcome is by no means certain: higher rainfall, humidity and cloud cover could off-set the effect of higher temperatures; and, as described above, higher carbon dioxide levels could reduce transpiration rates, making plants more drought-tolerant.

Temperature Impacts on Crops

Increased temperatures with global warming seem likely to have their main direct impact on crop production through a reduction in the length of the cool winter period. The nature of the impact would differ between crops.

Tropical and subtropical crops such as rice, maize, millets (e.g. *kaon*), sugarcane and bananas would benefit from a longer warm growing season (provided that moisture supplies remained adequate to support growth). For temperate crops grown in the cool dry season (e.g. wheat, potato, lentils and mustard/rapeseed), the impact of a shorter cool winter period would be negative. The size of the respective gains and losses in yield and production of these crops would depend on the number of days by which the cool winter period was actually reduced, assuming that moisture supplies were not limiting. This will vary in different parts of the country.

Rabi-season crops seem likely to suffer the greatest impact. Wheat, for instance, yields best when minimum temperatures are continuously below 15°C for more than 60 days between tiller formation and panicle initiation, and when day-time temperatures throughout the growing period are within the range 15°C to 25°C which is optimum for photosynthesis in that crop.

These conditions are currently met only in agroecological thermal zone T5 (FAO, 1988), the southern boundaries of which cross the north-western and north-eastern corners of the country. Thus, only Dinajpur and Rangpur regions, the western half of Rajshahi region and the eastern fringe of Sylhet region are presently rated very well suited for these crops (ignoring moisture and soil requirements). At the other extreme, the coastal fringe, in thermal zone T1, with an average of only 30 to 40 days with minimum temperatures below 15°C in the November-February wheat-growing season, is rated moderately or marginally suitable for wheat on thermal grounds, with potential yields only about 30 to 45 per cent of those attainable in thermal zone T5.

A reduction in the length of the cool winter period would move these thermal zonal boundaries northward. That would reduce the area in the north which is presently rated thermally very well suited for wheat and other *rabi* crops, and would increase the areas in the south rated moderately or marginally suited. Further studies are required to determine the extent to which present thermal zone boundaries might shift per degree increase in winter temperatures: see the final section of this chapter. Such studies should take into account the possibility that minimum temperatures might increase more than mean and maximum temperatures.

For *rice*, the impacts would vary between seasons. For *aman* paddy, the critical temperature is 20°C: minimum temperatures below this level at the time of flowering interfere with pollination and so reduce grain yields. Under present conditions, the average date by which night temperatures begin to fall below this level ranges from 24 October in the north (Rangpur and Srimangal) to 16 November on the coast (Cox's Bazar; even later on the Bay islands).

Postponing the onset of minimum temperatures below 20°C by one or two weeks would particularly benefit *aman* HYVs which are not photoperiod-sensitive. Provided that moisture or depth-of-flooding conditions were not limiting, that would make it easier to fit in an *aman* crop (HYV or local) after growing *aus* paddy or jute, especially in the north-western part of the country. The conjectured lengthening of the rainy season would reinforce this benefit.

For *boro* paddy, there are two critical temperatures: minimum temperatures below 15°C reduce or arrest plant growth; and maximum temperatures above 40°C at the time of flowering interfere with grain formation. Thus, higher winter temperatures might be beneficial, whereas higher pre-monsoon temperatures could reduce yields.

BOX 4.7: How will climate change affect crops?

The optimum range of temperatures for plant growth varies between crops. In general, with global warming, *kharif* crops (e.g. *aus* and *aman* paddy, jute) would benefit from the higher temperatures in their growing seasons; dryland *rabi* crops (e.g. wheat, pulses) would suffer reduced yields with higher winter temperatures; and yields of *boro* paddy could be reduced by higher extreme maximum temperatures in the pre-monsoon season. Projected increases in rainfall should benefit crop production through increased soil moisture and irrigation supplies (where it does not increase flood hazards). These benefits could be partially reduced by decreased light intensity resulting from increased cloud cover associated with higher mean temperatures.

Higher winter temperatures would benefit both local varieties and HYVs of *boro* paddy by reducing the length of the cool period. HYVs would be particularly benefited: an earlier end to the cool period would enable farmers to transplant earlier and thus either harvest earlier, before the onset of heavy monsoon rainfall, or alternatively grow higher-yielding, long-duration varieties. At present, the mean end-date of the period with minimum temperatures below 15°C ranges from 20 January on the coast (Cox's Bazar; even earlier on the Bay islands) to 3 or 4 March in the north (Rangpur and Srimangal). The conjectured increase in mean seasonal temperatures by about 1°C to 1.5°C by 2030 might advance these dates by one or two weeks.

On the other hand, increased temperatures in the pre-monsoon period could increase the number of days with maximum temperatures above 40°C. At present, the mean number of days per year with such extreme temperatures ranges from less than one in the centre and east of the country to around ten in the extreme west (FAO, 1988). The effect of global warming might be to move the boundaries of the present 0.5 to 5, 5 to10 and 10 to15 day >40°C zones eastward. It might also bring forward by a few days the dates in the pre-monsoon season when such damaging temperatures might occur. The conjectured earlier onset of the monsoon season, were it to occur, would reduce the present incidence of such extremes in late May and early June.

For *aus* paddy, the direct impact of increased seasonal temperatures seems likely to be small. Neither broadcast *aus*, grown as a dryland crop, nor transplanted *aus*, grown under wetland conditions, are likely to be affected by changes in the periods of occurrence of temperatures below 15°C and above 40°C. The sowing period of *aus* comes later than the end of the cool winter period, and the flowering period is after the pre-monsoon months when extremely high maximum temperatures are liable to occur.

However, both broadcast *aus* and transplanted *aus* might benefit from higher temperatures occurring throughout their growing period. *Aus* paddy seems more likely to be affected by the indirect effects of higher temperatures through increased evapo-transpiration rates, which may increase exposure to moisture stress during critical early growth periods in the pre-monsoon season. However, this could be off-set to some extent by a reduction in transpiration rates caused by higher carbon dioxide levels.

The agronomic implications of possibly increased seasonal temperatures need attention in two study areas: plant breeding; and crop suitability assessment. Recommendations are given in the final section of this chapter.

4.5 HOW MIGHT OTHER BIO-RESOURCE SECTORS BE AFFECTED?

Forestry, fisheries, and livestock are important resource sectors that contribute 11.8 per cent to the GDP of Bangladesh. In this section of the chapter, the potential effects of climate and sea-level changes on these sectors are explored.

4.5.1 Forestry

Forests officially cover about 1.88 million ha, but much of it in hill areas and on the Madhupur and Barind Tracts is severely degraded and comprises grasses, bamboos and scrub vegetation[14]. The forestry sub-sector contributes about 3 per cent to the national Gross Domestic Product (GDP), and employs about 2 per cent of the total labour force (Task Force, 1991a). The country's forest resources are under great pressure from the growing population. Accordingly, this sub-section only addresses the question whether projected climatic and sea-level changes might ameliorate or exacerbate existing trends.

Forests Under Pressure

Bangladesh's forests comprise moist semi-deciduous forests in eastern hill areas (0.46 million ha); dry *sal* forests on the Madhupur Tract (and, to a minor extent, on the Barind Tract) (0.12 million ha); the Khulna and Chakaria Sundarban mangrove forests (0.58 million ha); and the so-called village homestead forests of multipurpose trees and bamboos (0.72 million ha) (Task Force, 1991b).

All forest ecosystems in Bangladesh are under severe pressure from human interventions. Much of the forest in hill areas and on the Madhupur Tract is severely degraded and has been replaced by grasses, bamboos and scrub. Given the present rates of degradation, and still greater pressures likely to be exerted in the future by the burgeoning human population, it is very difficult to predict how much forest might remain and what its composition might be by the 2030 horizon being used in the present study scenarios.

Changes in Upland Forests

In principle, the conjectured changes in climate would be expected to increase tree growth in all forest ecosystems. Most trees are C_3 plants in which, as described previously, the increased carbon dioxide levels contributing to global warming appear to increase rates of photosynthesis and reduce plant water requirements. However, it needs to be stated that these benefits have not yet been demonstrated in trees as they have for a number of agricultural crops (Andrasko, 1990).

In upland forests (including homestead groves), tree growth would also benefit from the conjectured increase in annual precipitation and from a longer growing season (resulting, in turn, from a reduction in the cool winter period as well as the conjectured increase in pre-monsoon and post-monsoon rainfall). The extent to which such benefits might be off-set by increased pest and disease activity consequent upon a longer, warmer and more humid growing season can only be speculated upon at present. Similarly speculative are the extents to which the conjectured increased rates of tree growth might be off-set by increased waterlogging and increased rates of soil erosion and nutrient leaching consequent upon increased amounts and rates of precipitation.

Changes in Mangrove Forests

Mangrove forests seem less likely than upland forests to benefit directly from the conjectured changes in temperature and rainfall. For mangrove species, seasonal changes in levels of soil and water salinity appear to be more critical.

If salinity levels were to increase due to a rise in sea level or to a continuing diminution of dry-season flow in rivers passing through the forests, salt-tolerant species, such as *goran (Ceriops decandra)* and *keora (Sonneratia apetala)* presently living near the sea, would be expected to advance inland at the expense of the species (especially *sundri, Heritiera fomes*) more tolerant of freshwater conditions which presently occupy interior parts of the Sundarban. On the other hand, if increased river discharges were to push back the present isohalines, especially in the dry season, then the species succession would be in the reverse direction.

BOX 4.8: Climate change and forestry

> In principle, increased CO_2, temperature and rainfall levels should increase tree growth rates. Such benefits could be partially or wholly off-set if warmer, wetter conditions increased pest/disease incidence, soil erosion and nutrient leaching. Potential climatic benefits in mangrove forests (Sundarban) could be negated where a rising sea level increased soil salinity levels. However, all forest resources seem likely to suffer more severe impacts from increasing human population pressure than from climate change in the foreseeable future.

In the present state of knowledge, it is not possible to predict what the overall impact of climate change and a rise in sea level might be on the Sundarban. There are too many uncertain factors. These include both natural factors as well as those related to possible human inteventions in the environment. Apart from the basic uncertainties regarding the climatic and sea-level changes that might actually occur and the rates at which such changes might affect the region, there are uncertainties concerning environmental changes occurring in the Khulna Sundarban, independent of global warming, associated with secular and human-made changes in river flow through the moribund Ganges delta. (See Chapter 7, The Case of the Coast, for detailed discussion.) The rates of natural species succession that could occur under pressure of environmental changes are also uncertain. So is the outcome of the conflict between the forces tending to submerge or erode land as sea level rises and those tending to cause accretion of sediments on existing land during a rise in sea level and so maintain the status quo.

Studies needed to improve predictions about the possible impacts of climate and sea-level changes on mangrove forest ecosystems are described in the final section of this chapter.

4.5.2 Fisheries

Bangladesh's fisheries play a vital role in the country's nutrition, employment and economy. They provide about 80 per cent of human intake of animal protein, and give employment to an estimated 1.2 million people directly and to a further 0.9 million indirectly through informal subsistence fishing activities. They contribute about 6 per cent of GDP and about 12 per cent of foreign exchange earnings (mainly from shrimps). This sub-section examines the possible ways in which projected changes in climate and sea level could affect fish habitats and production.

A Multiple Resource

Bangladesh's fisheries resources comprise river, floodplain, beel, baor, pond and marine production areas. (Beels are perennially, or almost-perennially, wet depressions; in the north-east, they are called haors. Baors are ox-bow lakes in the south-west.) Floodplain fisheries largely overlap with agricultural land. Ox-bow and pond fisheries, most shrimp farms and some beel and haor fisheries form a separate land use, covering an estimated 444,000ha (Ahmed, 1991).

Official estimates give total fish production as 0.89 million tons in 1990-91 (BBS, 1992). Of this, inland fisheries of all kinds contributed 0.65 million tons and marine fisheries 0.24 million tons. Production has increased by only about 80,000 tons in the past 20 years, whilst the human population has grown by about 40 million. In consequence, per capita availability of fish has decreased from about

33gm per day in the late 1960s to around 20gm per day in 1991 (Task Force, 1991a).

Increased production has come mainly from marine fisheries and from cultured pond fisheries. Production from open-water capture fisheries is reported to have declined. This is of particular concern, because these common-property resources provide the major part of the fish consumed by poor rural people. Amongst the causes of the reported decline are: over-fishing; the construction of flood control embankments; a reduction in the naturally flooded area and the area of dry-season water bodies due to drainage and the use of water for irrigation; and the use of agrochemicals on floodplain farmland.

Given the increasing pressures on fish habitats likely to result from the continuing expansion of the human population, it is impossible to predict what might be the state of Bangladesh's fish stocks and industry by the year 2030 and beyond when the environmental impacts of global warming might be felt. Therefore, this sub-section considers only those changes in fish habitats and production which might be directly linked with the projected changes in climate and sea level resulting from global warming.

Carbon-dioxide and Temperature Effects

It needs to be made clear at the outset that there is little definite information on which to base predictions of possible impacts of climatic change on fisheries. Little seems to be known about the direct impact which increased carbon dioxide levels and temperatures might have on different fish and crustacean species. However, for all fish, metabolic activity (reflected in growth) is doubled for each 10°C increase in temperature, contingent upon other environmental factors and food supply remaining adequate (D. Deppert, pers. com., 1992). The conjectured 1°C to 1.5°C increase in average temperatures by 2030 could therefore potentially increase fish yields by 10 to 15 per cent.

There could also be considerable indirect impacts. In the first place, increased temperatures (especially increased winter temperatures) and increased carbon dioxide levels would be expected to increase biological activity in all fish habitats, especially in existing floodplain, pond, and shallow-water estuarine and tidal floodplain fisheries. In principle, this might be expected to increase food availablity, both at peak periods and over a longer annual growing season.

However, this potential benefit could be counteracted (though to unpredictable extents) by accompanying changes in the composition of aquatic flora and fauna in different fish habitats. Such changes could include, for instance, changes in the balance of edible and toxic algae, and the depletion of oxygen levels resulting from excessive algal blooms (although such consequences seem likely to be limited to stagnant water conditions in the dry season and in cultured fish ponds). These are subjects deserving early research so as to provide information for modelling the direct and secondary impacts of global warming and the identification of possible mitigatory measures.

Floodplain Fisheries

Floodplain habitats could also be enhanced by the conjectured increase in monsoon rainfall, especially if this were to be accompanied by increased pre- and post-monsoon rainfall. This might lengthen the period of seasonal flooding, increase the extent of dry-season water bodies, and push back the saline front in estuarine and tidal floodplains, as described earlier in the water resources section.

Associated with the conjectured increase in average temperatures, the lengthening of the flooding period might also increase the duration of the fish spawning season. This could possibly induce second and third spawnings in species that normally spawn only once a year under present conditions. On the other hand, a pushing back of the present salinity front could have a negative impact on post-larval shrimp nurseries.

A rise in sea level, by impeding drainage from areas immediately inland from the coastal zone, could also prolong seasonal flooding and extend perennial flooding beyond present limits. This could benefit fish habitats and production in the areas affected (mainly the Begumganj depression east of the lower Meghna, and areas in and to the south of the Gopalganj-Khulna *beels* west of the Meghna). Such potential benefits would, of course, be lost if, as seems probable, flood protection and drainage works were undertaken to protect agricultural land and settlements from increased flooding.

Coastal Fisheries

Near-coast fisheries in the Meghna estuary and on the continental shelf seem least likely to be affected (beneficially or negatively) by the conjectured changes in climate and sea level. These habitats are dominated by the huge river flows and sediment loads entering the estuary and the adjoining Bay of Bengal. Whether such river flows and sediment loads increase or not as a result of the conjectured increase in monsoon rainfall, the scale of any change seems likely to be insignificant in relation to existing conditions in these waters.

In theory, a rise in sea level might change the location of the present estuarine and island coasts. Whether a sea-level rise would enhance or offset large-scale morphological changes that take place in the estuary anyway can only be speculated upon at present. (See Chapters 3 and 7 for more detailed explanations.) However, fish habitats within adjoining coastal waters as a whole seem unlikely to change as a consequence of possibly increased rates of coastal erosion or estuarine accretion.

One caveat on the former interpretation needs to be expressed. The present composition of coastal fish and crustacean species could be changed if habitat changes in adjoining tidal and floodplain areas were to affect (whether positively or negatively) the spawning behaviour or fingerling success-rates of important coastal fishes and shrimps. This could be particularly relevant for the penaid prawns which rely on the freshwater-saltwater interface for their reproductive success. Any change in the location of this boundary could have an impact on the location and production of these shrimps in cultured situations.

BOX 4.9: Fisheries in a warmer world

> Indirect effects of climate change on fisheries may be greater than direct effects. Increased carbon dioxide and temperature levels could increase biological activity in all fish habitats, thereby increasing fish growth rates directly as well as indirectly through increased food supplies. Projected rainfall increases could extend the period of seasonal flooding of floodplain land which, combined with higher temperatures and increased food supplies, could extend fish breeding and growing seasons. Impacts on brackish-water shrimp production will depend on the net balance, in different areas, between salinity reduction because of increased river flows and salinity increase resulting from sea-level rise. Quantitative assessment of potential changes in production is made highly uncertain by potentially greater impacts on fish habitiats and stocks by continuing increases in human population pressure and interventions in floodplain, estuarine and tidal fish habitats.

Possible habitat changes in tidal floodplains appear to be the most conjectural amongst all of the fisheries environments. That is because, as described earlier under water resources, this is the zone where the nature, scale and direction of possible changes are least clear: i.e. whether dry-season river flows through the zone might increase or decrease; whether the salinity front in different seasons will advance or retreat in relation to existing positions; and whether land submergence or erosion by a rising sea level might be counteracted fully or in part by the continuing accretion of sediment on the land from tidal flooding as sea level rises.

One area where loss of present tidal floodplain and foreshore fish habitats does seem likely to occur is along parts of the Chittagong coast where a rising sea level would be expected to accelerate present rates of coastal erosion. Along sections of this coast where the present shore-line is backed by relatively higher piedmont plains (as along the coast north of Chittagong) or by hills (as, for example, south of Cox's Bazar), the effect of coastal erosion would be to reduce the width of the tidal zone. This could reduce the extent of habitats which often are important for fish and shrimp spawning and feeding. Along the Chittagong coast, too, the projected increase in monsoon rainfall might accelerate present rates of run-off and soil erosion in the adjoining hills, with consequential increased rates of sedimentation on tidal floodplain land. This could also lead to the progressive loss of valuable fish and shrimp habitats, including the areas in and around the Chakaria Sundarban presently used for shrimp farms.

The St Martin's Island coral reef fisheries will not necessarily be destroyed by global warming impacts (assuming that they have not already been destroyed by on-going human depredations on the reefs). Corals in this area are at the extreme limit of their tropical extent. Warming of the adjoining seas might therefore enhance conditions for coral growth in this latitude (provided that the waters are not contaminated by increased sediment flows from the neighbouring mainland). If higher sea temperatures were to increase coral growth rates, then fish habitats based on the reefs would remain unaffected, and might even be enhanced, by global warming.

Future Studies

The possible changes in inland and coastal fish habitats described above are entirely conjectural at present. Only when more precise indications become available of the probable rate, scale and direction of such environmental changes will it be possible to assess the probable impacts of global warming on floodplain, estuarine and marine fisheries resources. Studies designed to provide improved information in these respects will also need to take into account changes to the environment resulting from physical interventions such as river and coastal embankments, cross-dams, drainage schemes and coastal defence works, all of which might be taking place over the next few decades, irrespective of any changes due to global warming. Study needs are described in the final section of this chapter.

4.5.3 Livestock

Livestock play important roles in the national and the farming economies. Presently, the livestock sub-sector provides 2.8 per cent of GDP and nearly 10 per cent of foreign exchange earnings through exports of leather, leather goods, hides and skins. Additionally, it is estimated that it provides employment, directly or indirectly, to 20 per cent of the rural population (Task Force, 1991a). This sub-section explores how changes in climate and sea level might affect livestock, taking into account also the pressures which the growing human population imposes on livestock feed supplies.

Livestock in the Farm Economy

Livestock-raising rarely forms a separate land-use system in Bangladesh. Very nearly all animals are kept on farms, closely integrated into the agricultural production system. They play a vital role in the farming economy, providing the main source of farm draught power, supplying meat and dairy products, and producing manure for use as fertilizer and domestic fuel. At the time of the 1983-1984 Agriculture Census, there were approximately 22.5 million cattle and buffaloes, 14.5 million sheep and goats, and 76.5 million poultry (including ducks) (BBS, 1992).

The productivity of these livestock is very low. Current milk production is about 0.65 litres per cow per day; farm hens lay only 40 to 50 eggs per year; and average meat production from cattle is only 75kg (Task Force, 1991a). The main reasons for this low productivity are poor feed, poor protection against diseases, and over-work of draught animals. About 30 per cent of cows are used also for ploughing. Cattle feed supplies have been adversely affected by the intensification of agriculture in recent years. This has reduced the availability of dry-season grazing as well as of leguminous fodder, oilseed cake and good-quality straw: HYV rices produce unpalatable, stiff straw.

The increasing constraints on draught power are leading to an increased use of power tillers for cultivation. This trend must be expected to continue as intensification of agriculture continues in order to meet the demands of the increasing human population. Beneficial though this change may be for cultivation, it does have adverse implications for the supply of milk and of cow-dung for manure and domestic fuel, as well as for costs of cultivation for small farmers. However, a reduction in cattle numbers might have little impact on meat supply since the average per capita availability of animal protein from all meat and dairy sources is less than 3gm per day, less than one-half that derived from fish (7.6gm) (BBS, 1991).

If the cattle population is expected to decrease, it seems probable that the numbers of small livestock and poultry may increase. In part, this is because such animals are easier to keep in rural villages than cattle and have a higher turn-over rate. Goats and sheep are also less susceptible to diseases than cattle and poultry.

The growing urban population may also create an increasing demand for meat and dairy produce (Asaduzzaman, 1989), which could lead to the development of intensive dairy and poultry farms around major cities. While this intensification may not lead to an increase in total livestock numbers, it would be expected to lead to better-quality animals of greater economic value.

Impact of Greenhouse Warming

As in agriculture generally, greenhouse warming could have both positive and negative effects on livestock production. The direct effects of a 1°C to 1.5°C increase in annual and seasonal mean temperatures on different kinds of livestock seem likely to be small. Heat stress during an increased number of extremely hot days in summer could be off-set by a reduced number of days with cold stress during the winter.

In principle, livestock might benefit indirectly from global warming. As is described earlier under crop agriculture, higher carbon dioxide and temperature levels, and a longer plant growing season, could increase crop, weed and pasture growth, thereby potentially increasing the amount of food available to livestock. On the other hand, increased flooding due to the conjectured increase in monsoon rainfall could increase the frequency of *boro, aus* and *aman* paddy losses and consequently reduce fodder availability (including grain for poultry). Increased rainfall and flooding could also increase the incidence of humidity-related animal and poultry diseases.

BOX 4.10: How might climate change affect livestock

Higher summer temperatures could increase heat stress in farm animals, but higher winter temperatures could reduce cold stress. However, the temperature increases projected to occur by 2030 (0.5°C to1.7°C) seem unlikely to have serious direct effects on livestock health or production. Projected increases in rainfall and consequent flooding could increase the incidence of humidity-related diseases. In theory, higher CO_2 and temperature levels could increase plant food availability. In practice, this might be off-set by reducing grazing time due to an increased duration of flooding, reduced grain and fodder availability due to more frequent flood losses, and increased salinity in coastal areas. These negative impacts would aggravate existing problems for livestock production caused by cropping and environmental changes induced by the increasing human population pressure on land and water resources.

It is impossible to predict the magnitude of such potential benefits and losses. This is partly because of present uncertainties about the magnitude of possible changes in climate. In part, it is also because of the difficulty in predicting what might be the future size and condition of the livestock population itself, irrespective of any climatic change, in the face of the projected doubling of the country's human population by 2030.

There are similar uncertainties regarding the impact which a rise in sea level might have on livestock production. Loss of land to the sea or to perennial flooding inland would lead to a direct loss of land available to support livestock. However, the location and extent of such losses of land with the conjectured slow rise in sea level over the next few decades is unpredictable at present, as is the extent to which the government and people of Bangladesh might succeed in measures they might take to oppose such losses.

Study needs in the livestock sector are outlined in the final section of this chapter.

4.6 WHAT NEEDS TO BE DONE?

This section focuses on planning and research needs identified earlier in the water and bio-resource sectors. Topics are discussed under appropriate headings, following the sequence in which the subjects are dealt with in the previous sections. Research studies which are needed to assess the scale and rates of climate and sea-level change themselves, and the morphological and socio-economic impacts of such changes, are described in other chapters.

4.6.1 The Planning Context

The preceding sections have indicated some of the uncertainties which cloud attempts to assess the possible impacts of global warming on Bangladesh's vital land and water resources. Those uncertainties include the nature and time-scale of possible changes in climate and sea level in the Bangladesh region. They also include uncertainties (at least as great) concerning the environmental changes that might be taking place over the next 30 to 50 years as a result of increasing demographic pressure on the country's land and water resources, irrespective of global warming impacts. In effect, the possible impacts of climate and sea-level change have to be assessed against a shifting base-line. That makes prediction doubly difficult.

The most urgent need would seem to be to set up benchmark and monitoring field studies in order to assess environmental changes resulting from demographic pressures as well as from global warming impacts. If possible, such studies should aim to isolate the respective contributions of the two contributing factors. However, whether demographic impacts can be isolated from global warming impacts or not, such studies should seek to identify feasible remedial measures before adverse impacts affect the human populations dependent on threatened land or water resources. Human needs must drive all the research studies.

Difficult though it may be (especially at first) indicative predictions must be attempted in order to guide policy makers and planners on subjects where a long-term outlook is needed: e.g. the design of coastal defences and flood protection works; the location of new towns and industry (including agricultural and fish processing plants); agricultural, forestry, fisheries and livestock research programmes; and long-term national economic modelling.

Economic modelling will need to take into consideration possible future changes in regional and global economic conditions that could affect strategic imports and exports: for example, the possibility that foodgrains available for export from present major suppliers to Bangladesh could decline (and become more expensive) with climate change (Rosenzweig *et al.*, 1993).

4.6.2 Water Resources

The planning of coastal defence works and major river embankments, such as might be constructed under the Flood Action Plan, should take into account possible increases in flood discharges, sea level, and storm-surge incidence which might occur within the planning time-frame of projects. At the very least, the range of possible changes which might occur should be included in sensitivity analyses, based on present projections, and contingency plans should be prepared for possible design modifications which might become necessary if global warming impacts occur on the scale presently foreseen.

In support of such planning and design studies, hydrological modelling should include analyses of data for the range of possible increased river discharges which might result from the increased precipitation indicated by various general circulation models for the Bangladesh region: see Chapter 2. The implications which such increased discharges might have for flood levels inside and outside existing or proposed river embankments, channel stability, sediment movement, and seasonal salinity fluctuations need to be examined and assessed.

A number of institutions have roles to play in collecting and analysing the required data: Bangladesh University of Engineering and Technology, Bangladesh Water Development Board (including the Surface Water Modelling Centre), Flood Plan Coordination Organisation, River Research Institute, and the Water Resources Planning Organisation. The Ministry of Irrigation, Water Development, and Flood Control will need to allocate the lead role to one of these agencies to coordinate the studies and analysis of the findings. Linkages should be established through the United Nations Development Programme, the United Nations Environment Programme, World Meteorological Organization and United States Agency for International Development with on-going studies of hydrology in relation to climate change being made in the Mekong, Indus and other major river basins.

4.6.3 Crop Agriculture

Six kinds of studies are needed to improve knowledge about the probable impact of climate change on plant growth, yields and production, and to monitor environmental changes as they occur.

Basic research is needed to measure the response of a range of economically important crop plants to increased levels of carbon dioxide and temperature within the range of changes which could possibly affect Bangladesh within the next 30 to 50 years. Such studies need to be carried out in controlled climate chambers. The studies could be organised by the Bangladesh Agricultural Research Council (BARC) and allocated to appropriate research institutes and university departments. Association with international research institutions, such as the International Rice Research Institute and the International Maize and Wheat Improvement Centre, would be appropriate for exchange of data and experience within a regional network.

Phenological research is needed to correlate the response of crops to climate-related variables under field conditions. Observations are needed on farmers' fields and on research stations. Particularly needed is a system of continuous monitoring to relate crop growth, crop yields, and pest and disease incidence to differences in temperature, rainfall, insolation, etc., between seasons and between years. Such studies require simultaneous daily meteorological and field observations to be made, so they are best conducted on research stations and on immediately adjacent farmers' fields. They could also be integrated into farming systems research programmes as a routine activity.

Plant breeding research by Bangladesh's agricultural research institutes is already focusing on the selection or development of high yielding varieties, especially of rice and wheat, which are suitable for a wide range of environments, including tolerance of low temperatures, high temperatures, drought and, in the case of rice, salinity and deep flooding. Breeding programmes need to be reviewed periodically to ensure that the objectives cater for the range of environmental changes that could occur within Bangladesh with global warming. Particular attention is needed to the selection of varieties of temperate crops such as wheat, potato and pulses for better adaptation to high temperatures, and of *boro* paddy varieties to tolerate extremely high temperatures in the pre-monsoon season.

Agroecological studies are needed to extend the soil-climate-crop assessments carried out at the BARC with FAO of the United Nations assistance in the early 1980s. The existing BARC agroecological zones data base needs to be up-dated to include climatic data for the years since 1980. Then, using existing models, analyses should be made of the shifts in moisture and thermal zones which could occur with changed levels of temperature and rainfall derived from the various General Circulation Models. That would provide the basis for assessing possible changes in crop suitability and potential yields that could result from climate change.

Crop simulation studies now being undertaken by the Bangladesh Agricultural Research Council within a global network (described in Rosenzweig *et al.*, 1993) should be extended to include crops other than rice, a wider range of environmental conditions and a wider range of professional disciplines. They need to be supported by strengthened field and laboratory studies referred to elsewhere in this section.

In view of the many uncertainties presently involved in both with and without global warming scenarios, it would be unrealistic to expect dependable results in the near future from such simulation exercises. However, as with 'expert' systems generally, such exercises can be fruitful in focussing attention on critical factors and relationships that might otherwise be overlooked in projecting future scenarios. They have the further advantage that, to be conducted effectively, they require multi-disciplinary involvement which, through interactions and successive iterations, can lead to the identification of new study areas and the gradual refinement of the models and outputs.

Environmental monitoring needs to be organized in order to measure and assess changes in the physical environment and land use that take place over time. The first need is to establish benchmark sites where such monitoring can be carried out on a routine basis. Some of the problems associated with the selection of such sites are discussed in Brammer (1993).

Monitoring sites should be easily accessible so as to facilitate the work and supervision. They should represent specific physical environments. The agroecological zone regions and sub-regions (FAO, 1988) provide a basis for selecting sites on a rational basis. Preferably, sites should include a cross-section of the local relief from the highest to the lowest points in order to include a full range of soil, hydrological and land use conditions within each study area.

Criteria need to be established for selecting the parameters to be measured, determining the scale and frequency of sampling, and determining analytical and reporting requirements. The Environmental Impact Assessment Guidelines and Manual prepared for the Flood Action Plan (FPCO, 1992b; Irrigation Support Project for Asia and the Near East [ISPAN], 1992) could provide models for the recommended benchmark and monitoring studies. These studies should form an integral part of national environmental monitoring studies being planned as a part of the National Conservation Strategy in which the Department of the Environment and BARC have coordinating responsbilities.

4.6.4 Forestry

Two kinds of studies are required to improve predictions about the possible impacts of environmental changes on the Khulna Sundarban forest ecosystems. Such studies could involve the Forest Department and university botany departments. BARC could coordinate the studies and also establish linkages with international environmental research networks involved in studying mangrove ecosystems.

The first set of studies would form a part of the water resources studies recommended above. It would involve computer simulation of seasonal water flows and salinity levels in the rivers crossing the Sundarban under a range of scenarios: a continuation of present 'business-as-usual' trends; amendments of present trends by the diversion of water from the Ganges through the Gorai-Madhumati channels now being examined under the Flood Action Plan; river flow projected to occur with an average increase in monsoon rainfall of 12 per cent together with a rise in sea level of 14cm to 24cm by 2030.

Supplementing these simulations, field studies are needed to relate individual mangrove species (particularly those of commercial or ecological importance) to specific environmental conditions, especially salinity tolerance. Also needed are observations to establish the rates at which individual mangrove species can adapt to changing environmental conditions. For this purpose, evidence of recorded changes in forest composition and tree growth rates associated with environmental changes actually taking place would form an essential part of all such field studies. The findings would need to be used to refine the recommended computer simulation studies.

If natural rates of species succession were found to be inadequate to match existing or projected rates of change in environmental conditions, then consideration would need to be given to artificial restocking with appropriate species to supplement natural regeneration. Pilot studies with artificial restocking in different habitats might be warranted, taking into account changes in environments expected to occur with best-estimates of projected changes in river flow, precipitation and sea-level rise over the next 30 to 50 years.

4.6.5 Fisheries

Fisheries monitoring studies need to be initiated as soon as possible. The purpose would be, first, to describe a range of existing habitats with their associated fisheries resources, and then to record the response of important fish and crustacean species to any changes in those habitats resulting from projected changes in annual or seasonal water flow, sediment load, salinity, etc., whether 'natural' or attributable to human interventions.

Consideration should also be given to undertaking studies under controlled laboratory conditions of the responses of fish species to changes in carbon dioxide levels and temperatures projected to occur in Bangladesh over the next 30 to 50 years with global warming. The purpose would be to indicate the likely magnitude of possible changes in fish behaviour, growth and production as a basis for conducting field trials and designing any mitigatory measures that might be required.

The sooner information of these kinds is made available, the sooner might it be possible to predict the consequences of global warming and sea-level rise on Bangladesh's fisheries resources with greater reliability than is possible at present.

4.6.6 Livestock

Livestock considerations need to be specifically included in the planning of comprehensive studies addressing the possible physical, environmental, economic and social impacts of global warming on Bangladesh which have been described above and in other chapters.

Included in such studies could be the development, testing, and promotion of practical means to reduce the methane emissions by cattle, goats, and sheep which contribute to global warming. Amongst presently known techniques of achieving such an objective is the addition of urea or other suitable supplements to the diet of ruminants to reduce the amount of anaerobic digestion which produces methane.

The Livestock Research Institute could include the assessment of such techniques in its research activities.

BOX 4.11: Where do we go from here?

The insidious impacts of projected changes in climate and sea level will impinge on environments already changing under pressure of population growth. Planning, research and environmental monitoring studies need to consider both these factors together. Agricultural and biological research studies are needed to measure plant, animal and fish responses to higher CO_2 and temperature levels, and to select biotypes tolerant of expected conditions. BARC should use AEZ models to estimate possible shifts in moisture and thermal zones with climate change. Correlation of individual mangrove species with salinity levels should be the focus of forestry studies so as to improve predictions of probable changes with sea-level rise.

ENDNOTES

[1] Technically, the *monsoon season* is the period between early-June and August-September when a relatively stable, humid, air mass moves from the Bay of Bengal over Bangladesh and neighbouring India and Myanmar (Burma). However, heavy rain showers in the transitional pre-monsoon and post-monsoon seasons extend the period when rainfall exceeds evapo-transpiration rates, so the wet (or rainy) season lasts longer than the monsoon season proper: average 4 months in the west; 7 months in the north-east.

[2] One standard deviation indicates the range (+ or -) within which values may occur two years out of three on average; 2 x St D = 95% probability (i.e., 19 years out of 20); and 3 x St D = 99% probability (i.e., rainfall outside this value should occur only once in 100 years or more).

[3] Bari (1974) describes a Comilla farmer with 1 ha of land divided into 17 individual fields scattered over his village area who had 17 different rotations on his fields in the first year of study and a total of 61 rotations over five years. The rotations included different HYV and local paddy varieties, as well as dryland crops. The farmer's reasons for variations between years included differences in weather and flooding conditions, relative market prices and seed availability.

[4] The Bengali name for the Ganges River downstream of the Bhagirathi off-take (in India, roughly the site of the Farakka barrage) is Padma. However, hydrologists often restrict the name Padma to the section between the river's confluences with the Jamuna and the Meghna. The Jamuna is the name given to the 'new' channel of the Brahmaputra River, formed about 200 years ago, south of Dewanganj in Jamalpur district; the Old Brahmaputra channel flows eastward through Mymensingh District, then down the Sitalakhya channel east of Dhaka to join the Meghna north of its confluence with the Padma. The Meghna River occupies the former channel of the Brahmaputra southward from Bhairab Bazar. Only in exceptionally high floods does it receive water from the old Brahmaputra River; it receives water predominantly from the Surma, Kusiyara and other rivers draining the Sylhet basin. Between Bhairab Bazar and the confluence with the Padma, the channel is called the Middle Meghna; downstream, it is called the Lower Meghna until it grades into the Meghna estuary roughly south of the Chandpur Irrigation Project area.

[5] So-called 'public cuts' are unauthorized breaches in embankments made by individuals or groups, usually to relieve flooding of houses or land inside or outside an embankment, but sometimes to facilitate fish capture.

[6] The Flood Action Plan (FAP) was prepared following the devastating 1987 and 1988 floods when aid donors decided to support government efforts to find a lasting solution to the country's chronic flood problem. The first phase of the FAP (1990-1995) comprises a set of 26 regional, pilot and supporting studies costing approximately $150 million. The plan (really a programme) includes non-structural as well as structural elements.

[7] The term "functionally landless" is taken to mean household ownership of 0.0 to 0.2 ha of cultivable land. Note that Chapter 5 defines the functionally landless class of cultivable land ownership as 0.0ha to 0.4 ha, which gives 68.8 per cent.

[8] Different sources give different values for total basin area of the Ganges-Bramhaputra-Meghna/ Barak river systems, and the percentage lying within Bangladesh. For example: UNDP (1989) 1.5 million km^2 and 7 per cent; ISPAN (1989) 1.75 million km^2 and 8 per cent; and The French Engineering Consortium and the Bangladesh Water Development Board (1989) 1.55 million km^2 and 9.6 per cent, respectively.

[9] A 1 in 100 year flood has a 1 per cent probability (chance) of being equalled or exceeded in any one year, or an almost 10 per cent probability of occuring in a 10 year period.

[10] The total demand for cereals could, in fact, be somewhat below this figure if there is a substantial increase in per capita income during this period. Part of the cereals' demand could be replaced by an increased demand for quality foods (such as meat, fish, dairy products, vegetable oils, vegetables), purchases of which tend to grow faster than the rate of income growth. However, increased meat and dairy production would require additional grain production for livestock feed.

[11] In Bangladesh, 'paddy' is the unhusked grain; 'rice' is the milled grain, representing about two-thirds of paddy weight. Farmers report yields as paddy; so do international agricultural statistics. Production statistics in Bangladesh report yields as rice. In some countries, 'paddy' refers to the wet fields in which rice seedlings are transplanted. In Bangladesh, *aus* and deepwater *aman* are mainly broadcast sown as dryland crops in the pre-monsoon season. The main *aman* crop (transplanted *aman*), *boro*, and some *aus* and deepwater *aman* are transplanted into wet soils deliberately puddled to make them hold water on the surface.

[12] Irrigation by large-scale schemes diverting water from the country's main rivers which might occur in the future would largely displace existing irrigation provided by tube-wells and low-lift pumps, except possibly in saline coastal areas. However, the large-scale diversion of river water to the latter areas might have adverse effects on the salinity front elsewhere in the coastal zone, so the net benefit would need to be carefully studied before any project was implemented.

[13] Although the expansion of irrigation could benefit *aman* production through the use of supplementary irrigation, this increase would have to be off-set against production lost through the displacement of some *aus*, deepwater *aman* and wheat crops by *boro* cultivation.

[14] Different sources give different figures for the area under forest. The figures quoted are from the Bangladesh Bureau of Statistics (BBS, 1992, Table 5.05). The Task Force Report, Volume 2, gives 2.45 million ha, comprising 2.18 million ha of government-owned forest and 0.27 million ha of homestead forest land (Task Force, 1991a, page 139). Volume 4 of the Task Force Report states (pages 120-121) that the Forest Department controls 1.45 million ha, about half of which is reported to be under forest, and gives a figure of 0.716 million ha for village forests: i.e., a total forest area of 2.166 million ha (Task Force, 1991b).

REFERENCES

Ahmad, Q.K., Verghese, B.G., Iyer, R.R., Pradhan, B.B., and Malla, G.K. (1994) *Converting Water into Wealth: Regional Cooperation in Harnessing the Eastern Himalayan Rivers.* Academic Publishers, Dhaka.

Ahmed, M. (1991) *A Model to Determine Benefits Obtainable from the Management of Riverine Fisheries of Bangladesh.* International Centre for Living Aquatic Resources Management (ICLARM), Manila, Philippines.

Andrasko, K. (1990) Global warming and for sts: an overview of current knowledge. *Unasylva*, 163, Vol. 41. Food and Agricultural Organisation (FAO), Rome

Asaduzzaman, M. (1989) Feeding our future towns: an overview of urbanization and associated food policy issues. In Bangladesh Planning Commission, *Food Strategies in Bangladesh: Medium and Long Term Perspectives.* University Press Ltd, Dhaka.

Bari, F. (1974) *An Innovator in a Traditional Environment.* Bangladesh Academy for Rural Development (BARD), Comilla.

Bangladesh Bureau of Statistics (BBS) (1986) *The Bangladesh Census of Agriculture and Livestock, 1983-84.* BBS, Dhaka.

Bangladesh Bureau of Statistics (BBS) (1991) *Report of the Household Expenditure Survey, 1988-89.* BBS, Dhaka.

Bangladesh Bureau of Statistics (BBS) (1992) *Statistical Pocket Book of Bangladesh.* BBS, Dhaka.

Bangladesh Rice Research Institute (BRRI) (1983) *Annual Report* for 1982. BRRI, Joydebpur.

Brammer, H. (1993) Geographical complexities of detailed impact assessment for the Ganges-Brahmaputra- Meghna delta of Bangladesh. In R.A. Warrick, E.M. Barrow and T.M.L. Wigley (eds.), *Climate and Sea-level Change: Observations, Projections and Implications.* Cambridge University Press, Cambridge, England.

Brammer, H. and Khan, H.R. (1991) Bangladesh country study. In *Disaster Mitigation in Asia and the Pacific.* Asian Development Bank, Manila.

Catling, D. (1992) *Rice in Deep Water.* MacMillan Press, London.

Chowdhury, O.H. and Shahabuddin, Q. (1992) A study of food situation and outlook in Asia: country report for Bangladesh. Final report (unpublished). Bangladesh Institute of Development Studies (BIDS), Dhaka.

Downing, T.E. (1992) *Climate Change and Vulnerable Places: Global Food Security and Country Studies in Zimbabwe, Kenya, Senegal and Chile.* Research Report No. 1, Environmental Change Unit, University of Oxford, U.K.

Food and Agriculture Organization (FAO) (1988) *Land Resources Appraisal of Bangladesh for Agricultural Development*. 7 Reports. FAO, Rome.

Flood Plan Coordination Organisation (1992a) *Proceedings of the Second Conference on the Flood Action Plan*. Government of Bangladesh and World Bank, Dhaka.

Flood Plan Coordination Organisation (1992b) *Guidelines on Environmental Impact Assessment*. FPCO, Dhaka.

French Engineering Consortium (1989) *Prefeasibility Study for Flood Control in Bangladesh, Volume 1*. French Engineering Consortium, Paris.

Hunting Technical Services Ltd (1992) *FAP12 FCD/I Agricultural Study, Final report, Vol.I, Main Report*, Dhaka.

Irrigation Support Project for Asia and the Near East (ISPAN) (1989) *Eastern Waters Study*. ISPAN, Virginia, USA.

Irrigation Support Project for Asia and the Near East (ISPAN) (1992) *Manual for Environmental Impact Assessment, Volume 2*. Preliminary draft edition. ISPAN, Dhaka.

Master Plan Organisation (MPO) (1985) *Master Plan Report, Phase I*. MPO, Dhaka.

Master Plan Organisation (MPO) (1986) *National Master Plan, Final Report, Volume I*. MPO, Dhaka.

Parry, M. and Jiachen, Z. (1991) The potential effect of climate changes on agriculture. In J. Jaeger, and H.L. Ferguson (eds.) *Climate Change: Science, Impacts and Policy*, Proceedings of the Second World Climate Conference. Cambridge University Press, Cambridge, England.

Rozenweig, C., Parry, M.L., Fischer, G., and Frohberg, K. (1993) *Climate Change and World Food Supply*. Research Report No. 3, Environmental Change Unit, University of Oxford.

Serno, G. (1981) *A Reconnaissance Study of Changes in Settlement and Related Non-agricultural Land Use in Bangladesh 1952-1974*. FAD/UNDP Land Use Policy Project (BGD/78/014), Ministry of Agriculture, Dhaka.

Task Force (1991a) *Report on Bangladesh Development Strategies for the 1990's: Volume 2, Managing the Development Process*. University Press Limited, Dhaka.

Task Force (1991b) *Report on Bangladesh Development Strategies for the 1990's: Volume 4, Policies for Environment*. University Press Limited, Dhaka.

United Nations Development Programme (UNDP) (1989) *Agriculture Sector Review. Main Report*. Bangladesh Agriculture: Policies and Performance. UNDP, Dhaka.

United Nations Development Programme (UNDP) (1989) *Bangladesh Flood Policy Study* (Final report). UNDP, Dhaka.

SOCIO-ECONOMIC IMPLICATIONS OF CLIMATE CHANGE FOR BANGLADESH

N.J. Ericksen, Q.K. Ahmad and A.R. Chowdhury*

5.1 INTRODUCTION

Land and life are closely entwined in Bangladesh. Over 80 per cent of the country's 115 million people live in the rural sector. The land area of 148,393km^2 is mainly the deltaic plains of the Ganges and Brahmaputra River systems. The prevailing climate is monsoonal, and the dominantly agricultural economy is attuned to its wet and dry seasons. Lands are frequently flooded by heavy rains, over-full river channels, and sea surges associated with cyclones. Disasters are relatively common. Changes in climate in Bangladesh could, therefore, have serious implications for local economies and human welfare. This is the theme that is explored in this chapter. The main relationships between society and climate are identified so that the impacts of a changing climate, especially natural hazards, might have on society can be appraised. Understanding of the main relationships between society and climate will help assess the socio-economic vulnerability or resiliency of the country should it in future experience a period of rapid climate change. This theme is explored through four main questions: How does the current climate affect Bangladesh society and economy? What societal trends may influence the vulnerability of Bangladesh to changes in climate? What are the possible socio-economic impacts of climate change on Bangladesh in future? What alternatives are there for future adaptations to climate change? The chapter concludes with a summary of research needs and a framework for prioritising options. (Sea-level change with particular reference to the coastal zone is not highlighted in this chapter, as it forms the focus of Chapter 7: The Case of the Coast.)

* The authors are respectively: Director, CEARS, University of Waikato, Hamilton, New Zealand; Chairman, Bangladesh Unnayan Parishad (BUP), Dhaka, Bangladesh; and Director (Programme and Administration), BUP.

5.2 CLIMATE AND SOCIETY: AN INTERACTIVE PERSPECTIVE

Climate influences the water, vegetation, soil, and animal resources upon which people depend for food and other products (Chapter 4: Natural Resources). In seeking to exploit resources, a society — like that of Bangladesh — adapts to the natural environment, modifies it according to its needs and wants, and is in turn influenced by these changes: for example, by the embankment and pump drainage of floodplain polders, such as in the Chandpur Irrigation Project; the provision of coastal embankments to keep out tidal flooding with saline water; and the planting of mangroves to encourage *char* (low island) accretion. These are resource creating activities. However, occupying these areas puts populations and economic productivity at risk from hazardous events, like floods, erosion, and storm surges. For example, new *chars* that become quickly settled and brought under cultivation of crops adapted to the specific soil and hydrological environments are exposed to erosion.

The interaction between society and climate is ongoing. In the long term, both climate and society vary and change, as do relationships between them. In the short-term, a significant change in climate may impact on resource uses in an area, beneficially or detrimentally. Likewise, a significant change in society, such as moving from a traditional farming economy to a modern farming and industrial market economy, may result in both beneficial and detrimental effects on society, as well as on the environment. When rapid social change and climate variation coincide, the outcomes can be quite profound.

The interactions between climate and society are schematised in Figure 5.1. If considered in present-day terms, the diagram encapsulates discussion in response to the first main question dealt with in this chapter: *How does the current climate affect Bangladesh society and economy?* The diagram also tries to signify changes through time. This is initiated at the top of the diagram as trends in climate and society. As the solid arrows suggest, these trends directly affect natural and managed environments (such as the land, water, forests, and crops dealt with in Chapter 4: Natural Resources), which in turn affect socio-economic systems.

A review of socio-economic trends enables a response to the second main question: *What societal trends influence the vulnerability of Bangladesh to changes in climate and sea level?* Consideration of the secondary socio-economic effects allows the third question to be addressed: *What are the possible socio-economic impacts of climate change on Bangladesh in future?* These secondary effects feed-back (broken lines) to influence the natural and managed environments by, for example, introducing irrigation against droughts and embankments against floods, or, more directly, to influence social and/or climatic trends, such as enhanced migration flows. Review of these linkages enables a response to the fourth question: *What alternatives are there for future adjustment to climate and sea-level change?* Each of these questions will in turn be addressed.

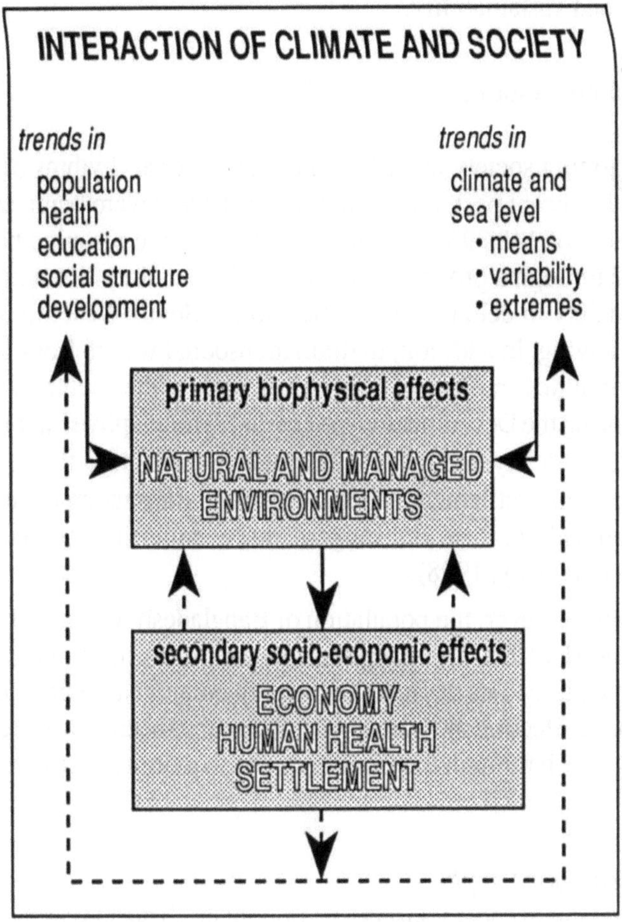

FIGURE 5.1: Interaction of climate and society. The solid arrows show the lines of influence caused by changing trends in society and climate on natural and managed environments, and consequently on people and their activities. Broken arrows show feed-back to environment and society.

5.3 HOW DOES CLIMATE CURRENTLY AFFECT BANGLADESHI SOCIETY AND ECONOMY?

The question of how climate currently affects Bangladesh society and economy is explored in this section through a discussion of two themes: climate as resource; and climate as hazard. The discussion focuses on the traditional rice-growing culture of Bangladesh, and concludes with observations about its resiliency, vulnerability, and sustainability.

5.3.1 Floodplain Resources

Bangladesh farming society has adapted to the seasonal rhythms of climate over millennia. In the recent past, a large number of people were being sustained on a limited land area of 148,393km^2 without destroying the resource base. This was made possible through a process of adaptation by a mainly rice and jute growing agricultural society to deltaic lands subject to the South-West monsoon (June to October) and having, in addition, to drain monsoonal waters from upper riparian countries. Difficulties are also posed in certain parts of the country by dry weather, particularly during the December to April period. The adaptive interaction that had evolved between society and the resource base helped shape the social organisation, kinship system, settlement pattern, economic transaction pattern, and belief systems of Bangladesh. Research suggests that even fertility and fecundity relate to seasonality (Maloney, 1988).

After 1965, however, the population of Bangladesh rose steeply, doubling to about 110 million by 1991. This has put severe pressure on the limited land resource leading in some areas to its degradation. Another major cause of land degradation in the recent past, affecting the Ganges basin areas, has been sharply reduced waterflows to Bangladesh through the Ganges due to withdrawal above and at Farakka in India.

Climate-Driven Food Supply

The traditional human endeavours on the floodplains have always been influenced by the impact of inundation and severe flooding from the major river systems and their numerous tributaries and distributaries. The proportions of land in relation to levels of normal flooding are shown in the pie chart in Figure 5.2.

What has evolved in response to this normal flooding is an integrated system of agriculture and aquaculture for food supply that is dependent upon the climate-driven water regime (Hossain, 1991; Khan, 1989). At the heart of the system is a cropping pattern, especially that of rice, that meshes with the climate and water regime, and a food and resource harvesting strategy that is dependent upon the flooding of large areas during certain months of the year. Cropping patterns are, however, influenced by annual rainfall, frequency and depth of flooding, and drainage patterns. These hydrological factors largely determine which crops can be grown and which rotations can be practised. Farming is not, however, a passive reaction to the forces of nature, and farmers have evolved many methods of water management that seek to control for periods of water deficits (through irrigation works and impoundments) and water excesses (through drainage lines and embankments).

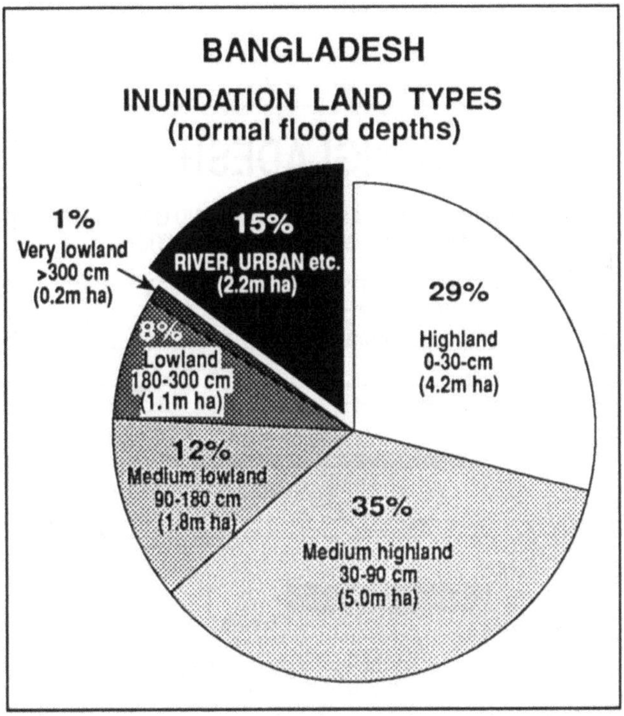

FIGURE 5.2: Land levels in relation to normal yearly floods. About 30 per cent of the total land area is normally inundated each year by river floods and impounded heavy rainfall. The type of land, depth of flooding, and the area affected as a proportion of the total flooded area are shown in the divided chart.

Source: Adapted from Task Force, 1991d, vol. 4, Table 5.1, p. 32.

Three separate growing seasons have emerged over time and four different crops have developed, each adapted to particular seasonal and hydrological conditions, and each accompanied by a distinct farming technology (Khan, 1990, 127-128). These are depicted in Figure 5.3. The time of summer monsoon is the *kharif* growing period wherein rice and jute are grown on seasonally flooded or wet land. About 85 per cent of all agricultural land is in rice and jute in the wet season. Broadcast and transplanted *aman* rice are the main crops. The dry winter is known as the *rabi* growing period wherein dryland crops like wheat and pulses are grown on land that drains quickly enough and has soils with good enough moisture retaining capacity. However, where land is low-lying and remains flooded throughout the year or where soils are impermeable and there is irrigation, *boro* rice is grown in the dry season. In the pre-monsoon and early monsoon or *kharif-I* period *aus* rice crop varieties dominate, along with jute and broadcast (deepwater) *aman*. (Chapter 4).

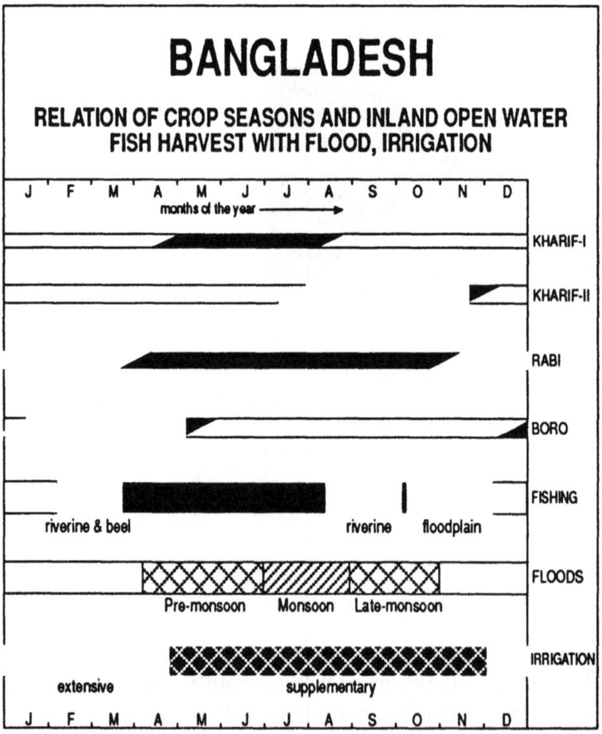

FIGURE 5.3: The relationship of crop seasons and fish harvest to floods and irrigation

Fragmentation and Subdivision

It is important to realise that very small changes in land elevation, and therefore flood levels, as well as in the water regime, such as the commencement of the monsoonal rains, are sufficient to evoke the most subtle of responses from farmers in terms of crop-mix, seed varieties, and appropriate technologies. These intricate responses are made on a seasonal and even sub-seasonal basis each year. They may be made in anticipation of changes or as a reaction to the impact of major events, such as recent floods or cyclones or to market prices.

These changes evolved out of the need to diversify cultivation so as to reduce the risk of drought or flood within the family farm. This was achieved by developing seeds that were adapted to the differing micro-environments within the farm with respect to drought, flood, water elevation, and soil types. This diversification of cultivation not only spread the risks genetically, but also spatially. As a consequence, family land holdings became geographically or spatially fragmented. In addition, because water management was critical in the early stages of wet rice growth, the size of individual plots was kept small. This enabled uniform standing levels of water in terms of terrain contours. But, increasing population and laws of inheritance have been causing subdivision and fragmentation of landholdings, which has, in the main, led to the progressive miniaturisation of landholdings. In the traditional system, these holdings were maintained through the application of local resources so that there was minimal contact with a national market. Capitalist penetration of the rural sector is still at an early stage (Khan, 1989; and Jansen, 1987)

5.3.2 Climatic Hazards

Because the economy and food supply are so closely linked to climate (Chapter 4), significant variations in climatic events have profound effects on society. In these circumstances, climate can be thought of as a hazard, rather than a resource. Floods, droughts and cyclones are examples of climatic hazards. Each of such hazardous events may be judged by characteristics such as magnitude, frequency, velocity, area of impact, speed of onset, and duration. These characteristics influence the nature of human response (Burton *et al.*,1978).

Floods, droughts and cyclones have occurred in Bangladesh over the centuries. Increased exposure due to growing population and development in hazardous areas has made recent disasters seem larger and more frequent. The frequency with which climate-induced disasters occur in Bangladesh is evident from Table 5.1. Only six of the last 32 years were disaster free. Events listed in Table 5.1 suggest

TABLE 5.1: Bangladesh droughts, floods and cyclones, 1960-91

Date	Drought (% area of country)	Flood (area affected 000 sq.km)	% of the Country	Cyclone (storm surge in metres)	Crop Damage (000s Mt)	Deaths Reported
1960				4.5-6 (09 Oct)		3,000
1960				2.5-3 (30 Oct)		5,149
1960	2.7	28.4	19.13			
1961	22.4	28.8	19.40			
1961				6-9 (09 May)		11,466
1962	11.3	37.2	25.06			
1963	8.6	43.1	29.04			
1963				4-5 (28 May)		11,520
1964		31.0	20.89			
1964				- (11 April)		196
1965		28.4	19.13			
1965				3.5 (11 May)		19,279
1965				4.5-6 (14 Dec)		873
1966	18.4	33.4	22.50			
1966				4.5-9 (10 Oct)		850
1967		26.7	17.99			
1968		37.2	25.06			
1969		41.4	27.89			
1969				- (17 Apr)		75
1970	9.1	42.4	28.57			
1970				- (23 Oct)		300
1970				6-9 (12 Nov)		220-400,000
1971	4.8	36.3	24.46			
1972	42.9	20.8	14.01			
1973		29.8	20.08			
1973				1.5-7.5 (09 Dec)		183
1974		52.6	35.44			28,000
1974				2-5 (28 Nov)		a few
1975		16.6	11.18			
1976	5.1	28.3	19.07			
1977		12.5	8.42			
1978	3.7	10.8	7.20			
1979	42.1	-	-			
1980		33.0	22.23			
1981		3.14	2.16	2 (10 Dec)		2
1982	N.A.					
1983	N.A.	11.1	7.48	321		
1984		28.2	19.00		1257	
1985		11.4	7.68	3-4.5 (25 May)	59	11,069
1986		4.6	3.09		140	
1987		57.3	38.61		1327	1,700
1988		89.97	60.62		2110	1,600
1988				1.5-3 (29 Nov)	575	2,000
1989		6.10	4.11			
1990		3.5	2.35			
1991		28.60	19.27	6-7.5 (29 Apr)		138,868
1992		2.0	1.35			

Another 18 damaging cyclones affecting Bangladesh did not result in reported deaths, making a total of 35 cyclones in 32 years. Cyclone information is from Table 1 in Haider *et al.*, 1991.

Source: Flood data from BWDB (1993), drought data from Mahtab, 1989 and BBS Agricultural Statistics (1991)

that droughts occurred on average every 2.3 years and floods and cyclones every 1.8 years. The spatial distribution of these events is quite extensive relative to the size of the country. It is the unique disposition of land, water, and people in Bangladesh in relation to climate that has resulted in this unusual pattern and frequency of disasters (Figures 5.4-5.7).

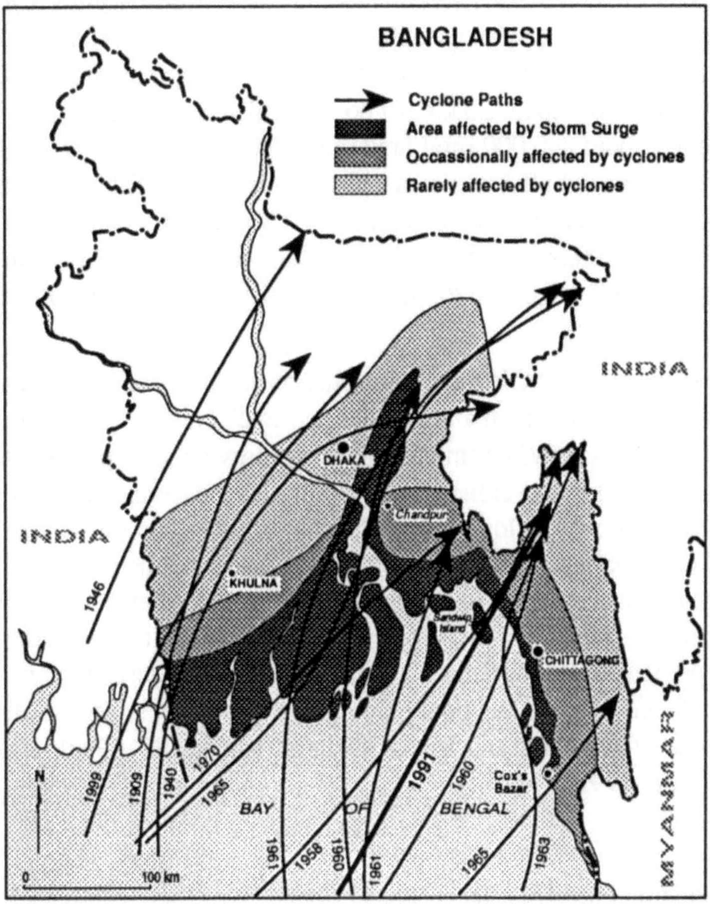

FIGURE 5.4: Areas of Bangladesh exposed to storm surge and affected by cyclones. The impact of the 1991 cyclone covered an area similar to that shown as "occasionally" affected on the map. Other selected cyclone tracks have been added for comparison .

Source: Various sources.

Cyclones

Cyclones bring severe winds, storm surges, and floods that impact on lives, crops and property. While cyclones tend to make landfall along a 710 km strip of coast, they also extend inland, sometimes reaching the north-east corner of the country (Figure 5.4). Cyclones appear suddenly out of the Bay of Bengal, and their paths are relatively unpredictable, as shown by the selected examples in Figure 5.4. In effect, cyclones pose multiple threats from severe wind, storm surge and heavy rainfall that result in both surface and riverine flooding. The flooding also accelerates the erosion of soils, riverbanks and coasts. Consequently, cyclones are very destructive of property and people and very disruptive of economic activities. For example, the severe cyclone of April 1991 had a surge height in places of over 7 metres and winds of up to 235 km per hour. It killed an estimated 138,868 people and destroyed about 840,000 rural houses in 16 districts. Another 910,000 or so houses were damaged affecting some 12 million people (Talukder and Ahmad, 1992). A cyclone in 1970, was even more severe with estimated loss of life ranging from 220,000 to 400,000. Some 280,000 cattle were lost, mostly draft animals, along with about 99,000 boats. Yet this was not the most severe recorded cyclone. That occurred in 1876.

People live in vulnerable places along the coast where cyclones strike, because a reasonable livelihood is obtained under "normal" conditions and pressures on land in inland districts. While embankments can provide some protection from the flooding associated with cyclones, they are much less able to cope with storm surge. Neither are traditional building materials able to withstand the severe winds. A system of embankments for cyclone protection was constructed in the 1960s and 1970s. They have been eroded over the years and/or exceeded in severe events, and are now in need of rehabilitation (World Bank, 1989, 16). After the disaster of 1970, a cyclone preparedness programme began, but has yet to reach all *thanas* in the high risk zone. Social customs and poor communications have, in general, limited the effectiveness of warnings and use of shelters (Islam, 1974; Paul, *et al.*, 1992; Talukder *et al.*, 1992).

Floods

Normal flooding (*barsha*) affects about 25 per cent of Bangladesh each year, but land use and settlement are well adapted to it. Abnormal flooding (*bonna*) can submerge more than 50 per cent of the land area damaging crops and property, disrupting economic activities, and causing injury and loss of life (Adnan, 1991; Ahmad, 1989; Siddique, 1989; Shahjahan, 1989; Hye, Solaiman, and Karim, 1986). Flood-prone land is basically of two kinds: active and stable. Active floodplains lie within and adjoined to the main river channels. These are marginal environments for human occupancy and thus highly vulnerable to floods and riverbank erosion. Stable floodplain land provides good crops in normal years, but *kharif* crops are vulnerable to untimely or unusually high floods (*bonna*). This vulnerability can be reduced by irrigation (dry *rabi* season) and flood control (wet *kharif* season). Early flash floods affect *boro* in the north-east. The severe flood of 1988 affected about 61 per cent of the country (Figure 5.5). More typically, however, it is the relatively high producing districts of Dhaka, Mymensingh, Tangail, Pabna, and Faridpur that are normally flood-prone (Adnan, 1991; Ahmad, 1989).

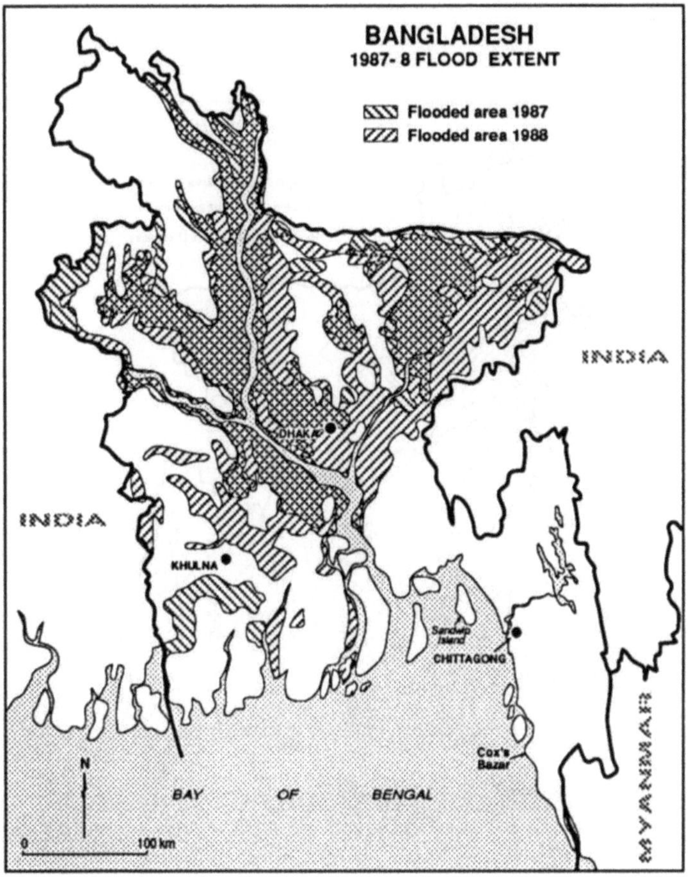

FIGURE 5.5: Areas of Bangladesh affected by the severe floods of 1987 and 1988.

Source: Elahi, 1991a, 3.

Droughts

Droughts are common in Bangladesh. They affect water supplies and plant growth leading to loss of production, food shortages, and for many people, starvation. In comparison with floods and especially cyclones, droughts are slow to manifest themselves and are relatively more pervasive. Typically, uncertainty of rainfall during pre-*kharif* and prevalence of dry days and lack of soil moisture during the dry season reduce potential yields of *B. aus*, *T. aman*, and *rabi* crops. Depending on the intensity of drought, estimated yield reduction of different crops varies from 10 to 70 per cent (Karim, *et al.*, 1990). A severe drought typically affects crop production in about 30 per cent area of the country, reducing crop yields by an average 10 per cent (Figure 5.6).

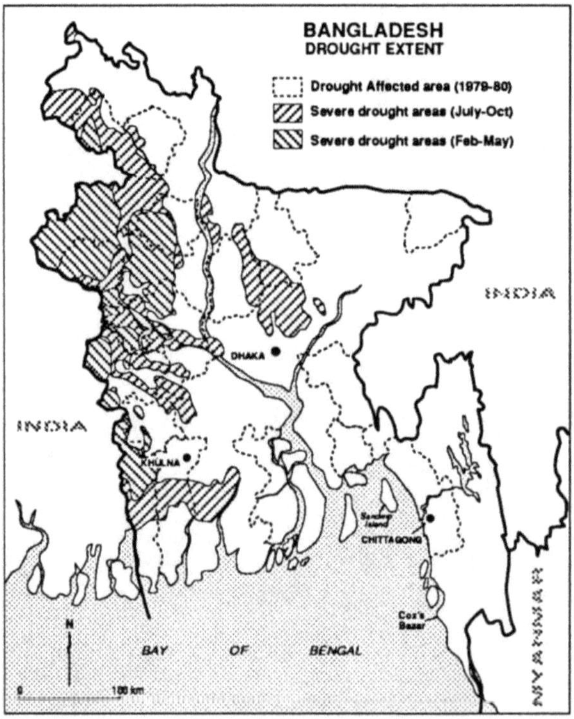

Figure 5.6: Areas of Bangladesh affected by severe droughts: *Kharif* (July-October) reference crop *T.Aman* and *rabi* and Pre-*kharif* (February-May). The extent of the area affected by severe drought 1979-1980 is given for comparison.

Source: Karim, *et.al.*, 1990.

Drought normally affects *kharif* crops (e.g. *aus* and *aman*), but sometimes *rabi* crops (e.g. wheat and mustard), as happened in the very severe drought of 1978-1979 to 1979-1980. This event directly affected about 42 per cent of the cultivated land and some 44 per cent of the population. Persistent drought is, however, relatively rare, but has the potential to cause famine. Drought tends to affect western districts more severely, especially when the monsoon is curtailed (Karim, *et al.*,1990; Mahtab, 1989; Task Force, 1991d, vol. 4, 64-67). Irrigation can help reduce drought effects, but HYV varieties tend to be more drought-prone than indigenous species (Hossain, 1990; Murshid, 1987).

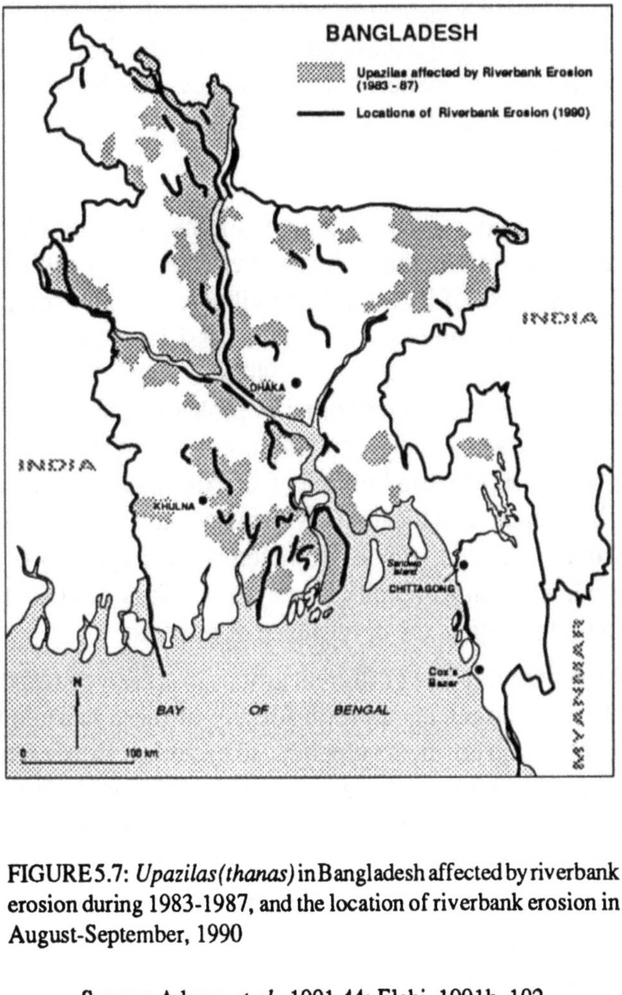

FIGURE 5.7: *Upazilas (thanas)* in Bangladesh affected by riverbank erosion during 1983-1987, and the location of riverbank erosion in August-September, 1990

Source: Adnan, *et al.*, 1991,44; Elahi, 1991b, 102.

Secondary Hazards

Important secondary consequences of climatic hazards include riverbank, *char* (river and deltaic islands), and coastal erosion. These are localised on-going processes, but tend to accelerate and become more severe during times of floods and cyclones (Figure 5.7). Erosional processes along the rivers render landless many of the one million or so people exposed annually to them (Elahi *et al.*, 1991). In badly affected districts like Faridpur, Barisal, and Noakhali, the proportion of landless households due to river bank and *char* erosion is 33, 37, and 42 per cent of total households, respectively, whereas the national average is 28 per cent (Elahi, 1991a; Rahman; 1991; Rogge and Coulter, 1991). Most of the affected households seem to move within 3 km of their original home, and become under-employed labourers. Only about 25 per cent of riverbank displacees move much further afield (Ferdous and Husain, 1988; Salaheen, 1991). Counter-balancing of the loss of land through erosion occurs through the deposition of silt and the creation of new lands for settlement. However, erosion-induced landlessness has a more immediate adverse impact than the positive impact of a deposition-induced settlement.

Capacity to Respond

When droughts and floods exceed the normally expected threshold, significant impacts begin to occur. In part, these impacts are dealt with by coping strategies that have evolved at the level of individual and village. At some point, however, the level of impact will trigger the need for external assistance, as was the case during the drought of 1979-1980 or the floods of 1987 and 1988. The multiple hazards of cyclones seem to present a different order of problem compared to droughts and floods. The magnitude and velocity of storm surge and wind cause such widespread destruction in a very short period of time, that external assistance is essential. In contrast, riverbank erosion is locationally focussed, and is primarily responded to at the level of individual and village. However, while crops may be re-sown after drought or flood, and homes rebuilt after cyclones, loss of land from erosion is irreversible and relocation is the only choice of response.

The ability of individuals or different groups in society to respond to extreme climatic events is by no means uniform for any of these hazardous events. For example, the people who are most vulnerable to cyclones occupy sea-side villages, low lying *char* lands, and unprotected islands. In addition, housing is destroyed selectively according to building materials and therefore wealth. Typically, almost all *kuchcha* houses of the poor are destroyed, along with a large portion of wooden and corrugated sheet-iron houses. Least affected are the reinforced concrete homes of the wealthy (Talukder and Ahmad, 1992).

External assistance to help cope with large-scale disasters involves the resources of the state, as well as relief funds and supplies from overseas countries and voluntary organisations. The scale of aid depends on the magnitude of the disaster. Over time, the Government of Bangladesh has been developing strategies for dealing with large-scale disasters. As economic development proceeds, it is assumed that the capacity of government to respond through disaster preparedness and emergency services and shelters will improve, and that the death toll from comparable events will fall. On the other hand, international comparative research clearly shows that as economic development proceeds, wealth at risk will increase and potential damages to property will rise (Burton, *et al.*, 1978).

5.3.3 Characterising Society

Various approaches may be used for characterising society. Of use in assessing response to climatic variations and change are:

- differentiating groups, livelihoods, subregions or activities by their potential vulnerability or resilience to climate change and variability;
- examining social factors, mechanisms, and trends that lead to greater or lesser resilience or vulnerability and;
- identifying extreme or unusual social events that are the equivalent of extreme natural events or climate anomalies (Kates,1985,14-21).

Resiliency

The process of long-term adaptation and short-term adjustment to the land-water regime suggests that the traditional rice-growing society of Bangladesh is quite *resilient* to adverse change. Coping strategies have been developed to deal with the environmental variations (like floods and droughts) that have always occurred in the local climatic and water regimes. For example, farmers in the Gopalganj district of the Faridpur region responded to the very serious floods of 1987 and 1988 by adjusting the acreage and timing of crops so that they were less exposed to the flood risk. This was done by devoting much more land to dry season *boro* rice and pulses than in previous years. The acreage of *boro* rice, especially HYV, and pulses almost doubled by 1989. This adjustment in response to having experienced severe floods would ensure a harvest that would be at least as good (even better) than a mixed *aus* and *aman* harvest in a good year. The farmers facilitated this change by installing increased numbers of irrigation pumps to ensure appropriate water conditions in the dry winter for the *boro* rice. This was achieved without outside intervention and in an area little affected by modernisation and the market economy (Sadeque, 1991).

The response in Gopalganj was repeated in many other flood-affected districts (Irrigation Support Project for Asia and the Near East [ISPAN],1992).

Obviously, the regional impacts of "abnormal" rains and winds on crop production are profound. This in turn directly or indirectly affects the livelihoods of millions of people. At the national level, however, evidence suggests that the overall effect of hazardous events on total crop production is less severe. This is because the shortfalls in production in affected regions are compensated by above normal production in unaffected regions (Hossain, 1990, 50; Montgomery, 1985). This outcome does not, however, necessarily result in the ready re-distribution of produce to needy areas due to transport problems and/or market rigidities. Another reason why the overall effect of hazardous events on national food production is less severe than suggested by the impacts at regional levels is that farmers affected by severe events (such as the flooding in Gopalganj) try to make up for a loss in one year by adjustments aimed at securing above normal production in the post-disaster season. However, there is evidence to suggest that recent ongoing disaster relief programmes can stifle this traditional productive behaviour (Hossain, 1990).

This view suggests that socio-economic systems in Bangladesh, particularly in rural areas, are based upon traditions that are resilient to climatic variations, which may, as has been the case in the past, also seek out ways of adapting to future changes in climate.

Vulnerability

Another perspective on the socio-economic scene of Bangladesh would suggest that the country is *vulnerable* to climatic variations and change. For instance, over much of the recent past, the country has been buffered from the worst effects of climate-induced disasters by liberal foreign aid, including: food-aid; food-for-work; post-disaster relief; project aid for irrigation and flood protection, etc. (Task Force, 1991a, vol.1). While other socio-economic groups perhaps benefit the most from aid that the country receives on a regular basis, it is the marginalised disaster-hit poor who are most reliant upon it for survival in the wake of a disaster. They have few resources with which to buffer themselves against adversity, whether induced by climatic variations, such as droughts and floods, or socio-economic change, such as escalating population and rural landlessness (Ahmed, *et al.*,1990; BRAC, 1984; Maloney, 1988; Sen, 1981).

As already noted, some geographic locations are more vulnerable to adverse impacts resulting from climatic variations than others, such as the coast and active floodplains. The rural landless and unemployed poor who are forced to relocate as a consequence of change are limited to siting in other risky rural areas or migrating to urban slums. It is, therefore, largely the poor who live in highly vulnerable rural and urban places. And, it is they who have fewest resources with which to respond to adversity. While emergency post-disaster assistance and structural food aid may help alleviate the worst effects of environmental and economic change, these do not empower the poor to achieve longer term improvement in resource accessibility and living standards. This is not to say that better-off groups in society are not adversely affected by environmental and economic change. But they have more power with which to take advantage of favourable economic changes and the resources to help shield themselves against adverse environmental impacts.

Sustainability

Important socio-economic and environmental perturbations have impacted on Bangladesh in the recent past. In responding to change, society demonstrates both resilience and vulnerability. Economic growth requires building on the resilient elements of human activity in order to reduce vulnerability. The challenge for Bangladesh is to provide for sustainable economic development in the face of a rapidly increasing population, a per capita GNP that is among the lowest in the world, limited natural resources, and frequent natural disasters (Government of the People's Republic of Bangladesh, 1992). While aquacultural production has markedly increased in recent years, this progress has been undermined by a high population growth rate. At the same time, per capita arable land has shrunk and the proportion of rural households that are functionally landless has increased. Poverty afflicts the majority of the population. Pressure from growing demands for fuel-wood and agricultural land has caused forested land to be halved in 20 years (Ahmad *et al.*,1986; Sarker, 1990). Urbanisation has been increasing primarily in response to in-migration from rural areas, and urban infrastructural systems cannot keep pace with demand (Laskar, 1983; Pernia, 1993). As a consequence of a rapidly increasing labour force in the face of sluggish economic growth, effectively one-third or more of the country's available labour time is unemployed (Ahmad, 1993a).

In these difficult circumstances, the challenge for Bangladesh is to pursue a development strategy which accelerates economic growth and equitably distributes the benefits towards alleviating poverty while at the same time sustaining its limited natural resources for future generations (Ahmad, 1992; Ahmad and Mirza, 1992; Task Force, 1991d, vol.4). The direction of climatic variation and change may well play an important role in meeting that challenge.

BOX 5.1: Climate-society interaction

> The effects of climate and sea-level change on society (e.g. incomes, health, and migration) arise from how interactions of climate and society affect the natural and managed environments. Traditionally, in Bangladesh, climate variations have provided opportunities (resources) and imposed costs (hazards), depending on how society adapted to the environmental behaviour and the nature and severity of particular variations. Thus, a bountiful floodplain rice-growing system, finely tuned to seasonal climate variations, is often disrupted by floods, droughts, and cyclones. In the future, the extent to which Bangladesh will be affected (whether adversely or beneficially) will depend on the future technological, demographic, and socio-economic trends and how they influence Bangladesh's ability to adapt in order to strike a new balance between resources and hazards.

5.4 WHAT SOCIETAL TRENDS MAY INFLUENCE THE VULNERABILITY OF BANGLADESH TO CHANGES IN CLIMATE AND SEA LEVEL?

The scenario for climate change outlined in Chapter 2, suggests that, for the BAU case, a 1.5°C to 1.8°C warming of Bangladesh could occur by the year 2050. In Chapter 3 on sea-level rise, it is suggested that, in response to climate warming, sea level could rise by 30cm in that period. What societal trends may influence the nation's resiliency or vulnerability to these changes? Is Bangladesh more or less vulnerable as a consequence of being in a state of socio-economic transition? In this section, selected socio-economic factors are used as indicators for assessing the vulnerability of Bangladesh to changes in climate and sea level. The chosen factors are: development, environmental interventions, industry, infrastructure, population, migration, health and education.

5.4.1 Society in Transition

When traditional farming societies are in transition to a modern state, many aspects of society are affected, both positively and negatively. For example, death rates fall in response to modern medicine much faster than birth rates because of the reluctance of people to adopt birth control due to social and economic reasons. Population quickly expands, but subdivision of family land escalates and uneconomic land holdings increase. Access to credit improves and new technologies result in improved production, but indebtedness strikes many small land owners. Although many of the landless move into new forms of employment, alternatives fail to keep pace with the expanding population. Industrial and service sectors expand, but this is typically at a rate that is insufficient to absorb surplus rural labour. Large-scale under-employment occurs and family incomes for some groups fall. For them, nutrient intakes deteriorate and rural to urban migration escalates. Migration to cities flows at a pace beyond the capacity of urban infrastructural systems to cope adequately. Thus, in the period of transition, a significant portion of the rapidly growing population becomes marginalised and vulnerable to social and environmental stresses.

This summary of socio-economic change by and large applies in Bangladesh, although the explanations for it may vary (Chowdhury, *et al.*,1989; Hossain, 1991; Jansen, 1987; and Khan, 1989). As already noted, human activity in Bangladesh revolves around climatic resources and hazards. With respect to climate change, it is important to know how socio-economic changes in traditional farming systems influence coping mechanisms.

It is frequently asserted in the scientific literature that a traditional farming economy is more resilient to environmental variations than a society in transition to a modern economy. Comparative research on how societies in different stages of economic development cope with extreme natural events has been carried out in many developed and developing countries (Burton, *et al.*, 1978; Hewitt, 1983; White, 1974). A major conclusion is that societies that are in transition from traditional to modern or industrial stages are more vulnerable to natural hazards (floods, droughts, and cyclones) than either traditional or industrial societies. This is because the mechanisms of traditional societies for coping with disasters are disrupted by the development process before being adequately replaced by mechanisms used in developed countries. One might surmise, therefore, that climate and sea-level change would have a more severe impact on a society in transition than one that is in a traditional or a developed state.

To what extent does this thesis apply to Bangladesh? Has the development process significantly separated people from their traditional means of production and as a consequence made them more vulnerable to natural hazards and social change than hitherto? If not, is this process likely to occur in the near future as transition proceeds? Since about 84 per cent of the total population is in the rural sector, this matter is an important one to investigate, for it may help indicate how well Bangladesh may cope should global warming significantly influence the farming sector.

Land and Wealth

Nearly three-fifths (61 per cent) of the total land area of Bangladesh is under cultivation — primarily with rice which occupies four-fifths of the area and provides three-quarters of agricultural produce. While irrigation and flood control projects allowed some expansion of cropped land since the 1950s, the limits of this are likely to be reached in the next 10 to 15 years. The significant increase in crop production that did occur over the last 30 years was due to the introduction of small-scale mechanised irrigation technologies, higher yielding varieties (HYV) of rice, wheat, and potato, and chemical fertilizers in the 1960s and 1970s — the 'green revolution.' This enabled the intensification of land use through multiple annual cropping of rice and increased production of wheat in the dry season. Consequently, foodgrain production has grown at a steady rate since the mid-1960s, and the food-gap has gradually narrowed due to a decreasing rate of population growth (Hossain, 1991; Rashid, 1991; World Bank, 1989).

Some 84 per cent of the population is wholly dependent upon rural landholdings as landlords, owner-operators, tenants, sharecroppers, and as landless labourers. Thus, any factor, such as climate change, that affects the resource base of soil, water, or forest will have important socio-economic consequences.

Bangladesh households have been divided into 10 socio-economic groups (Planning Commission, Ministry of Planning, (GOB) 1991; Task Force, 1991d, vol. 4, 190).

1. Landless agricultural labourers
2. Small farmers (0.5-1.49 acres of land)
3. Medium farmers, mainly tenant (1.5-5.0 acres)
4. Medium farmers, mainly owners (1.5-5.0 acres)
5. Large farmers (5.1-10 acres)
6. Very large farmers (> 10 acres)
7. Rural formal (households, mainly rich, in non-agriculture)
8. Rural informal (households, mainly poor, in non-agriculture)
9. Urban formal (households, mainly rich, in non-agriculture)
10. Urban informal (households, mainly poor, in non-agriculture)

The landless, small farmers, and rural and urban informal groups are poor and disadvantaged. Together they account for around 50 per cent of the total population.

Many of the medium farmers are also poor. Possessing few resources or assets, they are particularly vulnerable to the vagaries of the economic and physical environment (Ahmad, 1993b; 1993c; Chowdhury, *et al.*, 1989; Maloney, 1988; Sen, 1981). The proportion of rural households that are functionally landless rose from about 35 per cent (involving around 18 million people) in 1961 to 68.8 per cent (65 million people) in 1983-1984, and has since increased further. In the same period, average farm size declined from 1.45ha to 0.8ha and, under medium population growth, is expected to drop to 0.6ha by 2011. The net cultivable area per capita will have declined from around 0.12ha in 1975 to 0.045ha by 2011 (World Bank, 1989, 34). With unemployment in rural Bangladesh at 35 per cent or more (in terms of available labour time) and employment opportunities on the land and in other rural sectors limited, much of the rural population is subsisting below the absolute poverty level, a point to be discussed later in the chapter (Figure 5.8).

Without denying the obvious benefits that have accrued to Bangladesh and the rural sector from the 'green revolution' it is often argued that these adverse trends are a consequence of the highly skewed land distribution. However, capitalist transformation of agriculture has been limited. Land is still acquired more often by people for the more traditional reasons of subsistence, prestige and power. This means that while agriculture has experienced many important changes in the last 30 years, it is still very much in a state of transition (Khan, 1989). An important reason for the increasing number of families that are becoming landless and/or that are falling into poorer socio-economic categories is the increasing control of land by the rich. Other major reasons are the pressure on natural resources by a rapidly growing population and the subdivision of land holdings in successive generations. Repeated natural disasters, particularly river bank erosion, also play a part. However, the continuation of these trends into the future combined with a more fulsome penetration of capitalism into agriculture could lead to quite profound socio-economic changes. These changes could in turn significantly increase the proportion of people made vulnerable to climatic variations and extremes.

Economic Growth

Since liberation, the GNP (Gross National Product) of Bangladesh has grown at an average annual rate of around 3.8 per cent. This means that GNP per capita only grew at just over one per cent during the period. In 1990, the GNP per capita of Bangladesh was fourth to last in Asia, ahead of Laos, Bhutan, and Nepal. In addition, population continued to grow at an average 2.17 per cent per year. Because climate is the single most important factor in determining agricultural performance, sharp fluctuations in GNP or GDP (Gross Domestic Product) are not uncommon.

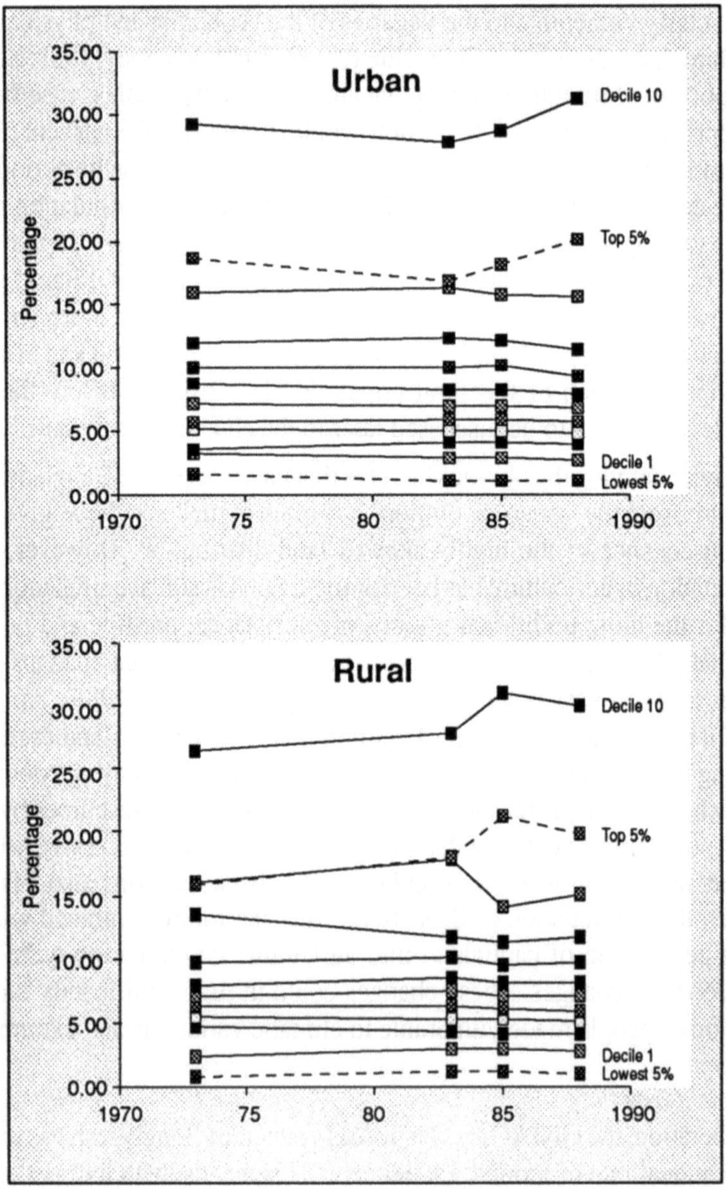

FIGURE 5.8: Trends in the distribution of the percentage of income to households in each decile for the years 1974, 1984, 1986 and 1989

Source: BBS, 1991d.

The changing pattern of the composition of the GDP over the past decade appears to reflect a society in transition. A significant proportion of the GDP has been emanating from sectors other than agriculture, which currently employs 57 per cent of the labour force (about 74 per cent if those, particularly women, engaged in expenditure saving activities in agriculture are also taken into account) and accounts for about 37 per cent of the GDP only. However, manufacturing has made little headway and accounts for only about 9 per cent of the GDP, while energy and construction account for about 8 per cent. It is the services sector that has grown rapidly and it now accounts for about 47 per cent of GDP.

In general, a more diversified economy is likely to be less subject to the vagaries of climate. However, while economic development has extended and diversified employment opportunities, it has not kept pace with population growth, especially in rural areas. In 1991, Bangladesh had a potential labour force of around 36.5 million, which was being joined by around one million new entrants each year. However, about one-third of the total labour time in the country was unemployed, and this translates into about 12 million workers (Ahmad, 1993a). With such a large pool of unemployed and underemployed workers, wages are very low, especially for women. As attitudes towards women in society are changing, increasing numbers of women are entering the workforce. The pressure this puts on overall employment opportunities is increasing (Ahmad and Mirza, 1992, 27-28). In addition, because of technological backwardness and since a large portion of the labour force is malnourished and poorly sheltered, productivity is relatively low in most sectors.

Another factor typical of a country that is undergoing economic development and transition is the very uneven distribution of wealth and income. As of 1988-1989, 20.5 per cent of the national income went to the top 5 per cent of households, and only 1.06 per cent to the bottom 5 per cent. In the same year the top 20 per cent of households acquired 47.2 per cent of the national income, while the bottom 20 per cent received only 6.64 per cent. The disparities have increased since 1973-1974 when the top 20 per cent of households received 44.4 per cent of the national income and the bottom 20 per cent 7.2 per cent (BBS, 1991c, 32). This trend may well continue for some time to come (Figure 5.8).

The geographical distribution of economic growth and employment is also uneven. This is clear from Figure 5.9 which, for 1990, shows the location of the most economically depressed of the 487 *thanas* in Bangladesh (administrative areas known as *upazila* between 1984 and 1991). The bottom 10 per cent of *thanas* were concentrated in regions to the north, north-west, and west of the country. People in these economically depressed *thanas* may be more vulnerable to the adverse effects of climate change than people elsewhere. A changing climate may also alter and/or exacerbate the pattern of uneven growth in Figure 5.9, but it is not yet possible to say how. The pattern in Figure 5.9 is, however, a static picture at a particular point in time. A more useful portrayal would show chronic 'depression' over a reasonable period of time. Research on regional disparities in economic growth and development over time, including the influence of environmental constraints, would help provide a basis for considering a future undergoing climate change.

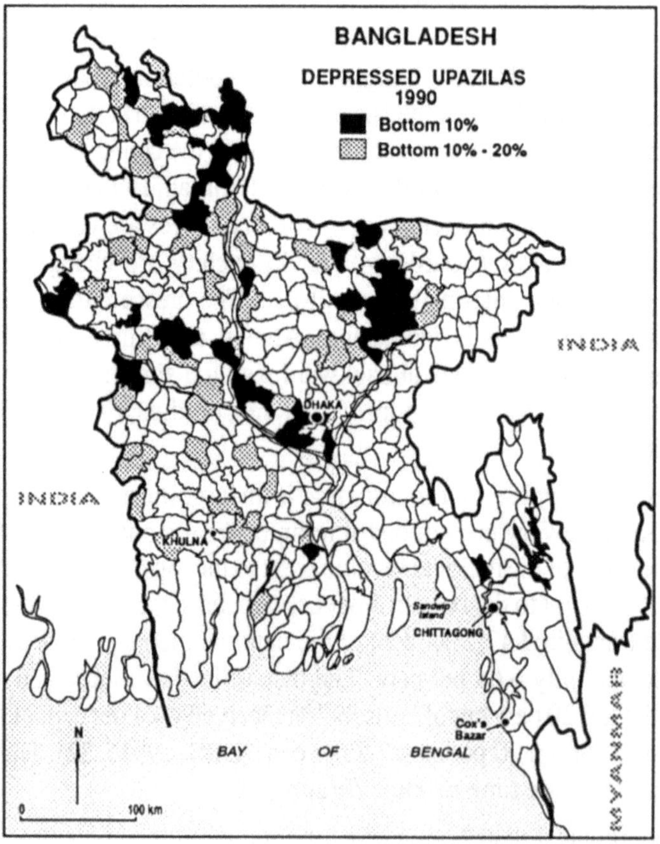

FIGURE 5.9: Economically depressed *upazilas (thanas)* in Bangladesh, 1990. The bottom 20 per cent of 468 *upazilas (thanas)* are shown on the map.

Source: Task Force, 1991a, vol. 1, 37.

Investment as a proportion of GDP declined during the 1980s from 15.9 to 11.7 per cent. This has eroded the economic base for future growth, which together with low domestic savings, uncertain foreign remittances, and dependence on foreign aid, for which prospects are no longer that bright, suggests that the task ahead in moving the economy forward is formidable (Ahmad and Mirza, 1992, 28-29).

Aid and Relief

Foreign aid has always been an important feature in Bangladesh economic management. Up to July1992, the country had actually received US$24.07 billion against a commitment of US$30.06 billion. Thus, as of July 1992 there was a substantial US$5.99 billion in the pipeline. About 49 per cent of the aid disbursed has been in grants. The outstanding debt is currently US$11.6 billion, about 50 per cent of the GDP. The debt servicing ratio was 17.7 per cent in 1991-1992, which may increase further in the coming years, thereby reducing the country's capacity to import unless exports can be increased (Economic Relations Division, GOB, Feb., 1993).

On a per capita basis, Bangladesh's aid receipt is one of the lowest at US$15. But, the outstanding debt burden on a per capita basis of US$99.5 is quite high for a sluggishly growing economy like Bangladesh's. Moreover, aid is the main source of public investment budget. It accounted for 77 per cent in 1991-1992, having been even higher in the 1980s. Also, a substantial portion of government revenue receipts come from duties and taxes on aid-financed imports. In fact, only about 40 per cent of the imports are paid for out of the country's export earnings; the balance is aid-financed to a significant extent, after foreign remittances have been taken into account. Hence, although quite low on a per capita basis, aid has a strong grip on the economy of Bangladesh. Donor induced and supported structural adjustment policies and programmes have further increased the donor influence in the economic management of the country.

On the other hand, aid has failed to help move the economy forward in any significant way so that the country remains trapped in a low growth-high poverty syndrome. For one thing, aid utilisation has been rather poor due to bad planning, managerial deficiency and corruption. Also, projects and programmes were often undertaken at donor insistence, which had little relevance to the country and led to large-scale wastages. For better results in the future, be it from aid or domestic resources, adequate emphasis should be placed on better planning, targeting and management.

Aid has, however, played an important role in relieving losses following natural disasters, and in helping the government to develop ongoing emergency preparedness programmes. It has, however, been suggested that aid has the potential to reduce resilience and the incentive to innovate and adapt, unless the aid is carefully targeted (Hossain, 1990).

Given the persistent low national income and savings, the exposure of Bangladesh to disasters, and the need to develop its resources to meet the demands of a rapidly increasing population, aid will be required by Bangladesh for a long time to come. It seems likely that the high level of international aid to date has cushioned Bangladesh to an extent against economic and disaster-related vulnerability. But when economic transformation in the future leads to a significant increase in the numbers of farmers and labourers moving off the land, the level of vulnerability may well dramatically increase, particularly if coincident with a changing climate that brings more frequent and intense natural events. One response should be a policy of maximum mobilisation of domestic resources supplemented by mobilising foreign aid as required and putting them to uses suited to the changing circumstances, ensuring their efficient utilisation.

BOX 5.2: Vulnerable in transition

Bangladesh is a developing country in *transition* from being a traditional farming society. In the drive for modernisation, evolving technologies and economic and social structures may alter existing systems and make many sectors of, and groups in, society *vulnerable* or more vulnerable to significant variations in climate. Environmental interventions notwithstanding, the country may become increasingly vulnerable should *climate extremes* (manifested in floods, droughts, cyclones, and erosion) worsen in future, even though certain benefits may accrue from climate change, such as improved crop production. In the long term, Bangladesh's vulnerability may depend more on the direction of technological, demographic, economic and social trends than on the rates of climate and sea-level change.

5.4.2 Environmental Interventions

Part of the modernising process should be to increase the scale and pace of environmental interventions. In a land that experiences both water excesses and deficits, this means enhancing traditional practices through the adoption of flood control, drainage, and irrigation technologies. Together, they aim to help to increase crop production, and protect land uses and populations at risk.

Irrigation

Irrigation expanded over the two decades to 1990 to cover 33 per cent of the cultivable area of around 9 million hectares (Task Force, 1991d, vol. 4, 79). Most ground water irrigation and potential irrigation is in the west and north of the country, while river-sourced irrigation is primarily in the south and centre.

Flood Control

There has been a steady expansion of flood control and/or drainage projects since the 1960s. There are over 190 functioning projects protecting 26,620km^2, about one-third of flood vulnerable land. Another 114 projects, discounting the proposed Flood Action Plan, are under construction or consideration aimed at protecting a further one-quarter of vulnerable land. Projects to date have resulted in nearly 10,000km of coastal and riverine embankments and nearly 3,400km of drainage and sluices (Safiullah, 1989, 173-187).

As resource creating projects, they have had benefits in terms of employment on construction, protection of lives and property, providing emergency refuges from rising waters, increasing agricultural production, and therefore income. However, with increased agricultural production comes the need for increased inputs of irrigation (from low to highland elevations) and of fertiliser and pest control. Projects may also adversely affect the open water floodplain fish catch, and poor embankment construction and lack of maintenance is cause for concern (Adnan, *et al.*, 1991; Aguero,*et al.*, 1989; Hunting Technical Services Limited, 1992; Safiullah, 1989; Thompson, 1989; and Zahurul, 1991).

FCD/I Projects

A recent study of six FCD/I (flood control and drainage/irrigation) projects showed that some have provided clear positive socio-economic impacts while others have resulted in negligible impacts or new problems (Hunting Technical Services Limited, 1991b, M-1). One project, Chalan Beel Polder D, showed no evidence of having reduced the environmental variability faced by farmers (Hunting Technical Services Limited, 1991b, M-15).

Across all projects, the distribution of income benefits from increased economic activity varied. Larger cultivating households were better off than smaller households, and they were better off than non-cultivating households. Worse off were poor labourers, fishermen, and boatmen.

Despite benefits from the FCD/I projects, they have been responsible for increased conflicts of interest between various groups both inside and outside of the project areas (Adnan, *et al.*, 1991; Hunting Technical Services Limited, 1991b, M-37; Naqi, 1990)). And, while some have provided protection from flooding to property and infrastructure, other project areas have experienced frequent flooding, some yearly, due to embankment failure, cuts, or greater than design standard levels of flooding. Generally, the incidence of damaging floods does not appear to have been much different between project and non-project (study control) areas. But in the large floods of 1987 and 1988, household losses were greater inside four of the project areas than in the study control areas; while crop damage was less in three of the projects compared with study control areas (Hunting Technical Services Limited, 1991b, M27-35).

Flood Action Plan (FAP)

The disastrous floods of 1987 and 1988 signalled renewed international interest in controlling the floods being caused by the major rivers of Bangladesh (World Bank, 1989). In 1989, a five year FAP was initiated with the support of many international donors and the World Bank as executing agency. Expert opinions varied as to the viability of such a massive project, which preliminary reports suggested could cost up to $US 5 billion. As ISPAN (1992) explains, opinions can be distilled into two opposing approaches for dealing with abnormal flooding:

- Bangladesh cannot be left to suffer disastrous floods indefinitely. All major rivers need to be progressively contained, so as to reduce the risk of abnormal floods and thereby enhance economic activity.

- It is technically and economically infeasible to prevent abnormal flooding, and embankments would create as many problems as they solve. A better approach is to build on the ability of Bangladeshis to cope with, and recover from, flooding.

The first approach has become part of long-term government policy (Task Force, 1991c, vol. 3, 363-410). However, as the second approach has merit, the FAP seeks to resolve the conflicting requirements of both approaches by conducting 26 studies over the 5 year period, 1990 to 1995 (ISPAN, 1992, 2). These component studies of FAP are estimated to cost around $US150 million. They have the major support of the World Bank, Asian Development Bank, United Nations Development Programme (UNDP), the United States, European Union, and Japan. The first six FAP studies are regional assessments. The remainder include: cyclone protection; Dhaka town protection; secondary town protection; flood forecasting and early warning; flood preparedness; agricultural study, operation, and maintenance; flood response; resettlement; environmental impact; fisheries; geographic information systems; compartmentalization; bank protection; floodplain management; floodproofing; river survey; flood modelling; and institutional development.

The spirit of FAP is therefore to examine the advantages and disadvantages of a range of alternatives for dealing with the abnormal flood problem and to combine the best options for various locations across the country. The long-term goals are: productivity (economic development); stability (insulation of incomes against minor disturbances); sustainability (continued growth over time despite flooding); and equity (gains evenly distributed over the population) (James and Pitman, 1992, 24). Many might well argue that the recent history of Bangladesh will sorely test the ideals of FAP (Chowdhury, 1992; Khondker, 1992; Mirza, 1991b; Parker, 1992; Rashid, S., 1991; and Zaharul, 1991).

5.4.3 Industry and Infrastructure

Industrialisation is an essential feature of a modernising society, and an adequate infrastructure provides an important ingredient for its success. Both are emerging features of Bangladeshi society.

Industry

The industrial sector in Bangladesh is relatively small, but growth in output has been at nearly 6 per cent per year during the past decade. Some 3 million people were employed in this sector in 1986 out of a total labour force of 31 million (BBS, 1991a).

Industries are largely based on agricultural commodities, such as jute, cotton, sugarcane, tea, and hides. However, Bangladesh also has some heavy industries, such as steel, pharmaceuticals, chemicals, machine tools, and diesel plants (Figure 5.10).

According to the Bangladesh Bureau of Statistics (1994, 241), the industrial sector accounted for 10.5 per cent of the GDP (constant 1984-1985 prices) in 1992-1993, of which 62 per cent was due to large-scale industries and 38 per cent to small-scale industries. However, there is a large number of rural (generally cottage type) industries dispersed throughout the country, most of which are not included in the statistics of the Bangladesh Bureau of Statistics, but significantly contribute to national income and employment, and have the potential to contribute much more to both.

In the 1980s, Government used various fiscal and monetary incentives to encourage industries to locate in less developed areas of the country (Figure 5.9), but with only limited results. Most industrial units tended to locate in the neighbourhood of the three metropolitan areas in order to reap the benefits of developed infrastructure, industrial agglomeration, and ready markets, while at the same time making the most of the economic facilities extended to underdeveloped regions (Mondal, 1989, 39-41).

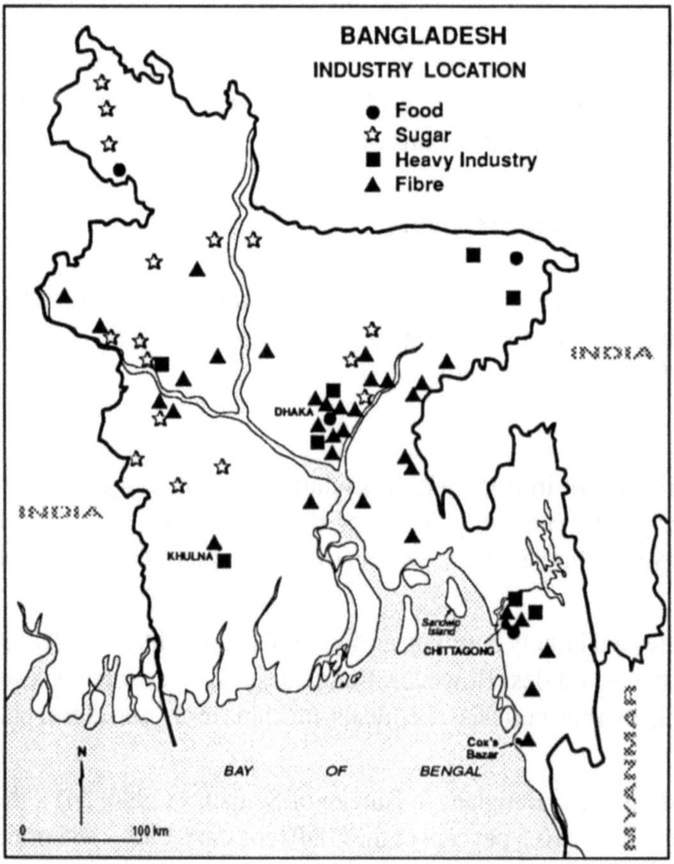

FIGURE 5.10: The general location of large-scale, fibre, and food industries in Bangladesh

Source: From maps in Rashid, H., 1991, 398 and 403.

Much of the industry is located on floodplains, although in the metropolitan areas it is often on elevated land or land that is protected by embankments. Nevertheless, whether concentrated or dispersed, much of the nation's industry is susceptible to severe flooding and/or cyclones. Exposure to the latter is greater in the coastal zone wherein lie the main industrial port cities of Chittagong and Khulna (Figure 5.10). As described in detail in Chapter 7: The Case of the Coast, these areas may also be at risk from sea level changes.

Infrastructure

Transport, communication, and energy reticulation are essential infrastructural elements for economic growth and development (Figure 5.11). Together with water

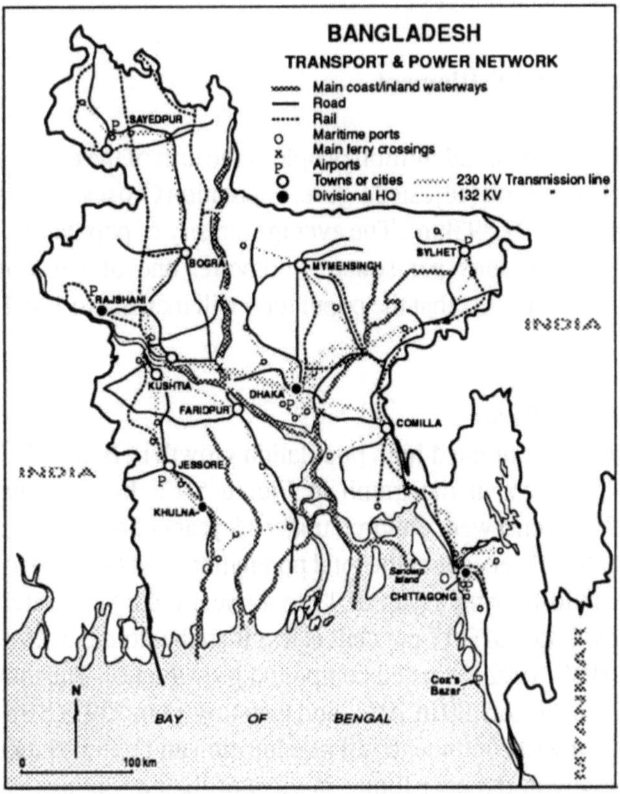

FIGURE 5.11: The general location of transport, communication, and energy infrastructure in Bangladesh

Source: Adapted from maps in Rashid, H., 1991, 377 and 392.

supply, they form the life-lines of a nation.

The deltaic system of tributaries and distributaries, together with flooding of around one-quarter of the land area during the monsoon, makes developing and maintaining infrastructural linkages across the country difficult. Periodic cyclones and severe floods result in some areas being isolated for days. While the waterways account for more than 50 per cent of inter-district movement, even navigation is hampered during the monsoon by river bank erosion and during the dry season by siltation. Steamers and launches, which rapidly expanded in the 1950s, declined significantly in the 1980s as road transport improved, in addition to the above reasons.

When important industrial nodes and their connecting lines of communication and supply are impacted by disasters, the disruptions that result are felt well beyond the immediately affected areas.

5.4.4 Population and Settlement

From a total population of 29 million people at the turn of the century, Bangladesh reached about 110 million (adjusted figure, Population Census, 1991) in 1991 living in a net land area of 107,893km^2. The average density of population, on this basis, is therefore 1,019 persons per square kilometre, one of the world's highest. Available estimates suggest that the population will increase to 136-140 million by 2000 AD.

Population Growth Factors
The main factor for continued high population growth in Bangladesh can be seen in the subdivided bars in the graph in Figure 5.12. In 1991, nearly half the population (47 per cent) were children under 15 years of age, poised to marry and have children, and 40 per cent of the total population (and therefore of females) are already in the child-bearing group of 15 to 44 year olds. Fortunately, the current population growth rate of 2.17 per cent is less than that found in 1981 when it was 2.35 per cent (BBS, 1992, adjusted compound growth rate). The total fertility rate fell from 6.78 in 1961 to 5.00 in 1981, and then to 4.33 in 1991 (BBS, 1991a). One can assume that it will continue to fall as education and living standards improve, but perhaps slowly in the near future in view of the present age-structure of the population and the continued socio-cultural resistance.

The future certainty is that population will continue to grow quickly in the foreseeable future. Low and high projections suggest that by 2025 AD the population will be between 213 million and 291 million, with a medium projection of 235 million (United Nations, 1989, 290-91). An extrapolation of the medium projection gives a total population of 305 million in 2050 AD — the end point for the climate change scenarios (Figure 5.13). These trends in population growth would imply aggravated vulnerability as increasing numbers of people place additional strains on limited resources, forcing higher unemployment and greater reliance on marginal economic pursuits that are susceptible to variations in climate and sea level.

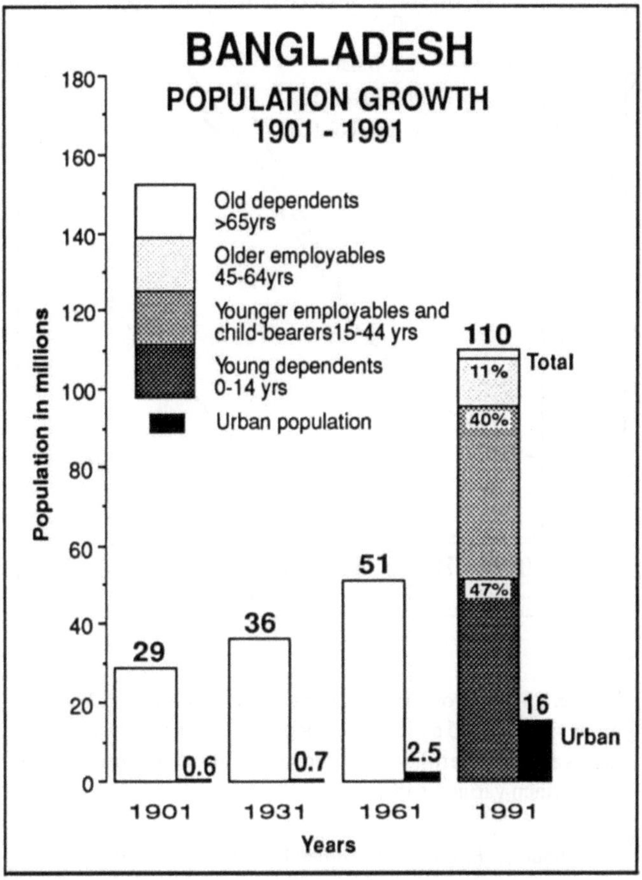

FIGURE 5.12: Historical growth in total and urban population in 30 year periods, 1901 to 1991

Source: Adapted from Faaland and Parkinson, 1976, 101; Task Force, 1991a, vol. 1, 120.

Rural Settlement

The population of Bangladesh is overwhelmingly rural, forming about 84 per cent of the whole (Figure 5.13). Rural land is densely settled, especially in the more fertile areas where alluvial soils support such crops as rice, jute, fruit and vegetables.

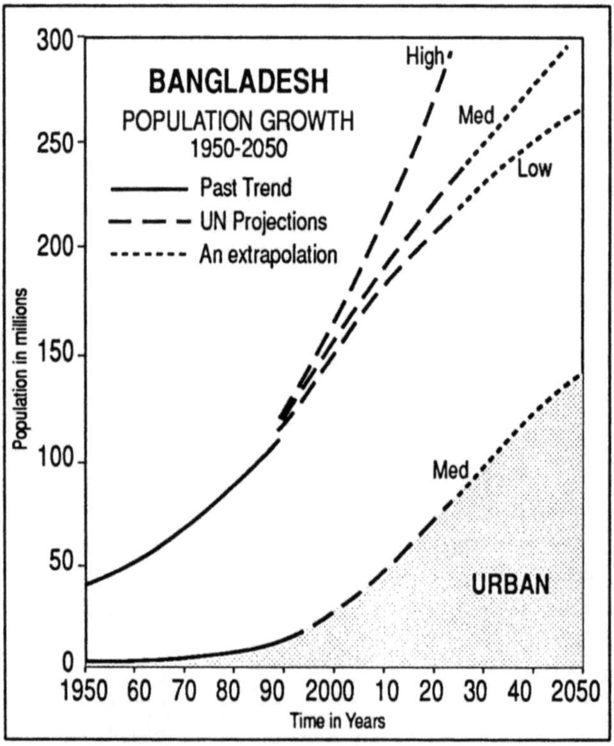

FIGURE 5.13: Growth in total population form 1950 to 1990; (UN projections to 2025 and extrapolations to 2050 for low, medium and high variants of fertility). The proportion of urban population for the medium variant is included.

Source: From data in United Nations, 1989, 290-291.

The map of 1991 population densities in Figure 5.14 shows that the densest population areas centre on Dhaka district, where it averages 3,000 people per km², and the nearby districts of Narayanganj and Narsingdi, where it is over 1500/km².

(These three districts also have large urban populations.) In a band extending south-east from Dhaka to Chittagong, the population averages over 1,000/km². The concentration in these areas probably reflects more stable agricultural production and less proneness to flood and drought than in many other areas of the country. However, in only seven of the 64 districts making up Bangladesh does the population density fall below 500/km².

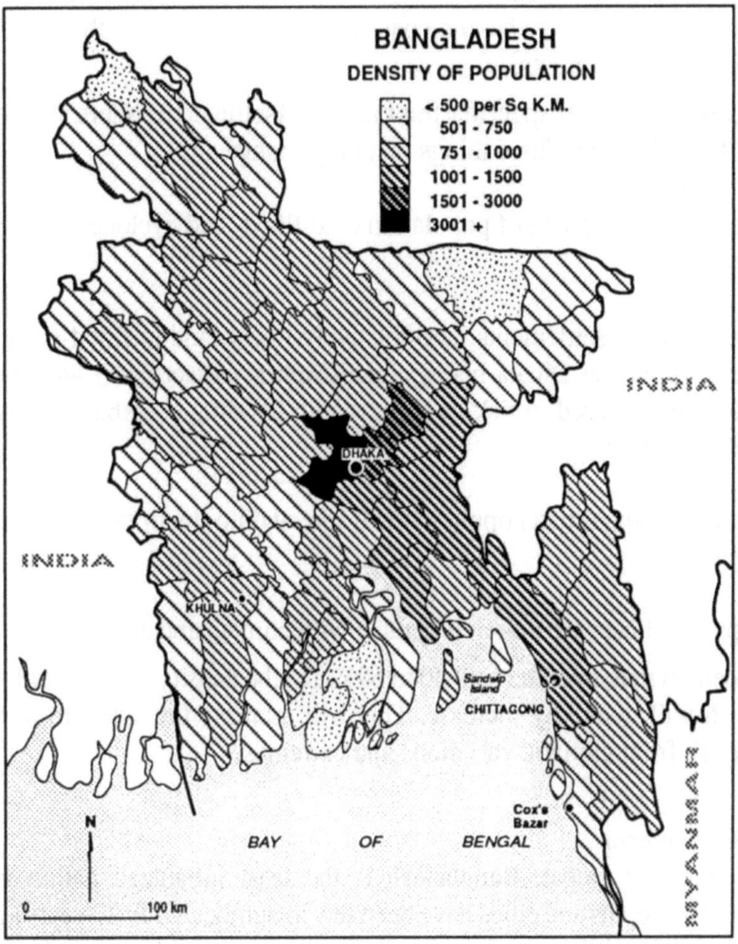

FIGURE 5.14: The distribution of population density in Bangladesh districts in 1991. The most dense districts are from Dhaka south-east to Chittagong.

Source: BBS, 1992.

Because of flooding in the rainy season, settlements in low basins, floodplains, and the delta are sited on natural or artificially raised land (ridges or mounds). Thus, linear settlements are the norm. About half the rural settlement in Bangladesh is of this type. The remainder — in areas of Medium Highland and Highland land types — the settlement pattern is either semi-nucleated or scattered. In low-lying basins, homestead mounds may be 3 to 5 metres high (Sultana, 1993). Also worth noting is the rather surprising prevalence of dispersed settlement in estuarine *char* areas, with homesteads built on plinths raised only above normal seasonal flood levels, not above experienced storm surge levels. The latter does not seem to reflect a misplaced sense of security behind coastal embankments, since the practice pre-dates modern embankments.

There seems little prospect that this basic rural settlement pattern will alter over the next 30 or 40 years. Three things that may change are:

a) the continued spread of population onto flood- and cyclone-prone *char* land;

b) the spread of settlements onto relatively lower land in flood-protected areas (as in the Dhaka-Narayanganj-Demra project area) where they would be exposed to risk of catastrophic flooding if embankments are breached and;

c) the expanding urban population will spread onto floodplain agricultural land.

All groups could become more vulnerable with climate and sea-level change.

It is certain that, over the next 40 to 60 years, the density of rural settlement in Bangladesh will markedly increase. This will increase the absolute number of people at risk from climatic variations and extremes.

Urban Development

After Bhutan and Nepal, Bangladesh is the least urbanized nation in Asia. Historically, its towns and cities have been few in number. Only 30 years ago, with a population of 51 million, there were only 78 towns containing a total of 2.5 million people (5 per cent of total). The annual average change in the level of urbanization between 1950 and 1970 was 3 per cent, and between 1970 and 1990, 3.9 per cent (Laskar, 1983; Pernia, 1993). The urban population has escalated during this period as rural poverty combined with economic development in and around urban areas encouraged people to move into towns and cities in search of work. By 1991, 95 municipalities, plus places with more than 5,000 people, contained 17.6 million people — about 16 per cent of the 110 million total (BBS, 1992). The four metropolitan areas held 9.53 million people (about 9 per cent of total national population), an increase of 59 per cent in 10 years (Figure 5.15).

The main towns and cities developed in response to trade and commerce, but many newer towns serve as local administrative centres. Growth in these between 1974 and 1991 is shown clearly in Figure 5.15. Dhaka municipal area is the main centre with around 3.64 million people, about 37 per cent of the total metropolitan population (BBS,1992). A ten-fold increase of people in Dhaka in 30 years has dramatically stressed services, such as sewerage and water.

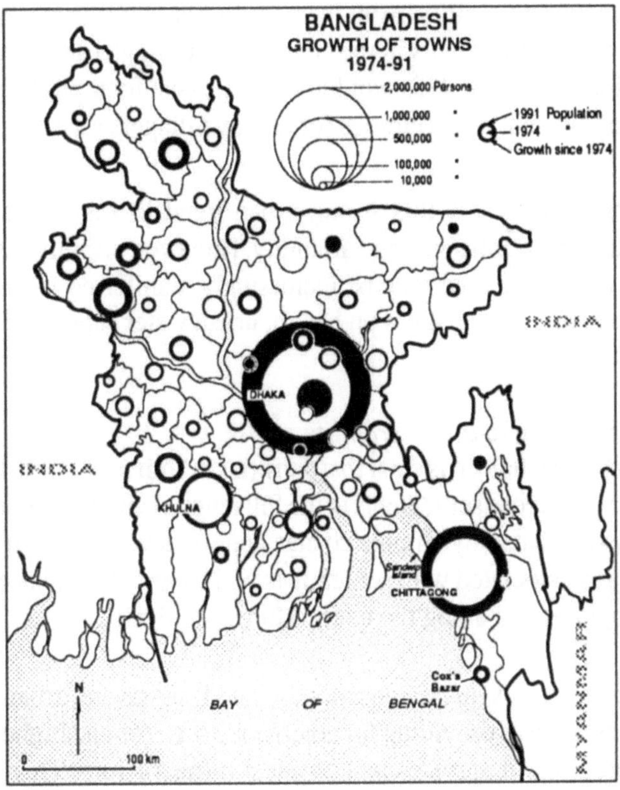

FIGURE 5.15: The cities and towns of Bangladesh and their growth in population, 1974 and 1991

Source: From data in BBS, 1992.

Similar problems are experienced elsewhere, such as in the industrial and commercial cities of Chittagong (1.57 million) and Khulna (601,000). A significant number of people in the main cities are unskilled labourers from the countryside living in poverty in crowded squatter settlements.

The urban population is projected to reach 84 million by 2025, according to the medium UN projection (United Nations, 1989), just over one-third of the total population. This implies a continuation of the high urban growth rate, at about 5 per cent per annum on average, and a total rural population increase of about 60 per cent over the same time period. Unless a strong interventionist policy dictates otherwise, it is likely that Bangladesh will follow the experience of other developing countries, and the few major centres will continue to dominate urban growth. Major expansion can therefore be expected in Dhaka, Chittagong, and Khulna, the last two falling in the high risk area for cyclones.

An alternative possibility to this increased urban concentration is to facilitate growth centres or market towns in the countryside through establishing food and other processing industries. This would not only assist food production, but also help create employment opportunities for nearby villagers (Asaduzzaman, 1989, 194).

It remains unclear how this strong urban trend in Bangladesh will affect vulnerability to climatic change and variability. However, one might speculate that the effects of climate change may be increasingly manifested in rather different ways, as through direct heat stress on health, urban water supplies or urban flood problems.

Population, Settlement and Vulnerability

For Bangladesh as a whole, it is clear that densities in both rural and urban areas will increase. This will expose settlements to the full range of climatic extremes. If these extremes are exacerbated by climate change and sea-level rise, then the exposure of infrastructures associated with settlements will be greatly enhanced, especially those on floodplains and along the coast. More specifically:

a) Greater frequency of serious floods could aggravate urban planning problems, such as providing for emergency services and higher levels of flood-proofing of services and physical infrastructure. This would increase capital and operating costs.

b) Higher sea levels would aggravate existing urban drainage problems in the port cities of Chittagong and Khulna. This would increase capital costs and probably health problems.

c) Increasing extreme events and sea-level rise could also accelerate rural to urban migration.

The shift of large numbers of people into the cities is in part due to a society in transition and the breakdown of traditional activities and ways of coping in rural areas as alternatives are sought in the cities. Under these circumstances, the vulnerability of people from social and environmental stresses increases, regardless of actual impacts of climate change and sea-level rise.

BOX 5.3: Population, settlement and climate change

After 1965, the population of Bangladesh rose steeply, doubling to 110 million by 1991. The medium UN projection gives 305 million by 2050AD. In rural areas, the population density will increase by over half. High density rural areas will continue to supply migrants to low density areas and to cities. The exposure of people to climatic extremes will persist and is likely to increase as more intense use is made of high risk areas. The urban population is projected to grow at a faster rate, about 5 per cent per year. This increasing concentration of people in large urban areas could increase the risk of catastrophe from rare climatic events and is likely to create additional risks of climatic variations more akin to those of other urbanised countries, such as heat stress, urban flooding, and urban water shortage. Overall, the trend of high population growth in Bangladesh should increase vulnerability to climate and sea-level change.

5.4.5 Migration and Employment

The permanent movement of people occurs for various reasons. Four main types of movement are apparent in Bangladesh: core to periphery; rural cycling; urban magnet; and international movements.

Core to periphery

Redistribution of people in Bangladesh has been due mainly to population pressure per unit area on the fertile alluvial plains and to labour saving agricultural innovations (Sultana, 1993, 43). Generally, the more densely populated mid-eastern districts are the main providers of migrants. This is due mainly to population pressure, and its corollary of land subdivision, as in Comilla, Mymensingh, and mainland Noakhali. However, less densely populated districts, such as Barisal and Faridpur, that experience severe flooding and riverbank erosion, also provide migrants to other districts.

The receiving or in-migration districts are mainly those with densities below the national average to the north and south west (or periphery), especially Rajshahi, Dinajpur and Rangpur and Kushtia, Jessore, and Khulna (Figure 5.16). The rationale for in-settlement in the north and west is the ability to exchange one unit of land in the over-crowded, more developed south-east for about 10 units of land in these less crowded, less developed areas. It is also suggested that people have been attracted to these areas since the 1970s by opportunities provided through new irrigated farming of rice and wheat based on deep tube wells and new HYV varieties, and because the areas were also less prone to flooding and river bank erosion (Sultana, 1993, 43).

Rural Cycling

The main reason for rural-rural migration has been overpopulation in relation to the capacity of the rural economy to absorb new entrants into the labour force. This means that there is large scale rural under-employment, which results in an unemployed labour force of an estimated 35 per cent or more. A recent Bangladesh Institute of Development Studies (BIDS) survey of 62 villages revealed that about 1.5 per cent of rural households *permanently* outmigrated to other rural areas between 1987 and 1989. It also found that nearly 60 per cent of movers circulated *seasonally* within the rural sector. Paradoxically, in the light of the above, a major reason for this type of back-and-forth movement is a growing shortage of labour in villages that have adopted new irrigation technologies for increasing crop production. Respondents in over three-quarters of the technology-adopting villages reported that labour shortages are experienced in the peak season, and that they rely upon an influx of migrant labour from other districts. This has the tendency to raise wages in both recipient and provisioning districts (Rahman, 1991). Another source of seasonal labour absorption is crop production on *chars* in the lower delta, where itinerant labourers move south from Faridpur, Barisal, and Noakhali. There is also northwards and eastwards permanent and seasonal labour movement into Sylhet, especially from Mymensingh, Comilla, and Faridpur districts (Figure 5.16). An increasingly reported reason for rural migration is the growing number of people moving into a state of landlessness, for reasons noted earlier.

Urban Magnet
Urban in-migration has become an especially marked trend in the last 30 years and is growing at around 2 million people each year. Many of the factors contributing to the rural cycling of migrants also apply to the movement of rural people into towns and cities. About 2 per cent of rural households migrate each year, around 25 per cent of them to urban areas.

Major recipients of in-migration have been the largest cities of Dhaka, Chittagong and Khulna (Figure 5.16). Many have come as a consequence of floods and riverbank erosion. In Khulna city, three-quarters of the squatter population is from Barisal (45 per cent) and Faridpur (30 per cent) where there has been recent flooding and riverbank erosion. These districts also supply significant numbers to the squatter settlements in Dhaka — 11 and 31 per cent respectively (Elahi, *et al.*, 1991).

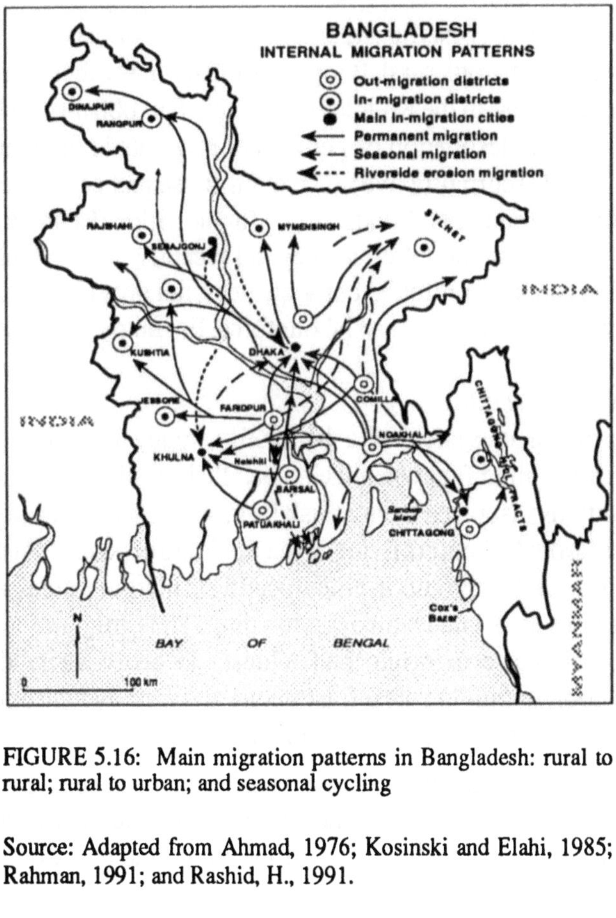

FIGURE 5.16: Main migration patterns in Bangladesh: rural to rural; rural to urban; and seasonal cycling

Source: Adapted from Ahmad, 1976; Kosinski and Elahi, 1985; Rahman, 1991; and Rashid, H., 1991.

During the past decade, an increasing number of migrants seem to have been attracted to larger rural towns (Sultana, 1993). This has likely been the result of improved road communications, investment of repatriated foreign worker funds, and deliberate government policy to develop rural growth centres. It is also likely to be the result of increased agricultural production and wealth that has accompanied the spread of modern technology. Analyses are needed to verify these trends.

Most migrants to the cities are adult males who are poor and in search of a better life. Where they are household heads, the consequences on family members left behind in the rural areas can be profoundly adverse.

In the cities, the rapid influx of people from rural districts has contributed to the pool of under-employed and urban poor and to a multiplicity of infrastructural problems. At the same time, it has relieved pressure on the rural labour market to some extent.

International Movements

A recent BIDS survey suggests that international migration accounts for 17 per cent of total out-migration from rural Bangladesh. The main recipient countries of temporary migration are in the Middle East (where 650,000 moved between 1976 and 1989). The main destination of permanent migrants is the United Kingdom (where some 300,000 currently live). The main consequence of international out-migration is the large amount of funds earned for the nation (Rahman, 1991). When conditions in the host nation falter, as they did recently in the Middle East, rapid return migration can have profound consequences, not only on remittances, but on those attempting to re-enter the local labour force. For many, it is poverty revisited.

Migration and Vulnerability

While people move in search of a better life, many have limited resources. Opportunities are hard to find when they reach their destination. Many are forced to live in conditions that place them at risk from social and environmental threats. For example, seasonal labourers living in the fields are exposed to sudden cyclones and storm surges, while migrants in cities locating in marginal areas are at risk from floods and diseases associated with overcrowding. Thus, migrants lose the support structures of their place of origin and relocate in areas where these, and job prospects, are poor. Improving the infrastructure and employment opportunities for migrants is critical for reducing their vulnerability.

There seems little doubt that significant movements of people will continue in Bangladesh. As population grows and the country modernises, the number of marginal farm units will increase, as will the rural landless. Continuing natural calamities will exacerbate this trend. It is difficult to predict whether migration rates will increase or decrease. But it does seem reasonable to assume that currently observed migration rates will continue well into the next century.

5.4.6 Health and Education

In a country where many people are poor, malnourished, and living in crowded conditions with inadequate waste disposal and water supply, many health problems occur. What are the trends in health care and education? What are the main relationships between climate and human health? How might climate change affect them? In this section of the chapter, these questions are explored through six interrelated topics: health care; educational opportunities; poverty; malnutrition; water-borne diseases; and vector-borne diseases.

Health Care

A recent report of the Asian Development Bank (ADB, 1989) documents that the health status in Bangladesh is low. Some of the main epidemic and endemic diseases have declined, but communicable diseases are still significant. The main causes of death are diarrhoeal diseases, malnutrition, and pneumonia. In the mid-1980s, respiratory and diarrhoeal infections accounted for 42 per cent of morbidity (disease) and 20 per cent of mortality (death). Life expectancy improved from 45 to 50 years between 1965 and 1985 and infant mortality from 153 to 121 per 1,000 births. The provision of doctors in 1981 was one per 9,690 people and of nurses was one per 19,370. By comparison, in India, it was 3,700 and 4,670 respectively (ADB, 1989).

However, the provision of health services has improved over the past decade. Life expectancy (about 55 years) and infant mortality (about 100) have consequently improved further. In 1990, there was one doctor per 5,498 people and one nurse per 11,861 people. Hospital bed provision improved from one per 3,740 in 1981 to one per 3,235 people in 1991. These statistics do, however, mask the fact that people in urban areas have much better access to health care than those in rural areas, but this is so for all social services. The extension of the Thana Health Complexes (which increased from 275 in 1980 to 352 in 1990), was aimed at helping rectify the imbalance. However, a recent survey showed 88 per cent had inadequate facilities, trained personnel, and equipment.

Only about 30 per cent of the population is estimated to be covered by primary health care. On the other hand, an extensive structure for delivery of services in family planning has been established, so that contraceptive prevalence rates have steadily increased (ADB, 1989).

The provision of medical care is made difficult, not only by limited resources, but also by difficulty of access, especially in rural areas. This is because of poor communications, especially during the monsoon.

Educational Opportunities

Access to education is also a problem in Bangladesh. Successive five-year plans have aimed at free and universal education with a view to eliminating illiteracy and improving trade skills. Progress has been slow. Only 24.8 per cent of the population was literate in 1991, virtually no improvement from 24.3 per cent in 1974. For men it was 31 per cent, and women 16 per cent. The uneven access to education is carried through as between urban and rural sectors, exhibiting literacy rates of 35 and 18 per cent in 1981, respectively. Primary schools increased around 20 per cent and teachers16 per cent between 1975-1976 and 1990-1991. By 1991, 81 per cent of boys and 76 per cent of girls were enrolled, although attendance has been much less. Unfortunately, some 60 per cent dropped out before completing primary school. This reflected mainly the fact that poor families could not afford to have children away from the tasks at home, especially girls; and a tradition that saw little room for formal education and the benefits that it can bring. Socio-economic status was also reflected in children who continue in technical training, and the types of trades chosen. (Data collected from Bangladesh Bureau of Educational Information and Statistics (BANBEIS), Dhaka; BBS, 1992; Ahmed, *et al.*, 1990.) Although rather slowly, educational opportunities have continued to improve, but very significant inequalities remain.

Obviously, large numbers of people do not receive formal education and this affects especially the rural poor. Schools are in poor condition and the curriculum content old, rigid, and irrelevant to every-day life. While recognised by educationists, these problems are difficult to rectify under existing socio-economic conditions and value systems (Task Force, 1991a, vol. 1).

For the education of primary school drop-outs and children who never went to school, small but important gains have been made through programmes of voluntary organisations. These aim at the disadvantaged and small farmer families with targeted programmes emphasising self-help, health, agriculture, enterprise, literacy, and population planning. For example, as of September1993, the Bangladesh Rural Advancement Committee (BRAC) ran 14,472 non-formal primary schools providing basic education in several districts — each teaching 30 students, along with several other development programmes. These schools are primarily focussed on the education of girls, as 70 per cent of these students are girls. There are many other organisations running similar and adult education programmes.

The consequences of this for the development of Bangladesh in future seem clear. Improvements in basic education and technical training are needed to effect changes in other aspects of social well-being, such as health, employability, and the alleviation of poverty. These are complex factors, and education will not by itself achieve the change. Some social transformation is required before the large number of poor people can perceive value in education and changes in the traditional ways of doing things (Maloney, 1988).

Poverty

The Bangladesh Bureau of Statistics estimates of the population below the poverty line based on energy-intake (calories and protein) show that the poverty ratio declined from 73 per cent in 1881 to 47 per cent in 1988-1989 (48 per cent for rural population and 44 per cent for urban population). But this sharp decline appears to be inconsistent with poor economic performance during the 1980s. Even so, the ratio in 1988-1989 was still 3 percentage points higher than the 44 per cent estimated for 1963-1964. However, if one were to consider the fulfilment of other basic needs (shelter, clothing, medicine, sanitation, clean drinking water, education, freedom of choice) at a minimum level, in addition to energy-intake, to measure poverty, the poverty ratio is found to be much higher. In fact, a UNDP estimate, on that basis — which may be called a poverty ratio with reference to the human dignity line — puts it at 86.5 per cent for 1990 (UNDP, 1992). This ratio would appear to be too high. Indeed, data and methodological differences produce different results, but food intake alone cannot be accepted as a basis for measuring poverty. Hence, the poverty ratio is certainly higher than 47 per cent, even though it may not be as high as 86.5 per cent. The controversial measurement of poverty aside, the fact is that the number of poverty-stricken people in Bangladesh is massive. The reasons for this include rapid population growth, low economic growth, unequal distribution of productive resources, large-scale unemployment, and domestic policy biases.

It is worth noting that while the overall poverty ratio has, according to estimates of the Bangladesh Bureau of Statistics, declined significantly during the latter half of the 1980s, hardcore poverty increased between 1986 and 1989 by 30 per cent in rural areas and 8 per cent in urban areas (BBS, 1992). It is highly likely the latter has involved mostly rural migrants in search of urban employment.

Malnutrition

As a consequence of poverty, malnutrition and hunger are widespread in Bangladesh, probably affecting over half of households (Ahmed, *et al.*, 1990; Maloney,

1988; Quddus and Ara, 1991; and Task Forces, 1991a, vol. 1). These include people who have very few resources at their command. Thus, in times of adversity, as when food supplies are affected by drought or flood, they may face starvation. In this sense, starvation is the result of entitlement failure (Sen, 1981).

A simplified diagram shows the main causes of malnutrition in Figure 5.17. Poverty lies at the centre of the diagram. There are diseases that stem directly from malnutrition, and others that can take hold because malnutrition weakens resistance to them. More fundamentally, to break the cycle, access to adequate income opportunities is needed. This would improve nutrition, and so improve health. The trends in household incomes shown earlier in Figure 5.8 show that for a large number of people, the prospects in the near future are not good.

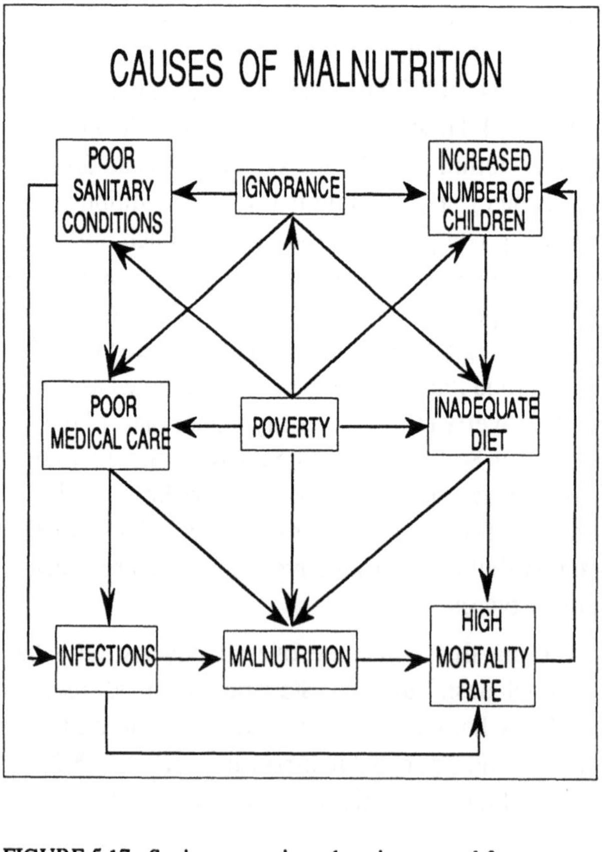

FIGURE 5.17: Socio-economic and environmental factors causing malnutrition in Bangladesh

Source: BRAC, 1979.

Several nutrition studies have been conducted in Bangladesh over the past 30 years, some of them nation-wide. The 1989-1990 child survey concludes that more than one-third of children aged between 6 and 71 months were stunted and about one-quarter severely underweight (BBS,1991b, 78). Rural areas were found to be worse affected (by about 10 per cent) than urban areas, due to lower incomes and poorer environmental conditions. Incomes and education, especially of mothers, had a significant influence on nutrition status. Thus, as economic growth and development improve in future, along with educational opportunities for girls, household incomes and expenditures, and food intake and nutrition status, should also improve.

A study by Quddus and Ara (1991) focused on the relationship between nutritional diseases and their seasonal variations to the socio-economic conditions of households in three geographically distinct villages. Whether in peak season (early winter) or lean (monsoon), 70 per cent of children five years and under suffered malnutrition. In both seasons, 12 per cent of the children suffered from third degree or advanced malnutrition (Figure 5.18a and 5.18b). While some were able to recover from second degree malnutrition in the lean season to first degree malnutrition in the peak season, almost none of the latter could recover to normal. Thus, many of the children had nutritional deficiency diseases, such as scurvy, stomatitis, night blindness, anaemia, rickets and beriberi (Figure 5.19a and 5.19b).

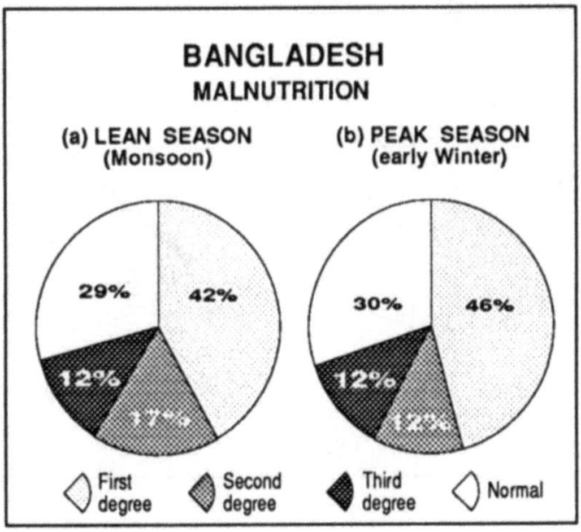

FIGURE 5.18a-b: Seasonal variation in malnutrition among 0 to 5 year old children in three villages

Source: Data from Qudus and Ara, 1991, Table 6.3

These diseases arise from a deficiency in protein, carbohydrates, fat, vitamins, and minerals. Improvement in disease incidence did, however, also occur in the peak season when more food was generally available, but many lacked access to it.

At a time of high population growth with consequent heavy pressure on the very limited land, the limited access to land, and limited employment opportunities outside agriculture to provide incomes with which to purchase food, the prospects for improved food production and intake for a majority of families does not seem bright. Improved education may help lead to a more balanced diet and improved sanitation, but may not in-and-of-itself significantly increase food supply. Aid programmes and/or monetarisation of the rural economy and/or technology transfers may help increase food production, or more importantly, employment, and therefore purchasing capacity.

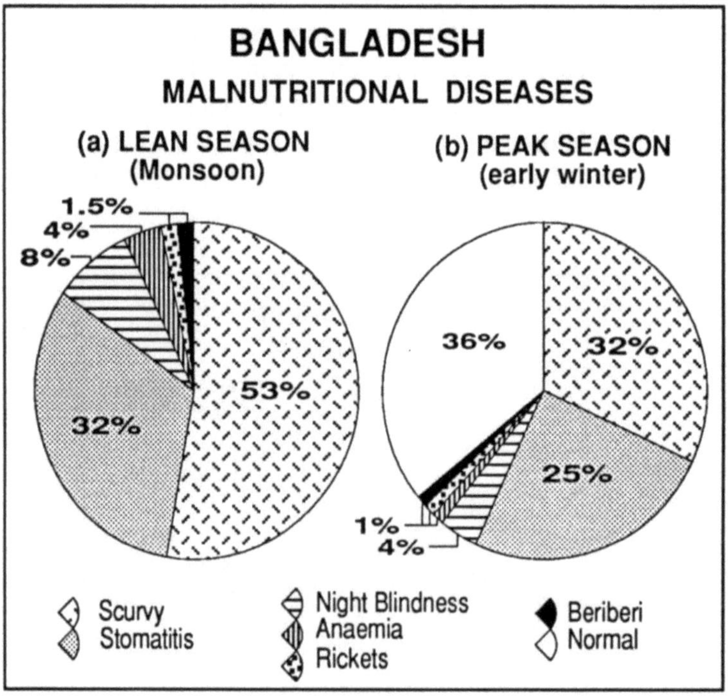

FIGURE 5.19a-b: Seasonal variation in nutritional deficiency diseases among 0 to 5 year old children in three villages

Source: Data from Qudus and Ara, 1991, Table 6.4.

These efforts need to be considered in the light of climate change. Food availability and malnutrition, like most other factors of life, are closely linked to seasonality and climatic variations. They are, therefore, likely to be vulnerable to climate change in future.

It is possible that a warming climate accompanied by more frequent and extreme events might act as a countervailing force to government and private efforts to improve farm development and alleviate poverty and malnutrition. This may suggest that policy should focus on creating secure employment opportunities outside of crop agriculture, that is in non-crop agriculture such as fisheries, livestock, poultry and forestry; and non-agricultural activities. This in turn would require policy focus on education and skills development for employment in professions and trades which could generate wealth and incomes, internally as well as internationally.

BOX 5.4: **Trends in health care and education**

> A population that is healthy and educated is better able to avoid poverty and the adverse effects of climate variations. While recent trends in improved health care and education in Bangladesh are encouraging, poverty and malnutrition remain rife, lowering the resistance of large segments of the population to disease. Even on a seasonal basis, the linkages between climate, nutrition and disease are apparent. Improvements in health care and education, as well as in security of food production, would help buffer Bangladesh against ill-effects of future climate change.

Water-Borne Diseases

In Bangladesh, some 80 per cent of all illness is linked to water-borne diseases. Surface waters are important sources of water intake, but are polluted by indiscriminate defecation practices and unsanitary disposal systems, as are surface soils. Water is the main means for spreading communicable diseases like diarrhoea and typhoid, since most water bodies are interconnected, especially during the monsoon. However, little research has been done on the relationship between surface water and diseases in Bangladesh (Haque and Hoque, 1990, 181). Cholera has been chosen from among the various water-borne diseases to illustrate some of the problems.

Factors Influencing Cholera. Cholera (*Vibrio cholerae*) is an important common cause of potentially fatal dehydrating diarrhoea, especially among adults. It occurs all year, but has a marked seasonal pattern. As the monsoon moves towards peak, the abundance of water dilutes the prevalence of bacteria that cause cholera, but as water volume reduces towards the end of the monsoon densities increase. Also, in the dry season, water tables eventually fall below the bottom of some tube wells and large numbers of people resort to bacteria infected pond water.

While different cholera strains have different peaks of incidence, natural disasters often facilitate cholera epidemics. For example, an explosive epidemic of cholera on Sandwip Island in 1985 was induced by a cyclone and tidal surge in late May near the time when the incidence of cholera is normally at its lowest (June and July) (Siddique, *et al.*,1989).

The precise mechanism of an epidemic is not clear, but on Sandwip the storm and floods caused overcrowding in an already densely settled area (1000/km^2). About three-quarters of the people normally depend on ponds for their drinking water. Flooding increased this dependence by limiting access to the few tube-wells, and by helping to contaminate pond water. This contributed to the outbreak and transmission of the disease, dominated by the *El Tor* strain (Siddique, *et al.*, 1989, 377-382).

Cholera also followed the severe floods of September 1988 (Siddique, *et al.*, 1991b). Sampling over 12 months revealed that, in the south of the country, four-fifths of cholera cases were of the classic strain and only one-fifth *El Tor*, whereas to the north the reverse applied (Figure 5.20). Changing ecological conditions in various regions of the country is given as a possible explanation for this pattern. For example Siddique, *et al.*, (1991a; 1992), claim that the hardier *El Tor* strain is more liable to be found in polluted water-logged areas of the north than the classic biotype, whereas the latter is more viable in the brackish and estuarine systems of the south. Further research is needed to clarify these sorts of relationships. Normally, the country is classified as cholera-free.

Climatic influences. Clearly, there are many social and environmental factors that influence the incidence and spread of diarrhoeal diseases, like cholera. Given these conditions, climate plays an important role. Both seasonality and climatic extremes are significant. Were social conditions to remain unchanged, would climate change significantly increase water borne diseases in Bangladesh in future?

Globally, diarrhoeal diseases occur in a wide range of temperatures under tropical to temperate climates. It has been gathered from International Centre for the Diarrhoeal Disease Research, Bangladesh (ICDDR,B) that an increase in mean

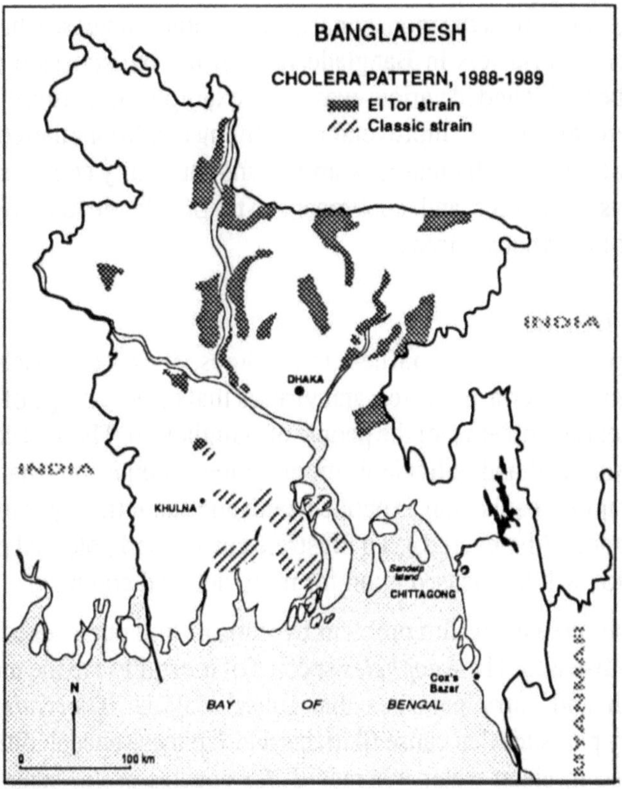

FIGURE 5.20: Clusters of V cholera isolations in the 12 months following the September 1988 floods appear spatially distinct. Mainly El Tor strains were in the north and mainly Classic strains in the south.

Source: Siddique, *et al.*, 1991a, 1126.

annual temperature in Bangladesh of up to 3°C may not in-and-of-itself be important in fostering epidemics, although an increase in extreme climatic events could be.

The main condition for the prevalence of diarrhoea is sanitation, which is tied to poverty. Improvements in economic growth and development ought to help alleviate poverty and improve living conditions. This could, however, be countered by other factors related to climate change. A change in the seasonal patterns of rainfall and temperature could adversely affect farm production and incomes thereby intensifying poverty and the susceptibility of farm families to disease. If climate change intensifies the monsoonal system and/or increases the incidence of flood and cyclone disasters in Bangladesh, then the transmission of diarrhoeal diseases may be facilitated in various ways. For example, increased disasters could intensify poverty and inhibit improvements in living conditions thereby facilitating transmission of disease. Increased warmth and humidity could cause disease-bearing insects, like flies and cockroaches, to proliferate thereby facilitating transmission of diarrhoeal diseases.

Vector-Borne Diseases

Vectors and pests, such as sandflies, mosquitoes, flies, cockroaches, lice, and rodents, transmit bacteria, parasites, and viruses that cause a range of diseases that pose serious health problems for the people of Bangladesh. These vectors and pests proliferate in polluted and unhygienic environments, conditions that may well be enhanced by rapid urbanisation as public health and sanitation systems fail to keep pace with demand (Ahmed, n.d.). To illustrate current and potential problems, the mosquito vector will be focused upon for more detailed comment.

Mosquitoes create a health problem by transmitting parasites to humans that cause various diseases. The *Anopheles* species of mosquito are the most prevalent in Bangladesh and carry parasites that cause malaria. The *Culex* species of mosquito carry parasites that cause filariasis and Japanese encephalitis. *Anopheles* mosquito breed in clean water whereas *Culex* species prefer stagnant, polluted water. Information on one of these types of mosquito is summarised in order to illustrate whether there are important relationships that might be susceptible to climate change.

Malaria. In the late 1950s, malaria affected over 1.5 million people and there were nearly 50,000 deaths per year. It was transmitted throughout the cultivated lowlands mainly by the *Anopheles philippensis* mosquito (Elias, *et al.*, 1987). Eradication using chemicals, like DDT, started in 1961, but stopped in 1970 by which time reported cases of malaria were relatively few. Within 5 years of stopping the eradication programme, malaria increased 10 fold, a trend shown in the graph in Figure 5.21 (Rosenberg and Maheswary, 1982a; 1982b).

A major reason for the resurgence of malaria in the lowlands was its transmission from the forested hills of the north eastern areas of the country by migrants to the lowlands during the political turmoil of the early 1970s. Malaria in the forested hills of Bangladesh has never been controlled. It is transmitted from person to person by *Anopheles dirus*, a highly efficient vector. Its infectivity rate is clearly tied to amount of rainfall (Rosenberg, 1982, 192). Because spraying of breeding areas under the forest is very difficult, the area remains a reservoir of intense, unchecked malaria transmission from which the thickly populated alluvial plains are being reinfected. This occurs by the return migration of plains people from hill areas carrying parasites acquired in the hills, and from re-infection of *A. philippensis* with parasites derived from malaria carriers who were originally infected through *A. dirus* bites while living in, or visiting, hill areas. Tribal hill people have resistance to the kind of plasmodium prevalent in the hill tracts, whereas people from the plains are highly susceptible to this strain (Rosenberg, 1982; Rosenberg and Maheswary, 1982a; 1982b).

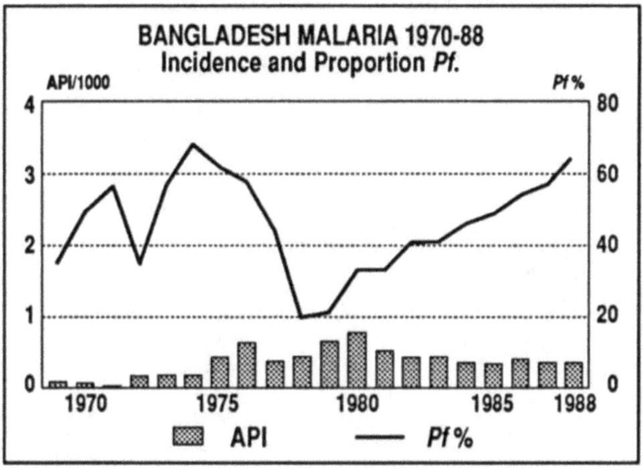

FIGURE 5.21: The incidence and proportion Pf of malaria in Bangladesh between 1979 and 1988. API is annual parasite incidence (the malaria positives per 1000 population under surveillance); Proportion Pf is the Pf (*Plasmodium falciparum*) infections per 100 malaria positive slides.

Source: International Assessment of the Malaria Programme: Bangladesh, 1989.

Invasion on the plains is selective, mainly in response to environmental factors associated with human landuse and settlement, including irrigation works, flood embankments, and overcrowded living conditions. Resurgence seems more pronounced in areas where the introduction of wide scale anti-malaria tactics preceded development of a primary health care system. Uncontrolled use of anti-malarial drugs increased parasitic resistance to them. In addition, mosquitoes developed resistance to control sprays (Elias, *et al.*, 1985).

The current distribution of the main malaria vectors appears in the map in Figure 5.22, together with the incidence of malaria. Incidence is most in the north

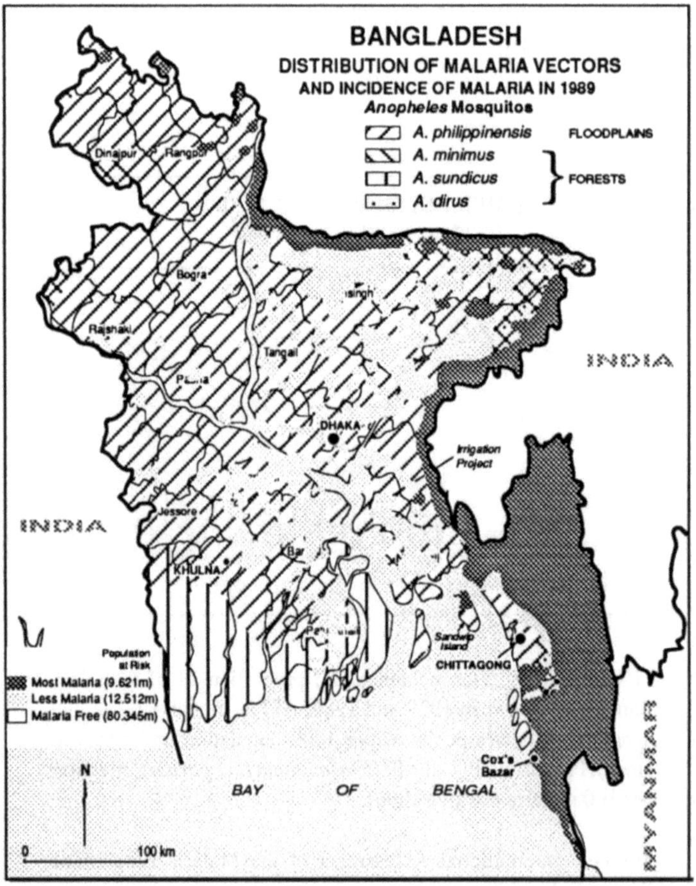

FIGURE 5.22: The distribution of Malaria vectors and the incidence of malaria in Bangladesh in the late 1980s. The incidence of filariasis is not shown, but is prevalent in the north-west and around the main cities.

Source: International Assessment of the Malaria Programme, Bangladesh, 1989.

and eastern forested hills and southeastern coast, and decreases towards the drier west — where filariasis is prevalent (Ahmed, 1990). Recently, both *An aconitus* and *An annularis* have been implicated in transmitting malaria on the floodplains of Bangladesh (ISPAN, 1992, 25; Ahmed, 1990).

In general, embankments for flood control, railways, and roads in the deltaic area have reduced the spread of silty water thereby facilitating vector breeding (*Anopheles*) and increasing the prevalence of malaria. However, poorly drained embanked areas can enhance stagnant and polluted water thereby facilitating vector breeding (*Culex*) and increasing the prevalence of filariasis (ISPAN, 1992).

The Chandpur Irrigation Project provides an example of mosquito invasion and resurgence on the floodplain. The project created approximately 2,830ha of water in the form of borrow pits and irrigation canals which carry water in the dry season, as well as during the monsoon. By 1990, about 60 per cent was covered in water-hyacinth. This, and the very favourable climate, created a permanent breeding ground for *Anopheles* mosquitoes (Mirza, 1991a, 160-162). The 104km of embankment enclosing 55,000ha of land meant that there was little chance of flushing out mosquito larvae by flood water, and localised rainfall runoff created pools within the project area that provided suitable breeding habitat. *P. vivax* and *falciparum* were the main parasites carried by the mosquitoes. Although ideal long-run records on infection were not available, analysis of records for 1986 indicated that the attack of malarial vector *P. falciparum* was significantly higher in the project area (200 per cent SFR) than in adjacent areas (Mirza, 1991a, 163).

Climatic Influences. Climate is one of five main epidemiological factors contributing to the transmission of malaria (Begum, *et al.*,1986; Elias *et al.*, 1987; Rosenberg, 1982; Rosenberg and Maheswary, 1982a; 1982b). Others are the habits and biology of the vector mosquitoes, environmental conditions, and the presence of high risk groups. While mosquito activity is linked to climate factors, the fluctuations in disease incidence in the past seem to be largely due to the activities of people.

Predicted temperature changes under a warming climate are small compared with the present annual range and year-to-year fluctuations. Nevertheless, it is generally agreed amongst researchers that the development rates of malarial parasites increase with warmer temperatures. It is possible, therefore, that changes in temperature, rainfall, humidity and storm patterns could directly affect the vectors' reproduction rate, biting rate, and the amount of time of human exposure. If increased warmth is associated with adequate rainfall for mosquito breeding, then, in the absence of an efficient eradication programme, the incidence of malaria could increase. In a warmer climate, mosquitoes may also move vertically into higher land (Rosenberg and Maheswary, 1982b). It can also be expected that irrigation works will continue to develop in future although the maximum areal extent may well be reached soon after the turn of the century, long before the impacts of climate change take effect.

Nevertheless, within irrigation areas, the conditions for mosquito breeding would remain high without suitable counter-measures, such as spraying, regardless of climate change. On the other hand, it could be argued that, if pressure on available water supplies were to lead in future to an expansion of irrigated dryland crop production (e.g. wheat) at the expense of (or instead of) flood-irrigated *boro* rice, this would be beneficial in terms of malaria control.

BOX 5.5: Water- and vector-borne diseases

> Temperature, precipitation and humidity influence the incidence of water-borne (and air-borne) diseases. Bacteria, parasites, and their vectors may breed faster and live longer in warmer, wetter conditions in Bangladesh. However, these climatic factors are necessary, but not sufficient for these diseases. Sanitation tied to poverty is the main condition for diarrhoeal diseases (like cholera). Drought and flood facilitate their transmission. Rainfall and poorly maintained human settlements facilitate breeding of mosquitoes, and people from infected forest areas is a main reason for its resurgence on the lowlands. Climate change in future could encourage such diseases, especially if economic development is impeded.

Information about the ecology of disease carrying mosquitoes in Bangladesh is limited. There is no base-line information from which assessments may be made as to the influence of environmental changes on the transmission of associated diseases, whether it be as a consequence of climate change or human alteration of drainage, such as through flood control and/or irrigation. These are areas in which further research is required.

5.5 WHAT ARE THE POSSIBLE SOCIO-ECONOMIC IMPACTS OF CLIMATE CHANGE ON BANGLADESH IN THE FUTURE?

The previous two sections posed two questions: How does the current climate affect Bangladesh society and economy? What societal trends may influence the vulnerability of Bangladesh to changes in climate and sea level? In this section, the question posed is: *What are the possible socio-economic impacts of climate change on Bangladesh in future?*

In addressing this question, emphasis is on adverse effects rather than on the benefits that might well be associated with an assumed change in climate. The appraisal is organised around climatic variations, or extreme events, because it is assumed that they may cause the most adverse impacts in the early stages of global warming. Running through the appraisal is consideration of *where* the impacts are most likely to be felt and by *whom*.

5.5.1 Current Vulnerability

It is often said that the best predictor of future possibilities lies in the pattern of past events. However, Bangladeshi society is in transition. As change proceeds, future socio-economic developments may prove to be significantly different from the recent past. Discussion in this section therefore begins by synthesising what has been made evident in the first two sections of the chapter. First, the notion of Bangladesh as a society in transition and its implications for future socio-economic change will be re-examined. Second, the present-day hazardousness of Bangladesh will be reviewed by bringing together the extreme climatic events (outlined in the first section) with the various patterns of human activities that have resulted from recent socio-economic trends (outlined in the second section). This synthesis may help indicate some future possibilities for a Bangladeshi society undergoing climatic change and sea-level rise.

Societal Change

A major element of social change in Bangladesh has been the rapid increase in population imposing on a limited resource base. Another has been the economic transformation along market economy lines. However, the adoption of small-scale technologies as part of the 'green revolution' has limited the degree of capitalist transformation of the agrarian sector. Thus, many of the traditional socio-economic structures of rural Bangladesh have been maintained. Nevertheless, the review of socio-economic trends in section two of this chapter suggests that the socially disruptive forces typical of the capitalist transformation of a traditional agrarian economy are emerging. It is therefore to be expected that this process will escalate as economic development proceeds into the future. A summary of socio-economic trends over the last 30 years, and their implications for increasing the vulnerability of Bangladesh to climatic variations and change in future, is provided in Table 5.3.

TABLE 5.3: Past socio-economic trends and vulnerability in Bangladesh

Past Socio-Economic Trends	**Direction of Socio-Economic Vulnerability** (+ positive - negative +/- uncertain)
Increasing population	+
Increasing numbers of landless	+
Increasing number underemployed	+
Increasing incidence of malaria	+
Persistence water borne diseases	+
Increasing permanent migrations	+
Increasing seasonal migrations	+
Increasing settlement on coastal lowlands	+
Increasing settlement on chars	+
Increasing settlement in other marginal locations	+
Decreasing absolute poverty	-
Increasing education opportunities	-
Increasing health care services	-
Increasing mid and upper classes	-
Increasing migration into western districts	-
Increasing urbanisation	+/-
Increasing economic diversification	+/-

Also provided is a general description of the direction of socio-economic vulnerability associated with each factor or trend. For example, improvements in health care, education, and training should help decrease the vulnerability of various groups in society to socio-economic change, whether or not it is climate induced. On the other hand, increasing population and landlessness should increase the vulnerability of affected groups to social and environmental changes. These factors are not of course unrelated. Economic growth ought to facilitate improvements in health and education making the landless more employable in jobs created by such growth. It is, however, of significance that most of the socio-economic trends in Table 5.3 point to enhanced vulnerability in future. This outcome is important and highlights the difficulties the government faces in developing policies for meeting the challenges of both economic development and environmental change. Central to both is slowing the growth in population and alleviating poverty.

It also seems reasonable to suggest from the review in sections one and two of this chapter that the sequence of large-scale natural disasters throughout the last 30 years has magnified some of the disruptive forces associated with economic development, while at the same time constraining the development process. This in turn has increased the level of foreign aid required for transforming the traditional economy of Bangladesh. It seems likely that social changes wrought by economic development and the social disruptions consequent upon natural disasters — whether or not exacerbated by climate change and sea-level rise — will dominate the future, while the more insidious socio-economic effects of a slowly warming climate on crops will pose a much lesser problem to resolve. Consequently, the remainder of this review will focus on extreme natural events and hazards.

Natural Events and Vulnerable Places

The areal extent of the four types of severe natural events and processes affecting Bangladesh was shown in maps in Figures 5.4 to 5.7. The events included cyclone, flood, drought and riverbank erosion. The patterns in each map are combined onto one map showing the composite of extreme events and processes in Figure 5.23. This map provides a generalised geographical distribution of the main hazardous natural events affecting Bangladesh.

Disregarding moderate and lesser events, the distribution in Figure 5.23 shows that only the north-eastern margins and part of the north-west of the country are free from severe natural events. The map also shows that about 25 per cent of the country experiences more than one type of severe event.

The socio-economic consequences of natural events are felt only in so far as they adversely affect human activities and well-being. Relating natural events to human activities broadly determines the degree of hazard or vulnerability of various places and the people therein.

A simple way to assess vulnerability is to compare the map of multiple natural events in Figure 5.23 with the spatial distributions of key socio-economic phenomena shown in maps in section two (Figures 5.8-5.11). To simplify the task, these

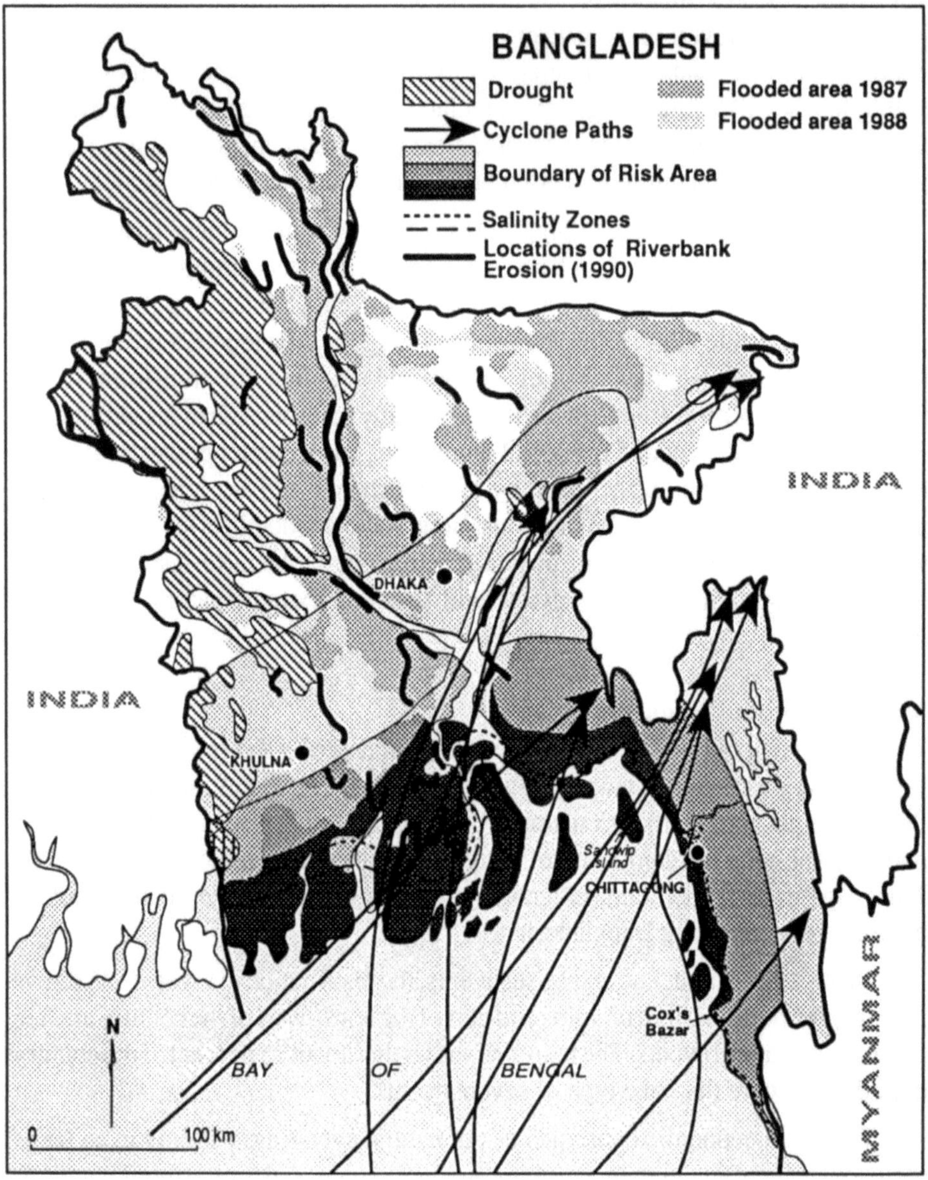

FIGURE 5.23: The spatial distribution in Bangladesh of five main types of severe natural events and processes: cyclones, floods, riverbank erosion, droughts, and salinity. A comparison of this with that of human activities (Figure 5.24) provides a generalised picture of the vulnerability or hazardousness of Bangladesh.

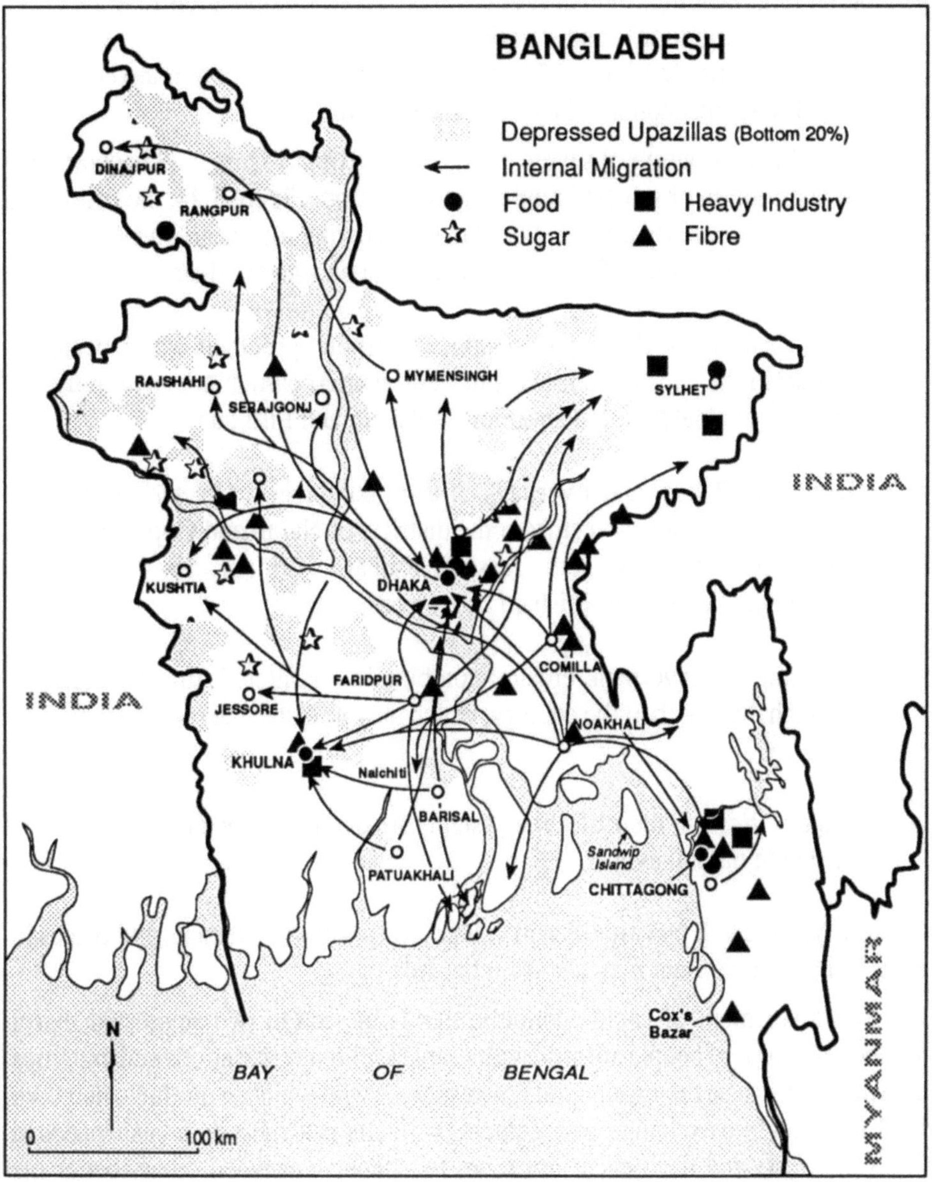

FIGURE 5.24: The spatial distribution in Bangladesh of selected human activities: settlements and migration, industry and infrastructure, economically depressed *upazillas (thanas)*. A comparison of this map with that of severe natural events and processes (Figure 5.23) provides a generalised picture of the vulnerability or hazardousnes of Bangladesh.

socio-economic phenomena have been re-grouped into three main elements: GDP generation (e.g. farming, industry, and infrastructure); population (e.g. density, settlement, and migration); and marginalisation (e.g, income, employment, and health). The activities in Figure 5.24 show: industry and transport, main cities and migration flows, and economically depressed *thanas*.

When the distribution of severe natural events from Figure 5.23 is compared with the distribution of selected human activities from Figure 5.24, the result is a generalised vulnerability map of Bangladesh showing the hazardousness of the country. The severe natural events and areas most exposed to them are:

- Cyclones, floods, riverbank erosion, and salinity problems which occur mostly in the *coastal zone* west of Feni district, particularly within the districts of Noakhali, Lakshmipur, Jhalakati, and Pirojpur.

- Droughts, floods, and riverbank erosion combine in the *mid-western zone,* reaching north-west from the districts of Narail and Gopalganj in the south through Magura, Jessore, Rajbari, Faridpur and Pabna to Sirajganj, Natore, and Rajshahi in the north.

- Flooding and riverbank erosion combine in a scattered, linear pattern through the length of the country, which may be called the *river-margins zone.*

- *Major cities* and their life-line systems are exposed to one type or more of severe natural events.

Obviously, there will be variations in vulnerability within each zone, and places in areas lying beyond them may also face hazards.

The assessment of socio-economic trends in section two suggested that a number of groups of people are more vulnerable to climatic variations and extremes than others, because they command few resources with which to produce their own food and enter into exchange arrangements with others to meet their basic needs in times of stress; and have even less room for manoeuver when others better off engage in economic exchanges that deprive them of their entitlements (Sen, 1981).

5.5.2 Future Vulnerability

The current vulnerability of Bangladesh summarised in the synthesis of Figures

5.23 and 5.24, provides a base for reviewing the question: *What are the possible socio-economic impacts of climate change on Bangladesh in future?* This is done by first, summarising changes in global warming that are expected to affect severe natural events, and then summarising the socio-economic impacts that may arise, particularly in terms of groups at risk.

Natural Events

How might climate change affect the pattern of natural events in Figure 5.23? First, warming of the oceans may lead to a rise in global-mean sea level of 18cm by 2030 and 30cm by 2050. While this may not sound like much, it is significant in terms of ongoing coastal erosion, inundation, salination, and the impacts that accompany the storm surges of cyclones (Chapter 7: Case of the Coast). Second, tropical cyclones could become more frequent and "ride" on a higher sea. (Chapter 3: Sea-level Rise). Third, changes to the monsoon could mean more rainfall in a longer season enhancing flooding and riverbank erosion. Fourth, changes in the timing of monsoonal rains could lead to decreased droughtiness. (See Chapters: 2: Climate Change, 3: Sea-level Rise, and 4: Natural Resources.)

In the absence of effective counter measures, to be described in a later section, these climate change prognoses suggest several likely outcomes for Bangladesh over the next 40 to 60 years. These are:

- The *multi-hazard core areas* (Figure 5.23) would, at a minimum, continue as such.

- Severe events and processes (cyclone, flooding, riverbank erosion, and salinity) in the *coastal zone* may intensify and become more frequent and spatially extended. A slowly rising sea level would exacerbate these effects along the coastal margin by: altering erosion rates; causing saline waters to intrude further inland; "shrinking" protective barriers; and increasing flooding by cyclone storm surges.

- Severe events (flooding and riverbank erosion) in the *river-margins zone*, may intensify and become more frequent. The design levels of existing protective barriers would, in effect, "shrink."

- Drought in the *mid-western zone* may reduce in frequency and intensity, although severe flooding and riverbank erosion would be maintained and may even increase.

- *Outside the multi-hazard core zones,* severe events may become more frequent and intense, except perhaps for drought.

Socio-Economic Prospects: the BAU Scenario

It is not easy to project with confidence the socio-economic circumstances of Bangladesh, say, 50 years from now. However, a broad outline of future prospects under a BAU (business -as-usual) scenario is provided below.

The major socio-economic goals in Bangladesh currently are — and have in the past been — concerned with economic growth, poverty alleviation, enhanced self-reliance, universal primary education, improved status and role of women in society, reduced population growth rate, improved primary health care, and improved potable water supply. The emphasis on different factors has varied from time to time. However, poverty alleviation has always been the top priority. There have also been shifts in strategies. The most important shift has in fact been the replacement of a planned approach by a market economy. However, even the planned development was primarily investment and GNP-focussed. Poverty alleviation and employment generation were not integrally built into the process. Of course, there were various poverty alleviation programmes; but they were often temporary, seasonal or experimental in nature, unsuited to prevailing conditions, and altogether rather marginal to the totality of needs. The outcomes have been low economic growth, persisting mass poverty, and burgeoning unemployment, accompanied by a failure to make much headway in respect of other socio-economic goals.

Beginning in the early 1980s, structural adjustment policies, such as those of privatisation, deregulation and globalisation, were adopted towards promoting a market economy and reducing the role of government in the economy. This process has since been carried forward very vigorously in recent years and it is the market economy philosophy that now underpins all economic policies and programmes in Bangladesh (Task Force, 1991b, vol. 2). But sluggishness in economic growth has persisted. The annual average economic growth rate achieved during 1991-1994 was 4.1 per cent, somewhat lower than the average rate achieved during 1981-1990 (4.8 per cent) or during 1973-1980 (4.9 per cent) (Ahmad, 1994).

If the poor in Bangladesh were left behind in the planned development strategy, they do not have the ability (in terms of resources, education, training and information) to participate in the market economy in a meaningful way. The essential argument in favour of a market economy relates to 'efficiency' promoted and nurtured by the competition that a market economy generates. Given that a large proportion of the population in Bangladesh is poor and unable to participate meaningfully in the market economy, the questions of competition and efficiency are irrelevant as far as they are concerned. Poverty alleviation and employment generation therefore remain outside the immediate concerns of this paradigm. The burden is left to what is known as the 'trickle down' effect. But it is well known from East Asian experiences that, even under conditions of high and sustained economic growth, strong government interventions are needed, particularly in the social sectors of education, training and health, for broad-basing the benefits. In Bangladesh, not only has high economic growth not been achieved so far, but also the social interventions have not been strong and comprehensive enough in practice. Thus, the country is at a very low point on the transitional trajectory. Very much the same situation, in both respects, may continue to be the reality in the foreseeable future under a BAU scenario. Economic growth may accelerate somewhat if the rich invest more and use upgraded technology; but, at the same time, poverty, inequality and insecurity in society will almost certainly increase.

Some likely outcomes in about 50 years from now, under a BAU scenario (indicated by the discussion in the above paragraphs and earlier relevant discussions) may be summarised, in more specific terms, as follows:

- population will have at least doubled;
- settlements will have intensified;
- migrations from high to low density rural areas and to cities will have intensified;
- urban expansion in major centres will have greatly increased;
- service sectors of the economy will have increased further; manufacturing may have become a relatively more important, but still small, component of the GDP;
- human development (in terms of economic and social progress) encompassing all citizens may or may not have made significant progress, depending on economic growth, population expansion, occurrence of natural hazards, and empowerment of the poor in terms of their access to resources and power to make decisions;
- the wealthy and middle class may form a larger proportion of the population than now; but the poor and disadvantaged will continue to

form the bulk of the total population, unless poverty-alleviating measures can be successfully implemented;

- traditional rural attitudes and traditional adaptive methods for coping with severe natural hazards should have undergone some change as modernisation should have proceeded some way;

- the number of landless would have increased as population will have grown, and agricultural transformation might displace significant numbers of farmers and labourers;

- a larger number of landless will have been absorbed into non-agricultural sectors, given appropriate policies and programmes; but their productivity will likely remain rather low;

- there will still be a large group of under-employed; and

- the extent and quality of education and health care will have markedly improved, but living conditions for the poor might not have improved much.

At Risk Groups

In summary, there are two ways in which the effects of global warming may register on a Bangladesh society in transition. In the first instance, a change in temperature and rainfall would, in time, affect biophysical systems (see Chapter 4: Natural Resources). Changes in the stock of natural resources, such as agricultural lands, forests, and water supplies, would in turn influence economic and social activities. Some changes may be beneficial, others adverse. These biophysical changes in response to climate change may be at a pace that will enable socio-economic systems to adapt. In the second instance, global warming may bring with it an increased incidence and scale of climatic extremes. And this may require improved adjustments so that society better copes.

If climate change and sea level rise occurred under BAU socio-economic circumstances, who would be most at risk, and who would be best able to take advantage of it?

In the past, severe natural events have not affected all groups of people in similar ways. It is certain that this will be so in future. Some people will possess wealth that cushions them from the effects of disasters; or at least enables them to recover more quickly after a disaster occurs. A climate that changes adversely is unlikely to affect them as much as it would people who possess very little. Thus:

- The *least at risk* will include: urban rich including absentee landlords, living in cities and towns; money lenders whose high interests almost guarantee that property will fall their way; and traders who artificially create scarcity by hoarding food before charging exorbitant prices for it.

- The *most at risk* will include people who already live in marginal conditions. For example: the marginal farmers, rural landless, urban squatters and migrants; the children suffering from malnutritional diseases; and women left to look after the household.

In other words, the adverse socio-economic impacts of extreme events in future will mostly fall on the weaker segments of the stratified society. In future, it is likely that, under a BAU scenario, there will be more people at risk than today, due to a rapidly growing population. There will also be more wealth at risk as economic development expands.

5.6 WHAT ALTERNATIVES ARE THERE FOR FUTURE ADJUSTMENT TO CLIMATE AND SEA-LEVEL CHANGE?

In this section, the focus is on how the potential for adverse effects of climate variation and change under a BAU scenario might be reduced in Bangladesh by adopting purposeful policies and interventions. In this context, adjustments or interventions may be of two main kinds, each complementing the other. One is the vision of an optimistic socio-economic scenario (as opposed to the BAU scenario outlined above) to be brought about by *socio-economic adjustments* aimed at accelerated economic growth and human development with equity, security and sustainability, regardless of adverse climate events. The other kind of intervention encompasses *hazard adjustments* aimed at reducing the impacts of, and controlling wherever feasible, disastrous natural events, such as floods, river bank erosion, cyclones and drought. Many traditional coping measures and government policies and programmes have evolved aimed at both controlling and reducing the impact of natural hazards. Under the BAU scenario, the prospects of success of these and other more appropriate measures (if formulated) will remain seriously constrained due to resource limitation and the limited ability of people and society to implement and maintain them effectively. Under an optimistic scenario, people and society should be better equipped and empowered to undertake appropriate hazard adjustments more efficiently.

BOX 5.6: BAU scenario

> If the recent past becomes BAU in future, low economic growth, mass poverty, and burgeoning unemployment will persist. If accompanied by a pattern of extreme natural events and hazard adjustments similar to those of the recent past, loss of lives and property will escalate. Should adverse climate and sea level changes occur under this BAU scenario, then the number of people at risk will increase, especially among the marginalised poor, and catastrophic losses will become more frequent.

5.6.1 Hazard Adjustments

Traditional practices that have evolved over the centuries, but reinforced through research and policy support in recent times, for coping with climatic variations and extremes include, for example, embankments, irrigation, seed selection, planting regimes, storing reserves, and so on.

Many flood control and irrigation schemes (FCD/Is) have been implemented over the past few decades. These have been aimed at moderating the floods and reducing their impacts on the one hand and enhancing the resource base on the other. A very ambitious approach to flood control across the country has been proposed and is being developed under the Flood Action Plan (FAP). It aims at providing a more secure physical environment for economic development by regulating the water regime. But, there are questions about its adverse socio-economic and environmental impacts. Also, the cost of implementing it will be very high. But if it can be ensured that the FAP's purpose will be achieved with no or negligible adverse social and environmental effects, that cost may not be too high when weighed against the resources (mostly foreign aid) that may be needed for recovery from disasters in the future that FAP will have moderated or controlled. However, it is important that lessons from recent experiences in Bangladesh relating to poor maintenance of embankments and other physical infrastructures, and the failure of embankments to cope with large floods, are adequately built into the design and implementation of various FAP components (Hunting Technical Services Limited, 1991b, 1992; Chowdhury, 1992; Khondker, 1992; Parker, 1992).

It must be remembered, in this context, that a major reason for severe floods in Bangladesh is the copious amounts of water that moves through this country from upper riparian countries.

A major emphasis is also being placed on creating awareness and preparedness among people to reduce their susceptibility to the adverse effects of extreme events. The measures adopted include: forecasting and warning systems, land use planning, and relief and rehabilitation. In the past, people have been reluctant to evacuate in response to warnings against impending disasters. It is also the case that, at times, warnings were not adequate or timely or did not even reach the people at risk. The government is seeking to strengthen the mechanism of providing adequate and timely warnings to the people concerned and preparing them to heed those warnings and seek shelter wherever possible. In response to a sequence of severe events in recent years, the government's programme of building shelters against cyclones and floods is being strengthened.

Since it is impractical to prevent settlement on unstable, flood-prone, and unprotected land, measures need to be adopted that help reduce vulnerability to floods and cyclones. 'Flood-proofing' settlements and services is one strategy that is being considered (Hunting Technical Services Limited, 1991a, 1992; James and Pitman, 1992).

In the next section, consideration is given as to how a more empowering socio-economic transformation (from the points of view of both society as a whole and people at large), may be brought about in Bangladesh, which would ensure better implementation, maintenance, and results in respect of both structural and non-structural disaster control and mitigation measures (being implemented or in the works, such as those outlined above, or others that may be developed in the course of time) than would seem likely under the BAU scenario.

5.6.2 Socio-Economic Adjustments

A strategy dictated by and suited to the prevailing socio-economic circumstances for accelerating economic growth and poverty alleviation is crucially needed in Bangladesh. These two goals are, by themselves, of top-most priority in this country. But progress in these directions will also enhance the ability of the people and society to respond better to other problems, including those arising from natural hazards.

Such a strategy must necessarily be people-centred in the sense that their full potential is used in conjunction with the best possible utilisation of available resources and that they benefit equitably from the outcome. If people at large are going to participate in the socio-economic transformation process in this way, activities must be planned locally, with adequate administrative decentralisation and political devolution to appropriate levels being essential prerequisites. Macro, meso, and micro policies must also be conducive to this process, supported by appropriate institutional networks.

The strategy may be based on policy adjustments within a market economy framework and not on centralised planning. A new market economy concept that mainstreams rather than alienates the poor needs to be evolved so that the latent creative energies of the people at large are released and mobilised in a competitive framework at the grassroots.

One way of evolving such a process would be to start by primarily focussing on employment, as opposed to investment as has hitherto been the case. In fact, the proximate cause of poverty in Bangladesh is unemployment in terms of unutilised time and low productivity. Instead of a certain rate of GNP growth, the target may be generation of a certain number of productive employment opportunities. Improvement in productivity should also be a part of the strategy. The annual employment target must be high enough to make a dent on the more than12 million jobless (on a labour-time basis), while the labour force is increasing by about one million persons every year. Once the target is set, it will be necessary to identify sectors, subsectors and activities where the jobs can be created; the public investment portfolio and policies to influence private investments may then be determined on that basis.

BOX 5.7: Optimistic scenario

> Rather than focus on BAU investment and GNP growth targets, an "optimistic" scenario would emphasise productive employment targets aimed at releasing the latent creative energies of the country's poor people at the grassroots level. In this future of Bangladesh, a new market economy may evolve in which the poor are mainstreamed through an employment-based strategy anchored on: basic education; skill training; and organisational support at local level. These social adjustments would be accompanied by an improved mix of structural and non-structural measures aimed at reducing the susceptibility of society to natural hazards — measures that would also prove helpful should climate and sea levels change in future.

Obviously, this approach places emphasis on the poorer segments of the society. Hence, when people who had no jobs or incomes before are productively employed and others who were employed, but at low levels of productivity, are more productive, the outcome will contribute directly to poverty-alleviation, as well as to economic growth. Equity will also be promoted by this process, as the focus is on the poor. At the same time, existing and potential opportunities in modern and export sectors should be identified and suitably promoted.

This employment-based strategy is anchored on the following crucial elements: basic education, skill training, and organisation. Basic education (literacy, numeracy and life skills) will enable the people to gauge the potential that they possess; skill training will equip them to undertake economic activities (on a self-employment or wage-employment basis); and organisation consisting of the provision of, for example, credit, information, technology, extension services, and marketing assistance, will help them undertake appropriate economic activities or find appropriate wage-employment. If assisted to make choices of activities with reference to existing and potential domestic (particularly rural) demand patterns, complementarity within the production pattern, and opportunities for export, demand may not become a constraint.

Adjustments in macro and meso policies and institutional arrangements will be necessary to make adequate resources, technologies and institutional support available where they are needed for this process to work effectively. In the institutional context, administrative decentralisation and political devolution are key factors toward enabling people in local spaces to find their rightful places in the process. An elaboration of this employment-based approach is provided in Ahmad (1993b).

This approach appears to be very relevant to the Bangladesh context; but there can be other mechanisms for moving the economy and society forward. Indeed, how a forward thrust can be achieved is constantly under review in Bangladesh. Hence, an optimism about much better outcomes than those outlined earlier as likely to materialise under the BAU scenario are justifiable.

5.7 KNOWLEDGE GAPS AND FUTURE RESEARCH NEEDS

The final section of this chapter addresses two questions: What lack of knowledge impedes the ability of Bangladesh to better adapt to environmental change and variability? and What research should be done to acquire the necessary knowledge? There are several general areas in which research could pave the way to improved adaptation.

First, in some areas, there is a lack of fundamental *knowledge concerning the relationship between climate variation and socio-economic effects*. For example, perhaps the most striking of these is the dearth of knowledge on the effects of climate on health: in particular, the role of climate in water-borne and vector-borne diseases could have in the future. Increased research in this field would appear to be well justified.

Second, *the socio-economic effects of climate variations in the urban environments of Bangladesh* is of concern. Despite the future uncertainties, two trends are quite clear: Bangladesh will be warmer and more urbanised. Large, dense urban settlements are a relatively new phenomenon in Bangladesh. Increased knowledge about the ways in which climate change and variability might affect urban areas could contribute towards the development of urban patterns and infrastructure better equipped to cope with existing climatic variability and extremes.

Third, there is a need to examine *the range of adaptive measures that are available for coping with environmental adversity*. Perhaps this is most urgent for traditional adaptive mechanisms. These include not only technical adjustments like seed varieties and planting dates, but also measures of social reciprocity that serve to share the burdens of loss and the benefits of bounty. Such traditional measures of coping with adversity are woven into the social fabric of Bangladeshi society. They require investigation and documentation lest they be forgotten in the transition to modern ways. Indeed, a primary goal ought to be developing modern equivalents out of the proven principles that underpin the traditional methods of coping.

Fourth, a key element in the development strategy of Bangladesh is its water control programme based on irrigation, flood, and drainage technologies. Studies are needed *to assess how, and to what extent, traditional technologies are being adapted to changing socio-economic conditions.*

Fifth, there is a need to *assess how customary behaviour is being modified in response to changing social and environmental conditions*. With the ongoing modernisation process in Bangladesh, new institutional arrangements, infrastructure, environmental conditions, and value systems have been developing, often being shaped by a strong western influence promoted through aid, western education, business contacts and so on. These impinge on traditional values and practices. Research into the impact of modernisation on customary behaviour may help identify how best to integrate traditional and modern systems so that vulnerability to environmental and social stress is minimised — with or without climate change.

Sixth, as a transformation of agriculture along capitalist lines proceeds, increasing numbers of displaced marginal farmers and labourers may be expected to migrate in search of employment opportunities. This process may be exacerbated by climate change and sea-level rise. Studies are needed *of various forms of migration and resettlement of the landless to help anticipate the likely dimensions of problems that may arise if climate extremes worsen and sea level rises.*

Finally, there is an urgent need to *develop means of empowering the landless and poor with entitlements to resources to ensure their resiliency in times of scarcity.* Ensured access to food, shelter, education and training, and health facilities surely lies at the heart of vulnerability to climatic variability and extremes. Of course, this also lies at the heart of poverty — the pervasive problem in Bangladesh. But if one message is clear, it is that poverty and vulnerability to environmental adversity are inextricably entwined. By attacking the problems of poverty and inequality, one is concurrently treating the issue of vulnerability and vice versa. The suggested employment-based approach merits further study.

5.8 WHAT SHOULD BE DONE?

Fulfilling the research needs noted above requires interdisciplinary research (integrating social sciences and natural sciences) aimed at developing an optimum strategy for reducing the vulnerability of Bangladesh to climatic extremes. This research would help ascertain the priorities that could be given to various kinds of activities.

A general framework for pursuing this research could take the form indicated in Figure 5.25. It aims to identify the main interactions between two main elements for reducing vulnerability. These elements include:

A. Policies aimed at socio-economic development through

- improving employment
- improving access to resources
- increasing incomes
- reducing poverty
- reducing population growth
- improving education
- improving technical training
- improving health care

B. Policies aimed at reducing vulnerability to extremes given present climate through

- embankments
- irrigation
- seed selection
- planting regimes
- storing reserves
- emergency preparedness
- landuse management
- disaster preparedness and management procedures

The nature and combination of the socio-economic policies (A) and the climate change policies (B) are likely to be different depending on whether climate changes at a slow or fast rate (Figure 5.25).

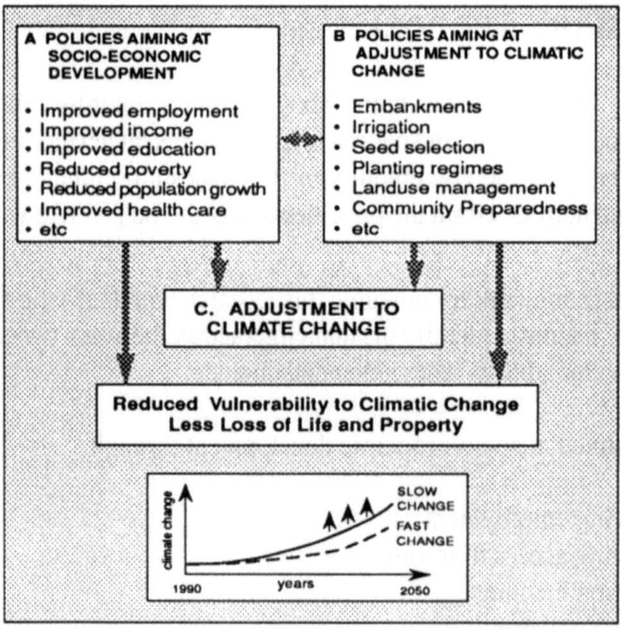

FIGURE 5.25: A schematic diagram showing the main relationships between three main elements for reducing vulnerability to global warming: socio-economic policies for alleviating poverty; hazards adjustments to severe events; and assessments of slow and fast rates of climate change.

Assigning priorities between the various activities will require reliable estimates with regard to not only the rate of change of climate, the occurrence of extremes, and the uncertainties of the projected frequency of extreme events, but also the rates of economic and social 'development'.

REFERENCES

Adnan, S. (1991) Floods, people and the environment. *Grassroots: An Alternative Development Journal*, 1 (1), pp.27-47.

Adnan, S. (with M. Ghani, S. Uddin, S. Khandaker, A. Dewan, S. Zaker, A. Sufiyan, S. Manik, A. Hossain, and S. Akhter) (1991) *Floods, People and the Environment: Institutional Aspects of Flood Protection Programmes in Bangladesh, 1990*. Research and Advisory Services, Dhaka.

Aguero, M., Rahman, A., and Ahmed, M. (eds.) (1989) *Inland Fisheries Management in Bangladesh*. International Centre for Living Aquatic Resources Management, Manila; Bangladesh Centre for Advanced Studies, and Department of Fisheries, Bangladesh.

Ahmad, A.J., Hossain, A., Mian, H.U., and Hossain, A. (1986) *Energy Crisis in A Bangladesh Village*, Rural Developement Academy, Bogra.

Ahmad, M. (ed.) (1989) *Flood in Bangladesh*. Community Development Library, Dhaka.

Ahmad, N. (1976) *A New Economic Geography of Bangladesh*. Vikas Publishing House Pvt Ltd., New Dehli.

Ahmad, Q.K. (1992) Policies and strategies for sustainable development in Bangladesh. *Futures*. November 1992, pp.879-893.

Ahmad, Q.K. (1993a) *An Assessment of the Unemployment Situation in Bangladesh and Towards Formulation of a National Employment Plan for the Country*. Paper presented to the National Conference on Bangladesh: Past Two Decades and the Current Decade, 29 January - 5 February 1993, Dhaka.

Ahmad, Q.K. (1993b) *Poverty and Environment: Breaking the Vicious Circle*. Meeting on the Occasion of World Environment Day, 1993, Dhaka.

Ahmad, Q.K. (1993c) *Population Policy, Poverty Alleviation and Environmental Protection in Bangladesh — An Integrated Approach Needed*. Paper presented at Seminar on Population Policies and Programs in Bangladesh, Dhaka.

Ahmad, Q.K. (1994) Anatomy of economic liberalization: efforts and results in Bangladesh. Paper presented at the International Conference on Economic Liberalization in South Asia, Australian South Asia Research Centre, The Australian National University, Canberra, Australia, 30 November - 2 December 1994.

Ahmad, Q.K. and Mirza, M.Q. (1992) Bangladesh's development: resources and constraints. *Bangladesh Development Dialogue*, 1, 20-36.

Ahmed, T. (1990) *Bangladesh Situation of Filariasis*. Institute of Epidemic Disease Control and Research (IEDCR), Department of Medical Antomology (mimeographed), Dhaka.

Ahmed, T. (no date) *Urban Vector and Pest Control*. IEDCR, Department of Medical Antomology (mimeographed), Dhaka.

Ahmed, M., English, P., Feldman, S., Hossain, M., Jansen, E., McCarthy, F., de Wilde, K. and Young, R. (1990) *Rural Poverty in Bangladesh: A Report of the Like-Minded Group*. University Press Limited, Dhaka.

Asian Development Bank (ADB) (1989) *Bangladesh Health and Population Sector Profile*. ADB, Infrastructure Department, Social Infrastructure Division, Manila.

Asaduzzaman, M. (1989) Feeding our future towns: an overview of urbanisation and associated food policy issues. In Bangladesh Planning Commission, *Food Strategies for Bangladesh*. University Press, Dhaka. pp.177-195.

Bangladesh Bureau of Educational Information and Statistics (BANBEIS) (1992), Government of the People's Republic of Bangladesh (GOB), Dhaka.

Bangladesh Bureau of Statistics (BBS) (1991a) *1991 Statistical Yearbook of Bangladesh.*. BBS, Government of the People's Republic of Bangladesh (GOB), Dhaka.

Bangladesh Bureau of Statistics (BBS) (1991b) *Report of the Child Nutritional Status Survey*. 1989-90. BBS, GOB, Dhaka.

Bangladesh Bureau of Statistics(BBS) (1991c) *Report on the Household Expenditure Survey (August 1991)*. BBS, GOB, Dhaka.

Bangladesh Bureau of Statistics (BBS) (1992) *Supplement No. 1 to the Preliminary Report on Population Census 1991*. BBS, GOB, Dhaka.

Bangladesh Bureau of Statistics (BBS) (1994) *Statistical Pocketbook of Bangladesh 1994*. BBS, GOB, Dhaka.

Bangladesh Institute of Development Studies (BIDS) (1992) *Multipurpose Cyclone Shelter Project*, (Annex G1: Population). BIDS, Dhaka.

Bangladesh Rural Advancement Committee (BRAC)(1979) *Ten Diseases: A Treatment Guide for Medical Paraprofessionals*. BRAC, Dhaka.

Bangladesh Rural Advancement Committee (BRAC)(1984) *Peasant Perceptions: Famine, Credit Needs, Sanitation*. BRAC, Rural Study Series, Dhaka.

Bangladesh Water Development Board (BWDB) (1993) *Annual Flood Report 1992*, BWDB, Dhaka.

Begum, A., Biswas, B. and Elias, M. (1986) The ecology and seasonal fluctuations of mosquito larvae in a lake in Dhaka City. *Bangladesh J. Zool.*, 14 (1), 41-48.

Burton, I., Kates, R., and White G. (1978) *Environment as Hazard*. Oxford University Press, New York.

Chowdhury, A., Hakim, M., and Rashid, S. (1989) *Changes in Land Ownership and Use in Rural Bangladesh: A Study of Seven Mouzas of Bogra District 1920-1987*. Rural Development Academy, Bogra.

Chowdhury, J.U. (1992) Flood action plan: one sided approach?' In J.L. Jr.Westcoat (ed.) *Five Views of the Flood Action Plan for Bangladesh*. University of Colorado, Institute of Behavioral Science, Natural Hazards Research and Applications Centre, Working Paper No. 77., Boulder, Colorado, pp.3-6.

Economic Relations Division (ERD) (February 1993) *Flow of External Resources in Bangladesh*. Ministry of Finance, GOB, Dhaka.

Elahi, K.M. (1991a) Impacts of riverbank erosion and flood in Bangladesh: an introduction. In K. Elahi, K. Ahmed and M. Mafizuddin (eds.) *Riverbank Erosion, Flood and Population Displacement in Bangladesh*. Jahangirnagar University, Riverbank Erosion Study, Dhaka, pp.13-29.

Elahi, K.M. (1991b) Riverbank erosion, flood hazard and population displacement in Bangladesh: an overview. In K. Elahi, K. Ahmed, and M. Mafizuddin (eds.) *Riverbank Erosion, Flood and Population Displacement in Bangladesh*. Jahangirnagar University, Riverbank Erosion Study, Dhaka, pp.95-110.

Elahi, K., Ahmed, K., Mafizuddin, M. (eds.) (1991) *Riverbank Erosion, Flood and Population Displacement in Bangladesh*. Jahangirnagar University, Riverbank Erosion Study, Dhaka.

Elias, Rahman, and Rahman (1985) DDT- susceptibility status of *Anopheles Philippinensis* - a mosquito vector of malaria in Bangladesh. *Bangladesh Med. Res. Counc. Bull.,*. XI (1), 1-7.

Elias, M., Rahman, A., Ali, M., Begum, J., and Chowdhury, A. (1987) The ecology of malaria carrying mosquito *Anopheles Philippensis Ludlow* and its relation to malaria in Bangladesh. *BMRC Bulletin*, XIII (1), pp.15-28.

Faaland, J. and Parkinson, J. (1976) *Bangladesh: The Test Case of Development*. C. Hurst & Company, London.

Ferdous, SK Z. and Husain, A.T.M. (1988) The utilization of and accessibility to accretion lands by displaced populations. *The Bangladesh Rural Development Studies*, I (2), .29-36.

Government of the People's Republic of Bangladesh (1989) *Study on the Causes and Consequences of Natural Disasters and Protection and Preservation of the Environment*. South Asian Association for Regional Cooperation (SAARC), Kathmandu.

Haider, R., Rahman, A. and Huq, S. (eds.) (1991) *Cyclones '91: An Environmental and Perceptual Study*. Bangladesh Centre for Advanced Studies, Dhaka.

Haque, B. and Hoque, M. (1990) Faecal pollution of surface water and disease in Bangladesh. In Rahman, A., Huq, S. and Conway, G. (Eds.), 180-187.

Hewitt, K. (Ed.) (1983) *Interpretations of Calamity: From the Viewpoint of Human Ecology*. Allen and Unwin Inc., Boston.

Hossain, M. (1990) Natural calamities, instability in production and food policy in Bangladesh. *The Bangladesh Development Studies*, XVIII (4), pp.33-54.

Hossain, M. (1991) *Agriculture in Bangladesh: Performance, Problems, Prospects*. University Press Limited, Dhaka.

Hunting Technical Services (1991a) *Draft Final Report, Volume 4, Appendix L: Gender Impacts*. Bangladesh Flood Action Plan, FAP 12 FCD/I Agricultural Study, Dhaka.

Hunting Technical Services (1991b) *Draft Final Report, Volume 4, Appendix M: Socio-Economic Impacts*. Bangladesh Flood Action Plan, FAP 12 FCD/I Agricultural Study, Dhaka.

Hunting Technical Services (1992) *Project Impact Evaluation of Chalan Beel Polder-D*. Bangladesh Flood Action Plan, FAP 12 FCD/I, Dhaka.

Hye, H., Solaiman, M. and Karim, M. (1986) *Floods- 1984*. Bangladesh Academy for Rural Development, Kotbari, Comilla.

International Assessment of the Malaria Programme, Bangladesh. Final Report, Extract, 1-28 October 1989.

Islam, M. (1974) Tropical cyclones: coastal Bangladesh. In G.White (ed.) *Natural Hazards: Local, National, and Global*. Oxford University Press, New York, pp.19-24.

Irrigation Support Project for Asia and the Near East (ISPAN) (1992) *Impacts of Flood Control and Drainage on Vector-Borne Disease Incidence in Bangladesh*. ISPAN, FAP 16 Environmental Study, Special Studies Program, Draft, Dhaka, December 1992.

James, L.D. and Pitman, K. (1992) 'Flood Action Planning for Bangladesh'. In J.L. Jr.Westcoat (ed.) *Five Views of the Flood Action Plan for Bangladesh*. University of Colorado, Institute of Behavioral Science, Natural Hazards Research and Applications Centre, Working Paper No. 77., Boulder, Colorado, pp.21-27.

Jansen, E.G. (1987) *Rural Bangladesh: Competition for Scarce Resources*. University Press Ltd, Dhaka.

Karim, Z., Ibrahim, A., Iqbal, A. and Ahmed, M. (1990) *Drought in Bangladesh Agriculture and Irrigation Schedules for Major Crops*. Bangladesh Agricultural Research Council, Dhaka.

Kates, R.W. (1985) The interaction of climate and society. In R.W. Kates, J.H. Ausubel and M. Berberian (eds.) *Climate Impact Assessment*. John Wiley and Sons Ltd, New York, Chapter 1.

Kates, R.W., Ausubel, J.H. and Berberian, M. (Eds.) (1985) *Climate Impact Assessment*. John Wiley and Sons Ltd, New York.

Khan, H.R. (1990) Water quality control. In Huq, S., Rahman, A., and Conway, R., pp.127-134.

Khan, S.A. (1989) *The State and Village: The Political Economy of Agricultural Development in Bangladesh*. University Press Limited, Dhaka.

Khondker, H.H. (1992) The political imperatives of 1988 flood in Bangladesh. In J.L. Jr.Westcoat (ed.) *Five Views of the Flood Action Plan for Bangladesh*. University of Colorado, Institute of Behavioral Science, Natural Hazards Research and Applications Centre, Working Paper No. 77., Boulder, Colorado, pp.9-20.

Kosinski, L. and Elahi, K. (1985) *Population Redistribution and Development in South Asia*. D. Reidal Publishing Company, Boston.

Lasker, S.I. (1983) *Urbanization in Bangladesh 1901-1981*. Bangladesh Institute of Development Studies, Dhaka, Research Report No. 34.

Mahtab (1989) *Effects of Climate Change on Bangladesh*. A report prepared for the Expert Group on Climate Change and Sea Level Rise, sponsored by the Commonwealth Secretariat, Dhaka.

Maloney, C. (1988) *Behaviour and Poverty in Bangladesh*. University Press Limited, Dhaka.

Ministry of Environment and Forest (1991) *Bangladesh Country Report for United Nations Conference on Environment and Development (UNCED)*. Ministry of Environment and Forest, Dhaka.

Mirza, M. Q. (1991a) *Environmental Impacts of Water Development Projects: A Case Study of Chandpur Irrigation Project*. Thesis: Master of Science in Engineering (Water Resources), Bangladesh University of Engineering and Technology, Department of Water Resources Engineering, Dhaka.

Mirza, M. Q. (1991b) Flood Action Plan of Bangladesh- the embankment issue. *Water Nepal*, 2 (2/3), 25-28.

Mondal, A. H. (1989) *The Pattern of Industrial Location and Its Determinants in Bangladesh*. BIDS Research Report No 114, Dhaka.

Montgomery, R. (1985) The Bangladesh floods of 1984 in historical perspective. *Disasters*, 9 (3), 163-172.

Murshid (1987) Weather, new technology and instability in foodgrain production in Bangladesh. *The Bangladesh Development Studies*, XV, 1, 31-56.

Naqi, S.A. (1990) Socio-economic impact of salinization and polderization: a case study'. In A. Rahman, S. Huq, and G. Conway (eds.) *Environmental Aspects of Surface Water Systems in Bangladesh*. University Press Limited, Dhaka, pp.109-119.

Parker, D.J. (1992) Social impacts of the Bangladesh Flood Action Plan. In J.L. Jr. Westcoat (ed.) *Five Views of the Flood Action Plan for Bangladesh.* University of Colorado, Institute of Behavioral Science, Natural Hazards Research and Applications Centre, Working Paper No. 77., Boulder, Colorado, pp.7-9.

Paul, S., Rahman, Z., Mirza, M.Q., and Mohammad, A. (1992) Towards development of effective warning system and preparedness programme for disaster management in the coastal region of Bangladesh. *Proceedings of the Workshop on Coastal Zone Management in Bangladesh.* Dhaka.

Pernia, E.M. (1993) *Urbanization, Population Distribution and Economic Development in Asia.* Report No. 58.

Quddus, M.A. and Ara, S. (1991) *Nutrition in Rural Communities with Seasonal Variations.* Bangladesh Academy for Rural Development, Kotbari, Comilla.

Rahman, A. (1991) Migration and food scarcity: a case study of Bangladesh. *Grassroots: An Alternative Development Journal,* 1 (2), 20-41.

Rahman, A., Huq, S. and Conway, G. (1990) *Environmental Aspects of Surface Water Systems in Bangladesh.* University Press Limited, Dhaka.

Rahman, M. (1991) Vulnerability syndrome and the question of peasants' adjustment to riverbank erosion and flood in Bangladesh. In K. Elahi, K. Ahmed, and M. Mafizuddin (eds.) *Riverbank Erosion, Flood and Population Displacement in Bangladesh.* Jahangirnagar University, Riverbank Erosion Study, Dhaka, pp.170-187.

Rashid, H. (1991) *Geography of Bangladesh.* University Press Limited, Dhaka.

Rashid, S. (1991) Flood Action Plan. *Grassroots: An Alternative Development Journal,* 1 (1), 8-11.

Rogge, J. and Goulter, I. (1991) *Riverbank Erosion Impacts in Bangladesh.* Paper presented to Hazards 91, Perugia Italy, August 1991. University of Manitoba.

Rosenberg, R. (1982) Forest malaria in Bangladesh (III. Breeding Habits of Anopheles dirus). *Am. J. Trp. Med. Hyg,* 31 (2), 192-201.

Rosenberg, R. and Maheswary, N. (1982a) Forest malaria in Bangladesh (I. Paristology). *Am. J. Trp. Med. Hyg,* 31 (2), 175-182.

Rosenberg, R. and Maheswary, N. (1982b) Forest malaria in Bangladesh (II. Transmission by Anopheles dirus). *Am. J. Trp. Med. Hyg,* 31 (2), 183-191.

Sadeque (1991) Flood plain agriculture: adjustment and household survival strategy— a case study. *Grassroots: An Alternative Development Journal,* 1 (1), 49-53.

Safiullah, A. (1989) Embankments for flood protection: success and failure. In M. Ahmad (ed.) *Flood in Bangladesh.* Community Development Library, Dhaka, pp.159-187.

Salaheen, M. (1991) The mobility characteristics of displacees: a case study from riverbank erosion hazard area. In K. Elahi, K. Ahmed, and M. Mafizuddin (eds.) *Riverbank Erosion, Flood and Population Displacement in Bangladesh.* Jahangirnagar University, Riverbank Erosion Study, Dhaka, pp.160-169.

Sen, A. (1981) *Poverty and Famines: An Essay on Entitlement and Deprivation.* Clarendon Press, Oxford.

Shahjahan, M. (1989) The devastating flood of 1988. In M. Ahmad (ed.) *Flood in Bangladesh.* Community Development Library, Dhaka,pp.141-143.

Siddique, A. (1989) Impact of flood on the economy of Bangladesh. In M. Ahmad (ed.)*Flood in Bangladesh.* Community Development Library, Dhaka, pp.245-253.

Siddique, A., Islam, Q., Akram, K., Mazumder, Y., Mitra, A. and Eusof, A. (1989) Developing world: cholera epidemic and natural disasters; where is the link? *Tropical and Geographical Medicine,* 41, pp.377-382.

Siddique, A., Baqui, A., Eusof, A., Haider, A., Hossain, M., Bashir, I. and Zaman, K. (1991a) Survival of classic cholera in Bangladesh. *The Lancet,* 337 (May), 1125-1127.

Siddique, A., Baqui, A., Eusof, A. and Zaman, K. (1991b) 1988 floods in Bangladesh: patterns of illness and causes of death. *J Diarrhoel Dis Res,* 9 (4), .310-314.

Siddique, A., Zaman, K., Baqui, A., Akram, K., Mutsuddy, P. and Eusof, A., Haider, K., Islam, S., and Sack, R. (1992) Cholera epidemics in Bangladesh: 1985-1991. *J Diarrhoel Dis Res,* 10 (2), 79-86.

Sultana, S. (1993) *Rural Settlements in Bangladesh: Spatial Pattern and Development.* Graphosman, Bangladesh.

Talukder, J. and Ahmad, M. (eds.) (1992) *The April Disaster: Study on Cyclone Affected Region in Bangladesh.* Community Development Library, Dhaka.

Talukder, J., Roy, G. and Ahmad, M. (1992) *Living With Cyclone: Study on Storm Surge Prediction and Disaster Preparedness.* Community Development Library, Dhaka.

Task Force (1991a) *Report on Bangladesh Development Strategies for the 1990's: Volume One, Policies for Development.* University Press Limited, Dhaka.

Task Force (1991b) *Report on Bangladesh Development Strategies for the 1990's: Volume Two, Managing the Development Process.* University Press Limited, Dhaka.

Task Force (1991c) *Report on Bangladesh Development Strategies for the 1990's: Volume Three, Developing the Infrastructure.* University Press Limited, Dhaka.

Task Force (1991d) *Report on Bangladesh Development Strategies for the 1990's: Volume Four, Policies for Environment.* University Press Limited, Dhaka.

Thompson, P.M. (1989) *The Impact of Flood Control on Agricultural Development in Bangladesh.* Middlesex Polytechnic, Flood Hazard Research Centre, Enfield, London.

United Nations (1989) *World Population Prospects, 1988.* United Nations, Department of International and Social Affairs, New York.

United Nations Development Programme (UNDP)(1992) Human Development Report, Oxford, New Dehli.

Westcoat, Jr., J.L. (ed.) (1992) *Five Views of the Flood Action Plan for Bangladesh.* University of Colorado, Institute of Behavioral Science, Natural Hazards Research and Applications Centre, Working Paper No. 77., Boulder, Colorado.

White, G. (ed.) (1974) *Natural Hazards: Local, National, and Global.* Oxford University Press, New York.

World Bank (1989) *Bangladesh: Action Plan for Flood Control.* Document of the World Bank.

Zahurul, M. (1991) Embankment failure in Bangladesh: causes and recommendations. *Grassroots: An Alternative Development Journal,* 1 (2), .20-41.

[illegible]

LEGAL IMPLICATIONS OF GLOBAL CLIMATE CHANGE FOR BANGLADESH

D. Freestone, M. Farooque and S.R. Jahan*

6.1 INTRODUCTION

The potential responses to climate change and sea-level rise will be governed to a large extent by the legal context within which Bangladesh functions. The Bangladesh legal system has a wealth of experience in dealing with problems directly relevant to the climate change issue, such as inundation, flooding and population displacement. However, the scale of such problems is likely to be exacerbated by global warming. It is therefore imperative to search for ways of strengthening legal institutions in order to assist Bangladesh in addressing current problems as well as potential future problems.

The purpose of this chapter is to examine both the national and international legal contexts that impinge upon decisions relating to climate and sea-level change. The chapter focuses on three critical questions. What are the characteristics of the national legal order, particularly with respect to land tenure and property rights, that may facilitate or constrain certain policy options or response strategies? What are the conventions or other rules of international law by which Bangladesh is bound that provide the framework for decision-making and action? Finally, what are the informational and research needs that would enhance the capability of Bangladesh to respond effectively to climate change and sea-level rise? This review suggests that there needs to be greater attention paid to ways of implementing international rules within the Bangladesh context in order to ensure coordination of internal policies with international norms.

* The authors are respectively: Professor of International Law, The University of Hull, Hull, U.K.; Secretary General, Bangladesh Environmental Lawyers Association (BELA), Dhaka, and former Consultant, BUP, Dhaka; and former Advocate, Supreme Court, Dhaka, and a former Consultant, BUP, Dhaka.

An analysis of the legal issues related to climate change in Bangladesh, as with any country or region, involves a myriad of different issues and perspectives. A traditional legal analysis of a problem would address the issues in terms of national law and assess the issues of damage (what harm has been suffered), causation, responsibility and redress. Although such an approach may have some place in the present context, it would be difficult to pursue in that the whole question of climate change remains surrounded with uncertainty, and most issues of responsibility (which would, in any event, be of an international or transnational character) must remain speculative.

Nevertheless, in broad terms it seems clear that the casual factors of the enhanced greenhouse effect, and thus global warming, are emissions of greenhouse gases, principally carbon dioxide, which are attributable largely to fossil fuel burning in the developed world (see Chapter 2). The worst impacts are, however, likely to be felt in the developing countries. While it may be true that the developing countries are contributing to carbon dioxide emissions through deforestation, principally in tropical regions, it must be borne in mind that most developed countries, particularly in Europe, felled most of their own forests generations ago, often as part of their own development process. There is therefore a basic inequity in the fact that the worst consequences of global climate change are likely to be felt by those countries, like Bangladesh, which have historically contributed least to, or derived least advantage from, the processes which have caused it.

The international community, and more recently international treaty law, has begun to recognise this issue. Some of the most worrisome aspects of climate change (possible changes in rainfall resulting in increased frequency of floods or droughts, sea-level rise, increased incidence and severity of cyclones) may most severely affect countries like Bangladesh with extensive coastal low-lying regions, with large populations and without the economic resources to respond quickly and effectively. The inequity of this situation, and the need for the developed countries to appreciate their disproportionate contribution to the greenhouse effect, and their consequentially disproportionate obligation in responding to it, is recognised by the 1992 Rio Declaration. Principle 7 of that Declaration, published at the end of the UN Conference on Environment and Development (UNCED) in Rio de Janeiro in June 1992, enunciated the principle of "common but differentiated responsibilities" of States "in view of their different contributions to global environmental degradation." This principle can also be seen in action in the 1985 Vienna Convention for the Protection of the Ozone Layer which permits developing countries a longer lead time in the implementation of controls on the production and use of ozone-depleting gases, in recognition of the costs involved and their historically low contribution to problem of ozone depletion. Similarly, the 1992

UN Framework Convention on Climate Change recognises "The specific needs and special circumstances of developing country Parties, especially those that are particularly vulnerable to the adverse effects of climate change..." These needs, and the differential obligations of the developed country Parties, will be further elaborated in the detailed Protocols to be negotiated under the auspices of the Framework Convention.

However, the fact that conclusive evidence is not yet available regarding greenhouse-gas-induced global climate change, or that legal responsibility cannot be attributed for the greenhouse effect, should not disguise the point that action needs to be taken before its first effects are felt — for by then the climate system will already be heavily committed to change. The approach that has been taken by the 1992 Framework Convention enshrines what has become known as the "precautionary approach". The precautionary approach or precautionary principle — which is arguably the most important new principle of international environmental policy and is emerging as international environmental law (Freestone, 1991) — is accepted by Article 3 of the Convention. It obliges precautionary action to "anticipate, prevent or minimize the causes of climate change and mitigate its adverse effects. Where there are threats of serious or irreversible damage, lack of full scientific certainty should not be used as a reason for postponing such measures..."

In the current state of knowledge, therefore, the role of legal analysis is most usefully directed towards the identification of impacts likely to have legal consequences and thus towards a view of law as a mechanism of social engineering — an essential element of planning and implementation of policy strategy — rather than as primarily a vehicle for the redress of wrongs. In such an approach, an appreciation of existing legal regimes and institutions as well as rights and obligations is important because these may also facilitate or obstruct possible response strategies.

The threats posed by climate change may also expose the strengths and/or limitations of existing legal regimes and institutions. The response strategy may similarly involve the strengthening or amendment of existing laws and institutions, or the development of entirely new approaches, or legal institutions and regimes. These issues are not simply national; they will also have international implications at both a regional and a global level.

As the other chapters indicate, global climate change may potentially have adverse impacts on Bangladesh. The most serious consequences may be from the possible increased frequency and severity of cyclones and from sea-level rise with consequent flooding. The cumulative impact of these possible effects will be complex and uncertain. In general terms, however, one could surmise that some low-lying areas are likely to be subject to the threat of permanent or enhanced seasonal inundation. Seasonal inundation is not a new phenomenon for Bangladesh, but with higher mean sea levels larger land areas would be threatened and some low-lying areas would be lost to existing uses. Similarly, patterns of sediment deposition may result in the emergence of new land areas; accretion and avulsion (land ownership changes by sudden movements of river courses) may occur as well as erosion and diluvion (land lost by flood). This process might be gradual, but as it occurred, it would have an immediate impact on human settlement, migration and land tenure. It would impose further burdens on the administrative institutions responsible for the management of coastal resources and for planning response actions as well as relief and rehabilitation of those affected.

The Intergovernmental Panel on Climate Change (IPCC) Coastal Zone Management Subgroup (IPCC, 1990, 1992) has suggested that national planned responses to sea-level rise can best be managed within a system of coastal zone management. Such a system would, at the least, involve the co-ordination of existing administrative activities in the coastal zone. However, the limitation of a purely coastal approach for Bangladesh is that the likely environmental impacts are not restricted to those associated with sea-level rise: the flood-plains cover some two-thirds of the country. An integrated response strategy must therefore involve more than simply the bodies responsible for coastal resources, and may indeed involve virtually all the existing national resource management institutions in Bangladesh.

Climate change will also have significance for Bangladesh at an international level in relation to issues covered by international legal regimes. Changes in sea level, current and wave patterns will inevitably affect the horizontal configuration of the coastline, which would affect geographical features from which maritime zones and boundaries are measured. Loss of arable land compounded by problems of land tenure and overcrowding in Bangladesh or its neighbours may exacerbate existing problems of international migration, even creating new classes of so-called "eco-refugees" — those seeking refuge from the adverse impacts of climate change across national boundaries.

These problems may not be addressed on a simple national basis. International law imposes constraints on the range of response strategies available, both for Bangladesh and for its neighbours. Actions by one country that have environmental impacts on neighbours are the subject of a rapidly evolving new body of international environmental law requiring, for example, prior consultation and environmental impact assessment. A network of international treaties also protects endangered species and areas of particular natural or cultural significance which may face new threats. The international law of state responsibility — by which states may be held liable for harm caused to their neighbours — provides basic concepts within which the whole issue of global climate change may be assessed. But perhaps of more significance is the rapidly evolving framework of international treaty law laying down new principles and obligations.

To address this wide range of issues, the rest of this chapter is divided into three sections. The next section addresses the national law issues. Measured and orderly response must take place within the constraints of the existing national legal order. An understanding of the system of land tenure and property rights is important to appreciate the ability of the legal system to accommodate changes in land holding which may be necessary to cater for the needs of temporary or permanent population movements. Similarly important is an appreciation of the existing administrative framework of the country. This framework itself may facilitate or constrain certain policy options or response strategies; previous experience of the sort of problems likely to arise under climate change may indicate the need for, or the viability of, proposals for innovation or reform. The main international law issues, already briefly addressed above, are discussed in a separate section. The final section addresses future needs for action and research.

6.2 NATIONAL LAW ASPECTS

6.2.1 Property Rights and Land Tenure in Bangladesh

Context and Issues
Bangladesh is the most densely populated country of its size in the world. The rapid growth of population has already pushed settlement to marginal and hazard-prone areas. Urban centres appear to offer the chance of prosperity to large sections of the rural population. Uprooted people also move onto land, often public land, encountering conflicts with relevant public agencies and amongst themselves. In the absence of any human settlement policy and equity in resource allocation, competition for acquiring, protecting and exercising titles over resources remains an active endeavour and struggle for all classes in Bangladesh society.

Climate change will exacerbate problems that already exist in Bangladesh. Migration out of areas threatened by increased flooding, by changed patterns of inundation or by cyclone impacts will put even greater pressure on the system of land holding and rights over property. The regulation of rights in property and ancillary interests involves a number of complex questions that would require policy changes as well as legislative and institutional development. Although the law allows the State, as "eminent domain", to acquire property, most of the rights which the State itself confers are designated virtually in perpetuity. Hence, careful allocation and regulation of rights and duties in property are needed now, as well as a contingency to accommodate the bio-physical changes that global climate change may bring.

Bangladesh is a disaster-prone country, susceptible particularly to floods and cyclones. People live in vulnerable areas either because they are confident that they can face the inevitable consequences, or in the knowledge that eventually they may be driven out by natural forces; often the choices are limited. Inundation, erosion and diluvion (loss of land through flood) of private lands and property are an integral part of the way of life of a large proportion of the population in Bangladesh, which has perhaps continued since the first settlements began on the deltaic silts and alluvial soil. Facing natural calamities with indigenous wisdom is a traditional and intrinsic response of the people. They are familiar with the processes of annual inundation and of erosion that compel them to re-negotiate and re-assert rights and interests over property, either temporarily or permanently. The experience of water covering land and belongings, causing landlessness and migration, is common place. Therefore, the physical fact of flooding or higher water levels changing the exercise of rights over property is not a new phenomenon in Bangladesh. The abundance of experience and traditional wisdom, at least psychologically, would help the people to accommodate the situation that sea-level rise might engender.

The structure of land holding will be an important constraint on governmental response to large-scale environmental change. The short-term interests of individual land holders (e.g. to maintain certain types of land use) may, in some circumstances, conflict with wider communal policy objectives. Land defence projects, retreat policies or even resettlement plans will require public land to be available, or may require compensation to be paid to individuals or communities for loss of land or rights such as grazing or fisheries.

Land tenure law has developed in accordance with experience and tradition. From the beginning of the codification of the law in the country, the fact that land does habitually change its physical characteristics has been reflected in various concepts, rights, duties and interests such as alluvial and diluvion law, the rules on rents of properties lost in natural processes, titles relating to *char* (temporary river islands) lands, fishery rights, the law on easements and other laws on land and water use. The law has had to address the question of the status of land left abandoned, whether for natural or other causes (such as political reasons), because of the number of times such events have arisen in the history of Bangladesh and its predecessor states. There is also experience of the need for legal measures for relocation and settlement of refugees because of the movement of a large number of families to and from India since the mid-40s prompted by the partition of India in 1947 and subsequent wars. Moreover, displacement caused by "development" programmes and other physical interventions are also facts of life in Bangladesh.

Historically, Bangladesh has already accumulated the experience of tackling, both legally and institutionally, the type of consequences that may result from climate change, particularly a rise in sea level. An in-depth investigation in the future into the interpretation of relevant laws and institutional procedures would also reveal that property rights, interests and duties have evolved to deal with such unforeseen and contingent consequences. Nevertheless, some appreciation of the system of land tenure and the way that rights may be exercised over property is central to an understanding of the means by which Bangladesh may be able to accommodate and respond to environmental threats such as those associated with global warming.

Historical Legacy

Historically, the system of property rights in Bengal had been regulated by traditions and customs. With the advent of monarchs and princely states, the traditional rights continued with payment of revenues to the immediate rulers, although often different tiers of intermediaries appeared in between the raiyats (tillers of the soil) and the sovereign ruler. However, often huge properties were transferred to wealthy individuals (*Zamindars*) not simply in return for revenues, but for feudal services, under a system institutionalised by the 1793 Permanent Settlement Regulation.

The "estates" vested in the *Zamindars* were known as *Mahals*. The pattern, form and conditionalities of resource settlement with the farmers or *raiyats* were not uniform, and *Zamindars* developed their own system as they pleased. To regularise the tenancy, the Bengal Tenancy Act was introduced in 1885.

After the withdrawal of British rule in India in 1947 creating two independent nations of India and Pakistan, a broad law was enacted to abolish the *Zamindari* system. The East Bengal State Acquisition and Tenancy Act 1950 (SAT) came into force in 1951. It provided for the acquisition of all "rent-receiving" interests of the *Zamindars* in favour of the Government. All the settlers or *raiyat* who held property under the *Zamindars* were made *Maliks* or tenants directly under the State. The properties that were retained personally by the *Zamindars* under *khas* (owner personal) possession, and other properties not settled with *raiyats* or tenants, were acquired or were vested with the Government as *khas* property. At the same time, the SAT declared certain categories of properties as non-retainable under private ownership. As such these were vested with the state as *khas*, for example, land consisting of forests, village *hats* and *bazars*, ferry *ghats*, fisheries etc.

Status of Existing Property Rights

To understand fully the legal implications of property rights, it is initially important to understand, in general terms, the system of property rights in Bangladesh. As with most legal systems, this is the most complex aspect of Bangladesh law. The primary explanation for this complexity, which results in a great deal of litigation, is largely historical, although it has been further aggravated by demographic factors. The fertile land of the Ganges-Brahmaputra-Meghna delta attracted spontaneous settlement with migrants coming from various surrounding regions. Such migration not only made settlement unplanned but also over-crowded, thus creating an agrarian community living in fluvial conditions. The land tenure system is therefore based on private ownership of agricultural land; it reflects the difficulties of regular seasonal inundation, which ensures the fertility of the soil, but which also results in physical changes to the land under cultivation. These sorts of changes would be exacerbated by climate change.

Property rights in Bangladesh can be very broadly classified into three categories, namely: private property rights, traditional rights and public property.

Private property rights of various types may exist over virtually all resources except those declared as "non-retainable". The right to fishery has been declared "non-retainable", but not the rights over water bodies, such as ownership of a village channel or even a seasonal river or canal. A right of navigation is also retainable.

Although in the Record of Rights no property is recorded in the name of a specific community as "common property", common use is maintained of resources such as *hats, bazars* and ferry *ghats* which are leased out to *Ijaradars* by the Government who then levy tolls for "maintenance". However, properties may be recorded in the name of religious or charitable institutions as private properties.

A private owner exercises and enjoys all rights in or over the property as understood in the common law concept of ownership. Such an owner also stands responsible for every activity on his/her land that creates liability. Apart from ownership, various other private titles may exist over immovable properties with various degrees of rights, duties and interests, e.g. lease, mortgage, licence, tenancies or rights acquired through easements, or prescriptions or adverse possession.

Customary or traditional rights over resources are another important aspect. These are often recognised by law through easements, prescriptions or adverse possessions. Although under various laws these rights are recognised to some degree (e.g. by the SAT, the Easement Act, 1882) the legal status of traditional rights is still unclear.

Customary rights over alleged Government property which is, in practice, used in a locality as common property are an essential component of the right to enjoy resources or to have access to them. Typical examples would be: rights of way, rights of grazing, rights over water courses, rights to fish in public waterways. Although title is generally regulated by law, the way that resources are actually used depends largely on customary practice. These traditions reflect the social and geophysical characteristics of each area and vary throughout the country. It may well be that customary practices, as dynamic norms of human behaviour, would become even more significant when statutory systems are put under stress by environmental change. There are traditional institutions such as *salish* which act as informal judicial bodies for adjudicating disputes arising particularly from traditional rights and customs. For many purposes the role of *salish* is more crucial than the formal courts and may acquire even greater significance in times of environmental stress.

With respect to public property, Article 13 of the Constitution of Bangladesh recognises, inter alia,

a) state ownership, that is ownership by the state on behalf of the people through the creation of an efficient and dynamic nationalised public sector embracing the key sectors of the economy; and

b) co-operative ownership: that is ownership by co-operatives on behalf of their members within such limits as may be prescribed by law.

Article 143(i) declares that, in addition to all other land or property being lawfully vested with the state, the following shall also be regarded as vested with the state.

a) all minerals and other things of value underlying any land of Bangladesh;

b) all lands, minerals and other things of value underlying the ocean within the territorial waters, or the ocean over the continental shelf of Bangladesh; and

c) any property located in Bangladesh that has no rightful owner.

Under the SAT, extensive properties including land, fisheries and forestry were vested as public property. The "non-retainable" category of property is all public property and anything that has no rightful owner is public.

Legal Protection of Property Rights

Laws have developed not only to protect the individual's physical well-being, but also to protect property from damage, destruction and other harmful acts. In Bangladesh, the right to property is a fundamental right. Clauses (1) and (2) of Article 42 of the Constitution of the People's Republic of Bangladesh states

1. Subject to any restrictions imposed by law, every citizen shall have the right to acquire, hold, transfer or otherwise dispose of property, and no property shall be compulsorily acquired, nationalised or requisitioned save by authority of law.

2. A law made under clause (1) shall provide for the acquisition, nationalization or requisition with compensation and shall either fix the amount of compensation or specify the principles on which, and the manner in which, the compensation is to be assessed and paid; but no such law shall be called in question in any court on the ground that any provision in respect of such compensation is not adequate.

Public property in its simplest form means anything owned by the Government. It means any premises belonging to, or taken on lease or requisitioned by, the competent authority under the law. The Government may acquire lands through acquisition and requisition wherever it feels necessary to acquire lands for the greater interest of the public (as under the Acquisition and Requisition of Immovable Property Ordinance 1982). It may become owner of the property abandoned by a private owner due to some political or social crisis. Examples can be found in the Law of Enemy Property passed after the War of 1965 and the Law of Abandoned Property passed after the Bangladesh War of Independence of 1971. Government may also acquire lands as excess lands by enforcement of a certain ceiling on individual ownership.

Administration of Government Lands

Newly acquired *khas* properties of the State come immediately under the authority of the Ministry of Land. It may be mentioned here that, under the prevailing law, the legal definition of other resources like water bodies or rivers, forestry, etc. are lacking. The legal definition of "land" includes land that is cultivated, uncultivated or covered with water at any time of the year, and includes benefits to areas of land, houses or buildings and also things attached to the earth, or permanently fastened to anything attached to the earth. This definition brings almost all the public land resources under the authority of the Ministry of Land, including fisheries, forests and water bodies. Often, appropriate agencies are deprived of their rights to develop management programmes unless the relevant land is transferred to them by the Ministry of Land. Another consequence of this definition is that, even after significant inundation, the rights of the land holder will not be extinguished, because the definition of land includes land "covered with water". Therefore inundated areas, however large, will not necessarily be classified as bodies of open water (and thus non-retainable) and will not vest in the state.

Ministry of Land operates through local-level revenue officers or Deputy Commissioners. It is the duty of the Ministry of Land to transfer specific resources to other relevant Ministries or Departments through notification in the Gazette. Although some of these resources have been allocated to appropriate agencies for management, many of these resources are still retained by the Ministry of Land. These amount to literally thousands of fisheries in both open and closed waters, and many thousands of hectares of forest land (often called Unclassed State Forest) which are still vested with the Ministry of Land. As a result, the local revenue officers (mostly Deputy Commissioners) often settle these lands or lease them out to various parties on widely different terms and conditions.

Rights Over Property

Fisheries, other than tanks owned by private persons, belong to the state, and the principles of management and settlement of these fisheries, were set out in the Government Estate Manual (1958), which subsequently was replaced by the Land Management Manual (1990). Management of fresh water fisheries is the responsibility of the Ministry of Fisheries and Livestock; these fisheries are leased out by Government for various terms from one to three years depending on type.

Changes in patterns of inundation and flooding seem likely to change the nature of individual fisheries. Given the short term nature of fishery leases these changes could be accommodated within the existing system if this were deemed appropriate.

With regard to Navigation, any river through which riverine vessels run throughout the year is considered to be a navigable river. All such rivers, connecting channels and river beds are held to be the property of the Government (Rule 333 of the Land Management Manual, 1990). The owners of the river lands shall not have any rights over the river beds or the foreshores (Rule 337), nor can any person or organisation remove sand or stones from the river beds without prior permission from the Government by taking lease (Rule 338).

BOX 6.1: Environmental change and property rights

> In the absence of any policies for human settlement and equity in resource allocation, the competition for acquiring, protecting and exercising titles over resources is the active endeavour and struggle for all classes in Bangladesh society. Flooding or permanent inundation have long affected the exercise of rights over property in Bangladesh. The abundance of experience and traditional wisdom would help the people, at least psychologically, to accommodate the immediate impacts of climate change and sea-level rise. Customary practices, as dynamic norms of human behaviour, would become even more significant when statutory systems are put under stress by environmental change.

6.2.2 Existing Law Relating to Displacement

If environmental changes bring about physical changes in the land form of the country or force movements in settlement patterns, large numbers of people are likely to be displaced — rendered homeless or deprived of their means of livelihood. These people will need to be resettled — a particular problem in a country which is already overcrowded. Although the scale may eventually be greater than anything previously experienced, Bangladesh has had to deal with similar problems in the past as a result of political upheaval and as a result of physical changes brought about by seasonal inundation and flooding. This section considers the way in which the existing law deals with issues such as rights over newly created (alluvion) or flooded land (diluvion), rights over abandoned property and previous experiences of resettlement of substantial groups. Future responses will be shaped by the structure of the existing system as well as by previous experience. If new policy approaches prove necessary the constraints as well as the strengths and weaknesses of the existing system will need to be fully appreciated.

There are no statutory obligations on Government to render assistance to the displaced due to any natural calamity (except in the situations discussed below) and the Constitution does not define the term "migration". There are certainly no constitutional guarantees for those who might have to be relocated due to inundation whether temporary or permanent. What the Constitution does, however, assure is the protection of the right to life and personal liberty (Article 32: "No person shall be deprived of life or personal liberty save in accordance with law"). It also assures freedom of movement in Article 36:

> subject to any reasonable restrictions imposed by law in the public interest, every citizen shall have the right to move freely throughout Bangladesh, to reside and settle in any place therein and to leave and re-enter Bangladesh.

This provision of the Constitution appears to prevent "compulsory" resettlement.

Section 54 of the Land Management Manual (1990) establishes the priority in the distribution of Government *khas* lands. It gives the highest priority to the families of farmers whose land has been washed away or eroded by rivers. According to Section 56 of the Manual, the first in the category of distribution of Government *khas* lands are those agricultural families whose land had been under cultivation. This provision would suggest that those in occupation of cultivated land rendered uninhabitable as a result of rising sea levels would fall into such a category and would get first preference to acquire Government *khas* lands.

Alluvion and Diluvion Law

Due to the frequent changes which take place in the channels of the rivers in Bangladesh, *chars* or small islands are often thrown up by alluvion in the middle of a stream or near the banks, and large portions of land are carried away by the encroachment of rivers on one side. Similar instances of alluvion, encroachment and dereliction also sometimes occur on the sea coast of Bangladesh. The Government has laid down rules for determination of claims and disputes relating to such alluvial lands.

Where land is lost (diluvion) by the same processes, property will be lost if the land has been deemed to be abandoned. The SAT Act (1950), (Section 86 — as amended by Act No.15 of 1994) made provision for the right to repossess immediately the land which re-emerged within 30 years, on payment of arrears of rent on the property on a fair and equitable basis but free of *salami*.

Status of Abandoned Property

The term "abandon" according to the Standard English Dictionary means "to desert or to give up all claims to". The act of abandonment therefore implies two factors namely, (i) desertion of the property and (ii) giving up one's right to the property. The word abandonment connotes the idea of the owner vacating the property with the intention of never returning to it. Mere temporary absence or absence in certain circumstances is not sufficient to warrant the inference of abandonment. To establish abandonment, some more positive act than mere non-residence has to be proved. The desertion of the property must have coincided with the positive intention to give up the rights vested in it. The owner of the property may not always be in possession of it and thus mere absence of the physical possession does not suffice to allow the property to be treated as abandoned (Farooqui, 1975).

The Bangladesh Abandoned Property (Control, Management and Disposal) Order, 1972 P.O. 16, section 2(1) runs as follows:

> In this order, unless there is anything [to the contrary] in the subject or content, "abandoned property" means any property owned by any person who is not present in Bangladesh or whose whereabouts are not known or who has ceased to occupy, supervise or manage in person his property ...

Therefore, under this definition a man may lose his property, even if he is in the country, if he is not "in possession of his property". This goes further than simple abandonment but obviously such a case will fall within this definition. If, for example, property became unfit for habitation due to flooding or sea-level rise, it would be deemed to have been abandoned.

Displacement and Resettlement

Displacement of people due to environmental degradation is prevalent in Bangladesh and encourages internal and transboundary migration of people. The causes of such environmental migration have been both natural and man made. Migration because of soil degradation is frequent. However, no clear data are available on the extent and magnitude of environmentally induced-migration in the country.

A large number of families have been displaced by the introduction of development projects. These displacements occur primarily for the following reasons:

a) acquisition of land for structural projects, especially in cases where a family does not retain enough land to sustain itself;

b) loss of professional opportunities as a result of the project (e.g. fishermen may be induced to migrate if a project gradually destroys the fishery of the area);

c) changed environment either within an area intended to benefit or outside the areas of the project (e.g. the execution of many flood control projects has created water-logging inside, or river bank erosion and heavy siltation outside, the project area leading to displacement).

Most of the sectoral laws governing development activities lay down conditions for compensation for land and other interests. For example, section 28 of the Embankment and Drainage Act 1952, states that "whenever any land, or any right of fishery, right of drainage, right to the use of water or other right of property shall have been injuriously affected by any act done or work executed under this Act, the person in whom such property or right is vested may claim compensation." Similar provisions can be found in other laws, like section 23 of the Agricultural and Sanitary Improvement Act, 1920.

Although provisions for compensation are incorporated in many laws, the institutional practice has been quite different. Compensation for consequential damage has hardly ever been paid to anyone.

Compensation and relocation for families displaced by environmental degradation has not yet developed as a matter of legal right. Nor has any law yet been enacted to ensure relocation if compensation is not paid in such cases. The Government policy of settling public *khas* lands on landless people through leases is a process that can be related to relocation.

The effort to relocate displaced people was given a legal mandate after the partition of India when Muslims from various parts migrated to the then Pakistan, including East Pakistan. The Displaced Persons (Land Settlement) Act and the Displaced Persons (Compensation and Rehabilitation) Act were adopted in 1958 to compensate persons for the losses suffered by them on account of expropriation by the Government of India of their rights in property in India or in any area occupied by India. Another example of rehabilitation was that done through the enactment of the Disturbed Persons (Rehabilitation) Ordinance, 1964, for those persons affected by the civil disturbances in the then East Pakistan in January 1964 and for the protection of immovable property of minority communities.

In the past, the Government of Bangladesh has undertaken several initiatives to resettle landless families. A recent example of resettlement of landless families took place in 1988 through the creation of a large number of "cluster villages" on Government lands. These villages were created with multi-dimensional objectives. The guidelines for effective organisation of cluster villages emphasise the optimum use of local resources and materials for ensuring effective utilisation of land and maintenance of hygiene and sanitary conditions.

There is also some experience in Bangladesh of formulating a policy to deal with the situations where natural disasters have forced people to leave or dispose of their land for reasons beyond their control. The question of property rights over such land was given special treatment in 1976 after the famine of 1974 had forced villagers to sell land at below the market price simply to buy food for their starving families. In 1976, the Alienation of Land (distressed Circumstances) (Restoration) Ordinance was passed permitting the repurchase of agricultural land sold for reasons of economic distress during the famine, at the price for which it had been transferred.

Current Provisions for Rehabilitation

The Government has awarded priority to landless people in the 1984 Land Reform Ordinance under which the rehabilitation of the landless in *khas* Government land yet to be utilised was declared to be one of the primary objectives of the Government.

One of the objectives of imposing an upward limit on land ownership was to enable the distribution of excess lands to landless and near landless families for cultivation. However, the priorities and policies regarding settlement of the excess lands frequently change.

Section 76 of the SAT provides that any land vested in the Government under this Act shall be absolutely at the disposal of the Government and that the Government shall be competent to make settlement of such land in accordance with rules made in this behalf or otherwise deal with such land in such manner as it sees fit.

BOX 6.2: Displacement and resettlement

> Compensation and relocation for families displaced by environmental degradation has not yet developed as a matter of legal right. Nor has any law yet been enacted to ensure relocation if compensation is not paid in such cases. The government policy of setting public *khas* lands on landless people through leases is a process that could be related to relocation. A precautionary, strategic public land settlement policy is needed to meet the relocation needs of those displaced through environmental changes.

It is arguable whether the Constitutional guarantees can be interpreted as an obligation on Government to protect both life and property from natural incidents such as those which might result from climate change. It is therefore difficult to argue that a strict legal duty exists to protect property and to resettle those made homeless by the effects of sea-level rise. However, the fact that such as policy has evolved in response to the problems of regular seasonal flooding and land movements indicates that basic mechanisms for such a response are already in place.

6.2.3 Legal Regulations Relevant to Climate Change

Bangladesh is a party to a number of international conventions responding to environmental and climatic change but has not yet ratified all of them. For those Conventions that it has not yet ratified, the provisions are not opposable to Bangladesh except in so far as the rules they contain have passed into customary international law. The Conventions to which Bangladesh is a signatory are the 1951 International Plant Protection Convention (ratified 01-09-78); 1972 Convention concerning the Protection of the World Cultural and Natural Heritage (ratified 03-11-83); 1985 Vienna Convention for the Protection of the Ozone Layer (signed 31-05-90; ratified 02-08-90); 1987 Montreal Protocol on Substances that Deplete the Ozone Layer (signed 31-05-90; ratified 02-08-90)); 1992 United Nations Framework Convention on Climate Change (signed at the Rio UNCED Conference, June 1992; ratified 15-04-94).

There are currently no legal controls of emissions of greenhouse gases in Bangladesh as no statutory environmental standards have yet been established. But the Environmental Protection Act, 1995 has set up environmental quality standards which include the control of emissions. Indeed, the extent of current emissions of such gases is uncertain.

National laws relevant for control of air quality are as follows:

1. *The Environment Pollution Control Ordinance, 1977:* This ordinance empowered the Department of Environment to require any person or commercial or industrial undertaking to adopt such measures as may be necessary for the prevention, control and abatement of existing or potential pollution of the environment. It further authorises the Department to inspect and test any wastes, air, water, etc.

2. *The Brick Burning (Control) Act, 1989* prohibits the use of firewood for burning bricks.

3. *The Factories Act, 1965* requires that every factory be kept clean and free from effluvia or other nuisance. Effective arrangements have to be made for disposal of wastes and effluents due to the manufacturing process.

4. *Motor Vehicles Ordinance, 1983*. Section 150 provides that: "Whoever drives or causes or allows or lets out a motor vehicle for use in any public place, the smoke of which would constitute a health hazard, shall be punishable."

5. *The Smoke Nuisance Act, 1905*. Section 6 provides:

   "The Government may, by notification in the Official Gazette, prohibit, within any specified area,

   a) the erection or use of any specified class of brick tile or lime-kilns or dumps for making bricks, or

   b) the erection or use of furnaces to be used for the calcining or smelting of ores or minerals, or for the casting, puddling or rolling of iron or other metals, or for the conversion of pig-iron into wrought-iron, or

   c) the manufacture of coke, in ovens, or with special appliances, or

   d) the making the coke without ovens or special appliances. Section 8 of the same act makes it an offence for smoke to "be emitted from any furnace in greater density, or at a lower altitude, or for a longer time, than is permitted by rules made under this Act."

6. *The Environmental Protection Act, 1995*. The Government has recently enacted the Environmental Protection Act, 1995. The law has set up environmental quality standards as well as guidelines for the Ministry of Environment. It has also made environmental assessment mandatory in each and every instance.

Most of the legal provisions listed above are not being enforced in Bangladesh due to various legislative and institutional shortcomings and incompatibilities. As stated earlier, the absence of statutory environmental quality standards deters their effective enforcement. Moreover, evidential standards of proof to establish legal liability for such offences have never been clarified.

BOX 6.3: Greenhouse gas emissions

> There are currently no legal controls of emissions of greenhouse gases in Bangladesh as no statutory environmental standards have yet been established. The extent of current emissions of such gases is uncertain.

6.2.4 Institutional Framework

The Intergovernmental Panel on Climate Change (IPCC) has stressed the importance of an integrated approach to coastal resource management as part of an effective response to possible climate change and sea-level rise. A first step developing such an integrated approach is to determine the basic existing organisational structure. The following section will therefore consider the way in which the component sectors of all the major national resources of relevance are organised.

Land

The Ministry of Land has overall control of government lands and the administration of all land and land-based resources. It is responsible for the lease/disposal of government lands and for transfer of land to other appropriate agencies for relevant purposes and projects. Unfortunately, this has not always been done. For example, forest areas have not been transferred to the Forestry Department for conservation and management.

The Department of Land Record and Land Survey prepares and maintains the Record of Rights which is not yet a definitive record of registered interests in land. For reasons explained above, ambiguities of title still persist. The local government authorities also have control over their own resources.

Water Resources

The Ministry of Irrigation, Water Development, and Flood Control has overall responsibility for inland water resources. The Bangladesh Water Development Board is responsible for the construction of polders and embankments for flood defence and management.

There are also engineers employed by the Local Government Engineering Department working with local bodies to design and execute local projects.

Sewerage within urban areas is the responsibility of Municipal districts — *thana*. There is no formal rural sewage disposal system. Environmental health aspects of water management are the responsibility of the Department of Public Health in the Ministry of Local Government, Rural Development and Co-operatives.

There is also a massive programme for flood control — the Flood Action Plan initially for a period of five years (1990 to 1995). The Plan is coordinated by the World Bank and funded by international donors. The Flood Action Plan is administered by the Ministry of Irrigation, Water Development and Flood Control. The Government has also enacted the Water Resources Planning Act, 1992 (Act 12 of 1992) for management of the water resources in the country.

Agriculture

Agriculture is the responsibility of the Ministry of Agriculture. The Bangladesh Agricultural Development Corporation has historically been responsible for arranging agricultural input supplies (fertilisers, seeds, irrigation equipment, etc.) but its functions have largely been handed over to the private sector in recent years. Agricultural extension services and technical assistance is the responsibility of the Department of Agricultural Extension (DAE), but more specialised aspects are covered by other specialist institutions (e.g. BWDB, cotton, tobacco, sugar cane and tea agencies). There are legal controls of pesticide manufacture and import but they are widely flouted, notably by unofficial cross-border trade.

The Bangladesh Agricultural Bank and some other Banks provide mortgages for agricultural land and loans for agricultural purposes.

Fisheries

This comes under the responsibility of the Ministry of Fisheries and Livestock, although some ten thousand closed fisheries are administered by the Ministry of Land. Marine fisheries are governed by the 1983 Marine Fisheries Ordinance which established a sea fishery licensing system for motorised vessels. The Ordinance is technically enforced by the Fisheries Department, although in practice this work is done by the Bangladesh Navy.

Navigation

Navigation is the responsibility of the Ministry of Port Shipping and Inland Water Transport (IWT). Inland navigation is regulated by the Inland Water Transport Authority. Maritime navigation is the responsibility of the Bangladesh Shipping Department. The Navy performs coast guard and shipping enforcement functions. There are two Port Authorities, at Chittagong and Mongla. The Bangladesh Shipping Department is also responsible for marine pollution control in Bangladesh waters.

Forests
The Ministry of Environment and Forest has overall control of such forest areas as have been transferred to it by the Ministry of Land. Forests are managed at District Commissioner level. The Forestry Department within the Ministry is responsible for general conservation and management of forests. It is responsible for:

- the coastal afforestation programme and the afforestation of embankments.

- protected and reserved forests including the Sundarban mangrove forest.

- designation of National Parks and Wildlife Sanctuaries under the 1974 Wildlife (Preservation) (Amendment) Act.

Some four parks and a number of sanctuaries have been designated. No effective management regime is yet in place for the administration of estuarine reserved forests.

Offshore Energy Exploration and Exploitation
This is under the Control of the Ministry of Energy which administers the continental shelf resources under the 1974 Act.

Environment and Pollution Control
The Ministry of Environment and Forest is responsible for the environment and pollution control. It executes its programmes through the Department of the Environment, operating under the Environmental Pollution Control Ordinance, 1977. However, other agencies are also empowered to control pollution through their respective laws, e.g. Department of Shipping (above), Road Transport Authority, Ministry of Industries, etc.

Relief and Rehabilitation
There is no permanent mechanism within the Government structure to deal with disaster response and relief. The Department of Meteorology issues forecasts, but these often use technical terms and scales of severity which are not well understood in the population at large. The Ministry of Relief and Rehabilitation is responsible for disaster response and relief. In the past it has not dealt specifically with disaster preparedness. It is, however, the government agency responsible for the cyclone shelter programmes, for the promulgation of cyclone warnings (handled through the Bangladesh Red Crescent Society) and the for the publication of Standing Orders for Flood, Cyclones and Relief — a codification of procedures for all

government agencies, including the Armed Forces, and Non Government Organisations (NGOs). Under the Flood Action Plan a Disaster Management Co-ordination Unit has recently been set up in the Ministry of Relief and Rehabilitation. Recommendations for the expansion and strengthening of this Unit as a permanent organisation are currently under discussion, as are arrangements for more effective inter-Ministerial co-ordination for disaster preparedness and response.

6.3 INTERNATIONAL LAW ASPECTS

International law governs the relations between states, including internal actions that can have an impact on other states. It can therefore provide either a constraint on policy options or an enabling mechanism for co-operative activities.

A large variety of international law issues come into play in relation to possible climate change in Bangladesh. Changes in temperature, rainfall, river flows, and cyclone frequency and severity seem likely to affect the present patterns of monsoon inundation and flooding problems. Changes in land and coastal form from sea-level rise are quite possible (see Chapter 7) and would undoubtedly precipitate further problems of population displacement and migration.

In general terms, sea-level rise could affect the horizontal configuration of the coast line, principally through erosion (indeed, some areas of the Bangladesh coast already have very high erosion rates). In normal circumstances, this would affect the delineation of baselines from which maritime zones (territorial sea, contiguous zone, exclusive economic zone, and continental shelf) are measured (the particular position of the Bangladesh baseline is considered after the general principles are discussed below). Movement of the coast and the erosion or deposition of islands may also affect the delimitation of putative maritime boundaries with neighbouring states, although boundaries settled by treaty will remain as agreed.

International law may also require co-operation from neighbours in certain situations or impose responsibility for failure to meet international obligations. It imposes a clear obligation on states not to cause direct harm to other states, but the situation is less clear in relation to indirect harm or to actions by neighbours which exacerbate the effects of harm caused by natural forces.

In relation to shared natural resources, such as rights to water (or other resources) from international water courses, case law makes it clear that the basic principle is that of equity. States through which an international river passes may use its renewable resources, but should treat it in an equitable and reasonable way. These principles will be relevant in considering the situation of damming or otherwise changing international water courses. Preventing water (or possibly even sediment) from getting through to the lower reaches may adversely affect land usage and habitation downstream but, similarly, increasing water or sediment flows may also have detrimental effects downstream.

A growing network of global and regional treaties protects critical ecosystems and wildlife habitats. Protection of forest areas and of marine ecosystems, such as coral reefs and mangroves, which may provide natural buffers and thus protection from cyclone impacts and coastal erosion, is an important response strategy for which international treaty law may provide constraints and opportunities.

International law does not currently possess a regime for dealing with populations forced to leave their native country for environmental reasons. This is an issue that may need to be addressed in the near future in relation to desertification and in the face of sea-level rise for low-lying islands as well as for coastal areas (Freestone, 1991b).

6.3.1 Maritime Zones

Article 143 of the Bangladesh Constitution vests in the Republic all lands, minerals and other things of value under the oceans within territorial waters or on the continental shelf. By the 1974 Territorial Waters and Maritime Zones Act, Bangladesh claimed a territorial sea of 12 nautical miles (nm), a contiguous zone of 6nm from the outer limit of the territorial sea and a continental shelf to the outer limit of the continental margin, as well as an Exclusive Economic Zone (EEZ) extending to 200 nautical miles from the baseline. Indeed, Bangladesh was the first state to claim such a zone. Bangladesh is a signatory to the Final Act of the Third United Nations Conference on the Law of the Sea (UNCLOS III) and to the 1982 Law of the Sea Convention (LOSC), which means that it accepts the principles contained in the Convention and intends to ratify it at some future date.

Coastal Baseline Delineation

Articles 3 to 16 of the 1982 Law of the Sea Convention set out the regime for drawing baselines from which maritime zones can be measured. Article 5, which is taken to represent customary international law, provides that:

> ... the normal base line for measuring the breadth of the territorial sea is the low water line along the coast as marked on large scale charts officially recognised by the coastal state.

Although in certain specified circumstances[1] the LOSC and customary international law permit coastal states to depart from the low water line and use straight baselines joining headlands, islands, rocks and other features (discussed below), the low water line remains the basic and most important baseline for measuring the breadth of all maritime zones[2]. The use of the low water line generally ensures that the maximum maritime area is included within the zones.

In broad terms, rises — or indeed any changes — in sea level which affect high and low water lines will obviously have a "knock-on" effect on the measurement of maritime jurisdictional zones. If the low water line shifts landward, the outward limits of maritime zones will similarly move landward and the coastal state's maritime zone limits will shrink proportionately, because all maritime zones are measured from that baseline. Where a broad coastline regresses rapidly (as in the case of areas of Bangladesh) the cumulative effect could quite substantial.

It was in recognition of this problem that, on 13 April 1974, the Bangladesh Government issued a Declaration announcing a baseline that consists of seven straight lines extending for 221 nautical miles (see Figure 6.1). None of the eight points that define these lines is located on land and, except for the sector immediately west of Cox's Bazar (where there is a steep landward indentation of the isobath), the lines are closely aligned with the 10 fathom isobath. These baselines, drawn according to what Habibur Rahman (1991) calls "the depth-method", were developed, he says, to take account of "the geological peculiarities of the coast and the peculiar topographical features of the coastal bay". On the 30 April 1974 the Governments of both India and of Burma (as it then was) protested these baselines. It might be noted in passing that Prescott (1985) considers the Burmese/Myanmar baseline, promulgated in 1968 and amended in 1977 to be contrary to both the "letter and spirit of the new Convention."

During the course of the Law of the Sea Conference, the Bangladesh delegation made a number of proposals — with the offshore waters of the Ganges delta clearly in mind — in an attempt to secure provisions within the text of the Convention addressing the particular problems of low water lines which move land-ward and seaward in unpredictable ways. The history of these proposals has been set out in detail elsewhere (Prescott and Bird 1989). Art 7(2) LOSC provides that:

> Where because of the presence of a delta and other natural conditions the coastline is highly unstable, the appropriate points (i.e. for straight baselines) may be selected along the furthest seaward extent of the low-water line and, notwithstanding the subsequent regression (sic) of the low-water

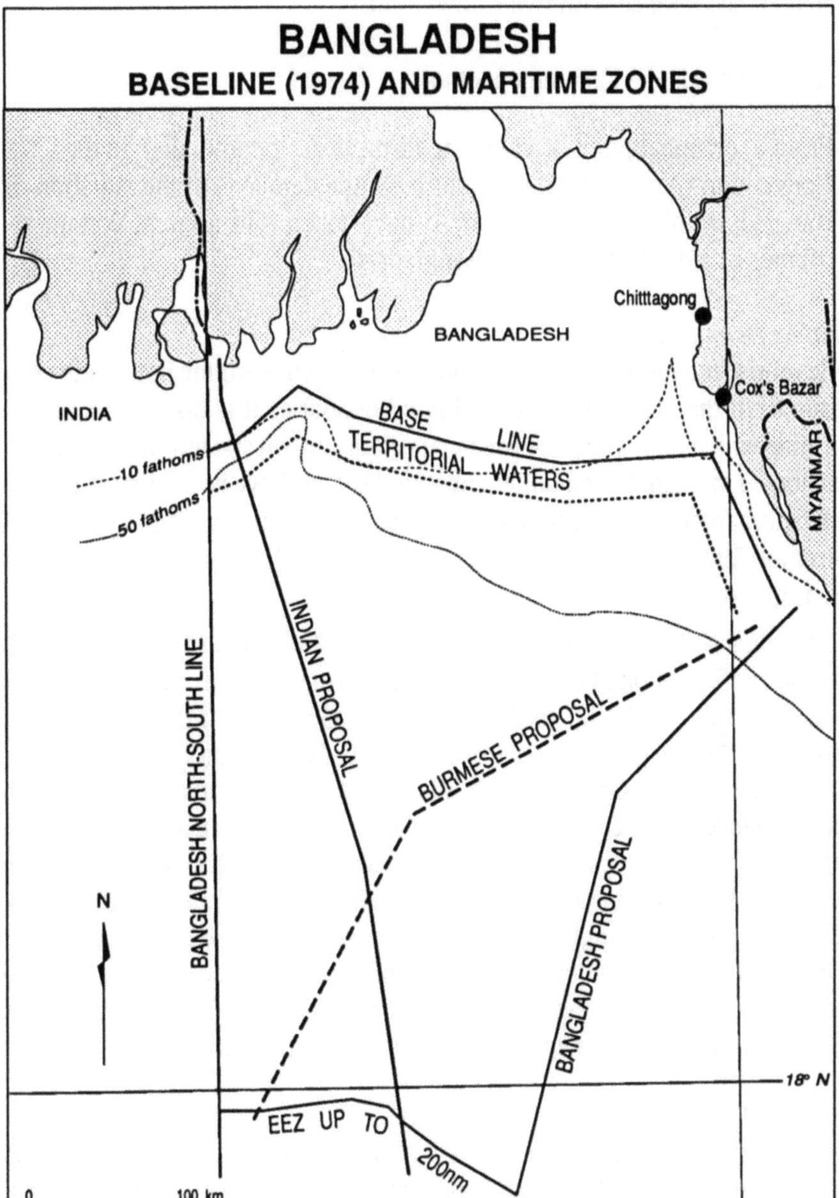

FIGURE 6.1: Map showing Bangladesh baseline of 1974 and maritime zones measured from it. Also shown are outstanding proposals for maritime boundary delimitation lines with adjacent states made by Bangladesh and its neighbours.

> line, the straight baseline shall remain effective until changed by the coastal state in accordance with this Convention.

Habibur Rahman (1991) comments that :

> From a practical point of view, the formula as incorporated in the LOS Convention (Article 7(2)) does not meet the demands of the situation of Bangladesh, although it appears to have taken into account certain attributes of a delta country like Bangladesh.

Maritime Boundaries

Sea-level rise may affect the base points from which maritime boundaries may be measured. In the negotiation of maritime boundaries it is open to the parties to determine which points shall be used in the delimitation and how much weight will be attributed to them. For example, inhabited islands have been given half effect in cases before international tribunals, (such as the Anglo-French Continental Shelf Arbitration 1979 and the Tunisia/Libya Case of 1982) and ignored completely in some treaty settlements.

Bangladesh has maritime zones adjacent to India and Myanmar (see Figure 6.2). The configuration of the Bay of Bengal and the position of Bangladesh within it is not entirely dissimilar to the position of Germany in the 1969 North Sea Continental Shelf cases. Hence, in a boundary delimitation agreement, it would expect to receive (like Germany in the North Sea) a larger zone than that calculated by equidistance alone. No maritime boundaries have yet been agreed with either of Bangladesh's neighbours, India or Myanmar. Both neighbours have refused to accept the 1974 baseline declaration. Therefore, despite the fact that the 1974 Declaration protects Bangladesh from the erosion of its baselines and thus of its maritime zones it is not axiomatic that neighbours would accept those baselines in bilateral talks. Hence de facto erosion and regression of the horizontal baseline might adversely affect its negotiating position in future talks.

The Case of South Talpatty Island

In 1971 a new island in the shape of a "U" was reported as having formed off the Ganges River delta. In 1978 this island, called South Talpatty in Bangladesh and New Moore or Purshaba in India, was reported to have an area of some three square nautical miles and to be growing. This geographical feature, is some five nautical miles offshore from the mouth of the Harinbhanga River, which marks the boundary between India and Bangladesh[3]. India is reported to claim the feature on the basis that the flow of the Harinbhanga River is to the east and that the area is

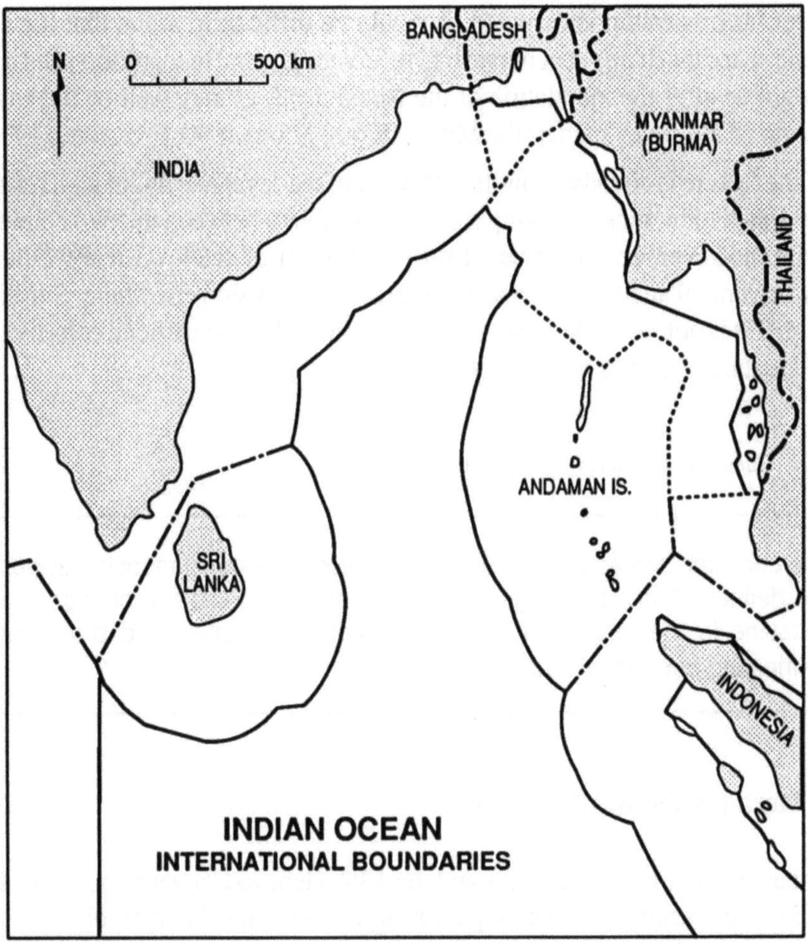

FIGURE 6.2: Map showing international boundaries in Indian Ocean

Source: Prescott, 1985.

within the natural prolongation of India. The Bangladesh view is that the river flows to the west of Talpatty and that it is not possible to distinguish the natural prolongation of India from that of Bangladesh in the Ganges Delta. As Talpatty is within the territorial sea/internal waters of either state (pending a determination of ownership) it would serve to extend the territorial sea of the state which established ownership. Obviously, its effect on outward limits of zones would be small but its potential effect in a maritime boundary negotiation could be disproportionate. However, as ownership is disputed, it would be difficult to argue that the extra maritime space that it might generate had received "general recognition" and could thus continue after the inundation of the island itself, as may well be the case in relation to island states such as the Maldives (see Soons 1990, Freestone, 1991b).

If the feature is already a low-tide elevation, sea-level rise might well result in its total inundation, thus removing a political problem between the two countries. It should, however, be remembered that the impact of sea-level rise would not simply be vertical and features may be nourished by erosion elsewhere. Similarly, although this is not settled law, Soons takes the view that the artificial conservation of such features "should be permissible" (Soons, 1990).

BOX 6.4: Climate change and maritime zones

> The coastline serves as the basis from which all national maritime zones are defined. Climate change and sea-level rise threaten to change the coastline in the future, causing jurisdictional difficulties and problems delineating borders with neighbours.

6.3.2 Marine Scientific Research

Under customary international law and the 1958 Geneva Conventions on the Law of the Sea, permission has been required for marine scientific research (MSR) by foreign vessels within the territorial sea and on the continental shelf. Research has suggested (Wooster, 1981) that prior to the Third UN Conference on the Law of the Sea (UNCLOS III), a number of pure research projects in maritime zones had been jeopardised, or even abandoned, as a result of the difficulties of obtaining requisite consents for projects in the territorial sea of other states, particularly in the zones of developing countries. The difficulties ranged from a complete failure to respond to requests, to permission being granted at too late a stage for the planned project to go ahead. It seemed that such difficulties and delays arose most commonly not for political reasons, but for purely bureaucratic reasons. It was also clear that research organisations avoided including within their ongoing research plans the waters of those countries with which they had encountered difficulties or, at an early planning stage, they excluded those states which had not replied to their requests for permission.

It was also clear at UNCLOS III that with the inclusion of control over MSR within the newly developed 200 nautical mile Exclusive Economic Zone, these problems were likely to continue and increase. The proposal was therefore made that, in the EEZ, the consent regime should be reversed so that a bona fide "pure" research project application made in proper form would be deemed to be accepted after a six month period if no reply had been received.

The 1982 Convention is not yet in force. Bangladesh has signed but not ratified the Convention; therefore it is not binding upon it. The 1974 Act — enacted before the finalisation of the text of the Convention — requires express consent for MSR projects in both the territorial sea and EEZ.

In the context of developing awareness of the problems of global climate and sea-level rise change, it is important that Bangladesh be included in any MSR projects seeking to collect data relevant to climate change. It is therefore important that the national legal regime — especially institutional practice and procedure — should be efficient and flexible enough to ensure that appropriate ocean-based international research projects can be carried out in Bangladesh waters.

There is considerable advantage to Bangladesh in having its waters included in such international research efforts.

6.3.3 Shared Riverine Resources

Shared riverine resources are of significance to Bangladesh as it lies in the delta of the Ganges, the Brahmaputra and the Meghna river systems, all of which rise in, and flow through, other states. Diversion of water by an upstream state may decrease agricultural output in areas consequently deprived of previous levels of water and supply of nutrients. The diversion may generate environmental hazards too. Other upstream changes, such as increases in volume and velocity of river flow, may also have deleterious effects by exacerbating problems of erosion and floodings.

In 1975, India commissioned the barrage at Farakka on the Ganges (18 km upstream from the Bangladesh border) to divert a larger part of its discharge in order to increase the flow of water through the port of Calcutta, which was silting up. The allocation of waters between the two countries resulting from this project was the subject of a detailed Agreement on Sharing of the Ganges' Waters between the two Governments. This agreement, which came into force on 5 November 1977, was for five years and renewable. The Agreement (without the "guarantee clause"4 of the 1977 agreement) was extended twice by Memorandum of Understanding for two years (signed in October, 1982 for the next two dry seasons and expired in 1984) and then three years (signed in November, 1985 for the next three dry seasons and expired in 1988). Since then, the alleged resultant water shortages in western areas of Bangladesh have been the source of considerable dispute between the two countries and are currently the subject of discussions between the them.

In a number of cases, notably the Diversion of Water from the Meuse case [1937] and the Lake Lanoux Arbitration [1957], international tribunals have held that there would a breach of international law if a joint watercourse were used by one riparian user in a way that is detrimental to the rights of another riparian state. In 1966 the International Law Association (ILA) adopted the Helsinki Rules on the Uses of International Rivers as a statement of then current customary international law. These rules propounded what they termed the "principle of equitable utilisation of the waters of an international drainage basin." Article IX defined "water pollution" as "any detrimental change resulting from human conduct in the natural composition, content or quality of the waters". States are in breach of international law if they increase existing water pollution, or introduce new forms of pollution into shared water resources so as to cause "substantial injury to a co-basin state"

The principles developed by the ILA are now contained in the new Helsinki Convention on the Protection and Use of Transboundary Watercourses and Lakes, signed in Helsinki, 17 March 1992. This imposes an obligation on parties:

> To ensure that transboundary waters are used in a reasonable way, taking into particular account their transboundary character, in the case of activities which cause or are likely to cause transboundary impact" (art. 2(2)(c)).

"Transboundary Impacts" are widely defined and would cover virtually any change in the conditions of transboundary waters (Art. 1 (2)). Although this Convention is not yet in force, international environmental lawyers widely regard it as a statement of existing customary law.

The 1991 Convention on Environmental Impact Assessment in a Transboundary Context, signed in Espoo (Finland), requires parties to conduct environmental impact assessments (EIAs) on projects likely to effect neighbours and to notify them of the results in Prior Consultation Procedures. This Convention, while not in force either, is also regarded as existing "good practice".

6.3.4 Legal Protection of Ecosystems

A growing network of global and regional treaties address the problem of the conservation of ecosystems. These regimes were initially developed for the protection of wildlife, but modern conservation thinking is directed increasingly towards the protection of habitats and critical ecosystems. Bangladesh has, since 20 November 1981, been party to the 1973 Washington Convention on the International Trade in Endangered Species of World Fauna and Flora (CITES), but does not as yet appear to have become party to all the key habitat protection conventions. In March 1992 Bangladesh became a party to the 1971 Ramsar Convention on Wetlands of International Importance; the Sundarban mangrove forest has been designated as a protected site under that convention. Bangladesh has also been a party (since 3 November 1983) to the 1972 UNESCO Convention concerning the World Cultural and Natural Heritage. Designation of sites under these conventions imposes legal obligations to provide protection and management of them; but such designation also opens opportunities for international funding to assist state parties in this process.

If environmental changes brought about by global climate change result in the displacement of people, there is considerable risk that existing forest areas would suffer from unregulated settlement by landless refugees. In the context of sea-level rise, wetland ecosystems, particularly mangrove systems, can provide important natural buffers to assist coastlines to adjust naturally (see Chapter 7). Participation in the international treaty network protecting both terrestrial and marine ecosystems does impose obligations to provide effective management regimes for threatened ecosystems, but conservation of these systems may as well be an important aspect of a response strategy to global climate change. The treaty frameworks can provide important advantages in terms of financial assistance and management expertise.

6.4.5 State Responsibility and the Issue of Climate Change

International customary law imposes a clear obligation on states not to use their own territory in a way that will cause harm to their neighbours. This principle is established in the much quoted Trail Smelter Arbitration (1938 and 1941) and the 1949 Corfu Channel Case before the International Court of Justice. As discussed above, there is also increasing evidence that international law requires prior assessment and consultation before a state undertakes actions that may have environmental impacts on other states. Violation of any of these duties may give rise to the liability — or as international lawyers term it, responsibility — of the offending state under international law with a commensurate obligation to make reparation. In addition, the 1972 Stockholm Declaration Principle 21 (which many international lawyers now think represents customary international law) provides that states have:

> the responsibility to ensure that activities within their jurisdiction and control do not cause damage to the environment of other states or of areas beyond the limits of national jurisdiction.

Indeed this Principle was recently reiterated in the 1992 UNCED Rio Declaration (Principle 2).

Can it therefore be argued that, if a low-lying state such as Bangladesh suffers harm as a result of global warming and sea-level rise, that a claim could be mounted for compensation against those states which contributed the most (or overwhelmingly) to the emissions of greenhouse gases?

Such an argument was discussed within the context of the Intergovernmental Negotiating Committee and there is clearly a strong moral argument involved. However, in the current state of international law, it is difficult to sustain an argument of liability. Primarily, climate change has still to be definitely proven. The 1992 Climate Convention, like the 1985 Vienna Convention on the Ozone Layer, is based on precautionary action. Secondly, the harm suffered is indirect, in that emissions of greenhouse gases (the majority of historic emissions of which took place before the greenhouse effect was recognised) have altered the composition of the atmosphere (a shared resource) and it is this altered atmosphere which may cause global warming. However, what was proposed at the INC by the Association of Island States (AOSIS) was the conclusion of an insurance protocol, against sea-level rise. Under such a protocol the states most vulnerable to sea-level rise would be insured against damage to, or ultimately the loss of, their coastal resources or low-lying territory, the costs of the premiums being met by levies pro rata on the major greenhouse-emitting states in the developed world. The insurance industry apparently regard this as economically feasible, and the proposal was discussed at the IPCC Coastal Zone Management Subgroup and is still on the table for consideration in the forthcoming negotiations on possible protocols to the 1992 Convention, discussed below.

What the Rio meeting did accept, however, was the principle, which is contained within both the Rio Declaration and the 1992 UN Framework Convention on Climate Change, of the "common but differentiated responsibilities" of States. This expressly recognises that the developed world has played a much larger role in contributing to global environmental degradation by their contribution to the greenhouse effect. Thus, it imposes upon them a disproportionate obligation to respond, both by curbing their own emissions of greenhouse gases and by providing finance and technology to developing countries and "countries whose economies are in a state of transition" for assistance in reducing emissions and in responding to the problems caused by climate change. In this sense, the general principle of strict legal responsibility is avoided but a new legal mechanism is devised to recognise the moral responsibility of those involved. In the fullness of time, the details of the financial implications of this "differential responsibility" will have to be negotiated within the context of protocols to the Framework Convention. In this respect, the example of the way that the Montreal Protocol on Substances that Deplete the Ozone Layer (discussed below) has been put into effect is likely to be highly significant.

BOX 6.5: Climate change, state responsibility and international cooperation

> Many environmental problems will be felt across national borders. Treaties and customary law impose obligations to co-operate with neighbours in facing common hazards and to avoid uses of shared resources that are detrimental to neighbours — as in the case of shared riverine resources.

6.3.6 The "Atmosphere Conventions"

The Vienna Convention for the Protection of the Ozone Layer

The Vienna Convention for the Protection of the Ozone Layer was concluded in March 1985. Bangladesh ratified, and became a party to, the Vienna Convention and its 1987 Montreal Protocol (discussed below) on 2 August 1990. Article 2 of this Convention imposes general obligations on each party "in accordance with the means at their disposal and their capabilities" to:

(a) Co-operate by means of systematic observation, research and information exchange in order to better understand and assess the effects of human activities on the ozone layer and the effects on human health and the environment from modification of the ozone layer;

(b) adopt appropriate legislative or administrative measures and co-operate in harmonising appropriate policies to control, limit, reduce or prevent human activities under their jurisdiction or control should it be found that these activities have or are likely to have adverse effects resulting from modification or likely modification of the ozone layer;

(c) co-operate in the formulation of agreed measures, procedures and standards for the implementation of the Convention, with a view to the adoption of protocols and annexes;

(d) co-operate with competent international bodies to implement effectively the Convention and protocols to which they are party.

The Vienna Convention was supplemented by the 1987 Montreal Protocol on Substances that Deplete the Ozone Layer, itself adjusted and amended by the London Meeting of 1990 and most recently by the Copenhagen Meeting of November 1992. Article 2 of the Protocol requires parties to impose controls on ozone-depleting chemicals, namely CFCs (Art. 2A); Halons (Art. 2B); other fully halogenated CFCs (Art. 2C); Carbon Tetrachloride (Art. 2D); and 1,1,1-trichloroethane (methyl chloroform) (Art. 2E). However, under Article 5, special exemption is given to any party which is a developing country and whose annual calculated level of consumption of controlled substances in Annex A is less than 0.3 kilograms per capita from the time of the coming into force or until 1 January 1999. Such parties may delay the implementation of these controls "to meet its basic domestic needs" for up to ten years, subject to certain provisos set out in Art 5(2)(3). Such parties are not, however, exempted from the duty to report data to the Secretariat (Art. 7). The Protocol also provides for technical and financial co-operation for developing country parties claiming exemption under Art. 5. A Multilateral Fund was agreed to assist such parties to research and to meet their obligations (Art 10) but did not come formally into existence until the London Amendments themselves came into force — which was not until 10 August 1992, some seven months later than expected (Gehring and Oberthur, 1993). The Fund has, however, begun to operate on an interim basis and has already financed some 40 country studies.

In the light of the 1991 Scientific Assessment of Ozone Depletion which found that there was new evidence of ozone decreases in both hemispheres at middle and high latitudes, a number of new measures were agreed in Copenhagen in November 1992. These are subject, of course, to the general regime out lined above (delays under Article 5 are still possible for developing states). All parties were required to cease using Halons by 1994, and CFCs, HBFCs (added latter in the list of the substances of the Montreal Protocol that deplete the ozone layer), carbon tetrachloride, and methyl chloroform by 1996. The use of HCFCs should be frozen by 1996, reduced 90 per cent by 2015 and eliminated by 2030. Methyl bromide — still used as a fruit and grain preservative — was subjected to voluntary controls; contracting parties agreed to "make every effort" to freeze its use by 1995 at 1991 levels. As Figure 6.3 shows, the measures agreed in Copenhagen aim to stabilise atmospheric chlorine loading by the year 2000 and then reduce them to below critical levels by about 2060.

Developing countries were concerned that, even with the ten-year period of grace built into Article 5 of the Montreal Protocol, they would have difficulty meeting these new obligations if the relevant technology and finance was not made available to them. The Copenhagen Meeting agreed to add an extra provision to Article 5 (Art. 5(1)*bis*) allowing the Meeting of the Parties to decide appropriate reduction targets and dates before 1996 without an additional formal amendment to the Protocol. This gives more flexibility to countries such as Bangladesh in meeting the stricter obligations of the amended Montreal Protocol.

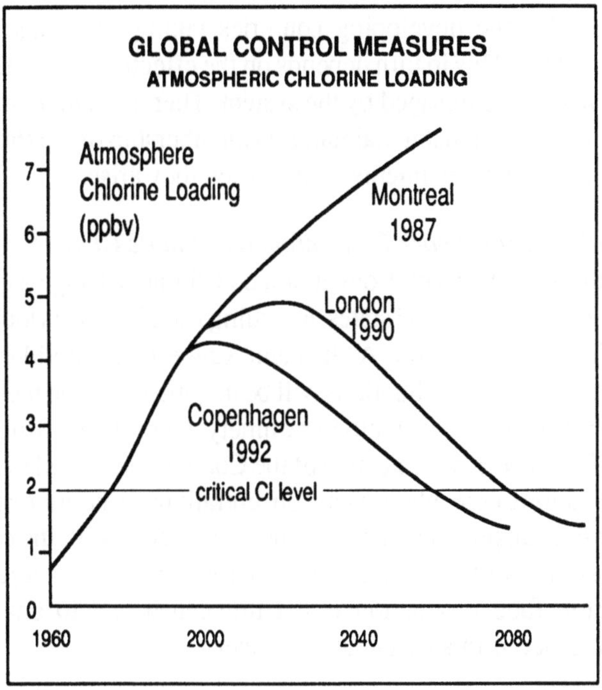

FIGURE 6.3: The impact of control measures on the atmospheric chlorine loading

Source: World Meteorological Organisation.

The Copenhagen Meeting also introduced a novel and significant procedure for monitoring and supervising compliance with these technical obligations. This procedure may be initiated by complaint of non-compliance made by one or more parties against another, by any party temporarily unable to meet its obligations, or by the Secretariat itself. A ten-country Implementation Committee will then review information submitted, request further information and even conduct information-gathering itself in the territory of the state. The Meeting of the Parties will then consider this evidence in the absence of the parties concerned and may take appropriate measures. These might include appropriate extra assistance, a caution or even suspension of rights and privileges under the Protocol.

This new enforcement regime is one of the strictest devised for a global treaty of this kind. Commentators (Gehring and Oberthur, 1993) seem clear that one of the most important items for the future will be strengthening and enforcing control measures in the so-called "countries with economies in transition" (the old eastern bloc states) but also the developing countries taking advantage of Article 5, including Bangladesh. This in turn depends on the effectiveness of the transfers of finance and technology envisaged by the system. Therefore, there is an advantage to Bangladesh in ensuring that it maintains its membership of the Protocol fully up to date by ratifying the amendments as and when they arise.

The United Nations Framework Convention on Climate Change

The United Nations Framework Convention on Climate Change, signed at Rio de Janeiro in June 1992 by some 155 states, including Bangladesh, is loosely modelled on the 1985 Vienna Convention. As its name suggests, it provides a framework within which more detailed obligations will be negotiated by the means of detailed Protocols. The Convention came into force ninety days after its fiftieth ratification on 21 March, 1994. The basic objective of the Convention, stated in Article 2, is to achieve "stabilization of greenhouse gas concentrations in the atmosphere at a level that would prevent dangerous anthropogenic interference with the climate system ... within a timeframe sufficient to allow ecosystems to adapt naturally to climate change, to ensure food production is not threatened and to enable economic development to proceed in a sustainable manner."

Article 3 lays down certain basic principles to guide the parties in meeting these objectives, namely: the need to protect the climate system for the benefit of humankind as a whole, on the basis of equity and in accordance with their "common but differentiated responsibilities"; the needs and special circumstances of developing country Parties — especially those particularly vulnerable to the adverse effects of climate change or those that would have to bear a disproportionate or abnormal burden under the convention; and the need to take precautionary measures to anticipate, prevent or minimize the causes of climate change and to mitigate its adverse effects. These policies and measures should be cost-effective ("to ensure global benefits at the lowest possible costs"), should take into account different socio-economic contexts, should be comprehensive in order to cover all relevant sources, sinks and reservoirs of greenhouse gases and adaptation, and comprise all economic sectors. The Article recognises the right to promote sustainable development, the need for counter-measures to be appropriate and integrated into national economic development, and the need for parties to co-operate in promoting a supportive and open international economic system which, by leading to economic growth, would allow developing states to address the problems of climate change. However, climate change measures should not be used as a means for unjustifiable discrimination or a disguised restriction on international trade.

The detailed commitments of the parties are laid down by Article 4. Two primary duties are imposed on all Parties:

a) to develop, periodically update, publish and make available national inventories of anthropogenic emissions by sources and removals by sinks of all greenhouse gases not controlled by the Montreal Protocol using comparable (and yet to be agreed) methodologies;

b) to formulate, implement, publish and regularly update national and regional programmes of measures to mitigate climate change by addressing anthropogenic emissions by sources and removals by sinks of all greenhouse gases not controlled by the Montreal Protocol and measures to facilitate adequate adaptation to climate change.

In addition to these clear obligations — which are, however, subject to some conditions for developing countries (discussed below) — all parties agree to a number of general obligations which are hortatory only (that is, they lack the precision to make their provisions clearly enforceable as such). Precision will be added by more detailed protocols. Parties commit themselves to the following: to undertake to promote, and co-operate in, the development of technologies, etc.; to control, prevent or reduce anthropogenic emissions of greenhouse gases; to promote sustainable development and the conservation and enhancement of sinks and reservoirs including biomass, forests, oceans and other terrestrial, coastal and marine ecosystems; to cooperate in adaptation to impacts of climate change, by elaboration of plans for integrated coastal zone management, water resources and agriculture and for protection and rehabilitation of areas affected by, inter alia, floods; to promote and co-operate in the exchange of scientific, technological, socio-economic and legal information relevant to climate, climate change and response strategies; and to promote and cooperate in relevant education, training and public awareness.

Parties are also obliged to communicate information about their implementing measures,in the form of a National Communication (Article 12), to the Conference of the Parties when it meets. All Parties are also obliged to take climate change considerations into account "to the extent feasible" in their relevant social, economic and environmental policies and actions and employ "appropriate methods" to minimise the adverse affects of projects or measures undertaken by them to mitigate or adapt to climate change.

Developed countries are obliged to take the lead in meeting this and the remaining commitments of Article 4. In a further qualification, which will be of particular relevance to Bangladesh, developing country Parties are only required to implement these commitments to the extent that developed countries have met their commitments to provide financial resources and to transfer technology. Even then the Convention recognises that for developing country Parties economic development and poverty eradication are the first and overriding priorities (Art. 4(7)). More significant for Bangladesh is the further provision in Article 4(8) that in implementing these commitments the Parties shall "give full consideration" to the actions necessary (including funding, insurance, and technology transfer) to meet the specific needs and concerns of developing country Parties arising from the adverse effects of climate change and/or impacts of response measures. The specific groups of developing countries singled out for this special treatment include two groups which would apply to Bangladesh, namely ...

(b) Countries with low-lying coastal areas; ...

(d) Countries with areas prone to natural disasters;

It should be noted that at this stage all the obligations in the Convention are phrased in general terms and are aspirational or conditional. In order to become effective they need to be made more specific; this process will be undertaken through the negotiation of detailed protocols. However, by ratifying this Convention, Bangladesh and other state Parties commit themselves to the objectives of the Convention and the principles which it contains. It should be clear from the summary above that the onus of the obligations to develop and finance measures to control greenhouse gas emissions and respond to the challenge of global climate change has at this stage been placed squarely on the shoulders of the developed world.

BOX 6.6: Framework Convention on Climate Change

> The Framework Convention on Climate Change, signed by Bangladesh in 1992 at Rio de Janeiro and ratified in 1994, establishes the framework for the negotiations of future protocols for stabilising greenhouse gas concentrations. It recognises that some developing countries, including Bangladesh, are particularly vulnerable to climate change. It also commits the developed countries to provide technical and financial assistance to developing countries in meeting their obligations under the Convention. By signing and ratifying the Convention, Bangladesh has committed itself to prepare national inventories of anthropogenic greenhouse gas sources and sinks and to embark on a national programme of mitigation and response measures.

By ratifying the Convention, Bangladesh has committed itself to a number of important obligations, not least of which is to undertake national inventories of anthropogenic greenhouse gas sources and sinks. It will also be required to embark on a national programme of mitigation and response measures. However, the Convention also makes it clear that it will not have to shoulder the burden of this single-handedly — technical and financial assistance is available from the developed world to assist in this process. Bangladesh is currently receiving support from the Global Environment Facility (GEF) as part of a region-wide programme.

6.4 LOOKING FORWARD

One theme emerges from the various issues discussed in this chapter: the problems that climate change is likely to pose for Bangladesh are not necessarily new problems; it is the scale which is likely to be different.

The Bangladesh legal system already has a wealth of experience in dealing with the problems posed, for example, by inundation, flooding and population displacement. The scale of these problems is, however, likely to be exacerbated by global warming. Legal institutions, like other organisational frameworks, are likely to be put under increasing, perhaps intolerable, stress. There is still time to look more closely at the way the current system operates to ensure that it does not crack under this new strain.

Planning for change at a legal and institutional level should be similar to that in other areas. New policies, strategies and legislation to address anticipated problems should be ones which will be of general benefit to the community even if the worse fears of climate change do not materialise. Legal changes to facilitate advance planning which, wherever possible, assist in addressing current problems, as well as possible future ones, are likely to be the most desirable.

Similarly, at an international level, the emphasis on international co-operation that new treaty law appears to be placing on responses to the threat of climate change will be advantageous to Bangladesh in addressing current, problems as well as possible future threats. This applies as much to issues such as the protection of threatened natural eco-systems (so-called life support systems) as it does to co-operation in scientific research. The emphasis of the Climate Convention, like the Ozone Layer Convention, is on providing assistance to particularly vulnerable countries and regions. Bangladesh is known to be one of the most vulnerable; it needs to be able to ensure that its internal institutions are in a position to take advantage of the technical and financial assistance planned to become available. Furthermore, Bangladesh must clarify for itself, at a strategic level, the way in which these initiatives can assist it in addressing its short to medium term difficulties as well as in its longer term responses to the environmental changes that global warming may bring.

6.5 FURTHER RESEARCH

There is a place for legal, as well as physical or social science research in this effort. More needs to be known about the way that legal rules and institutions actually work in practice. Changing the law is not a solution in itself unless the means exist for putting it into operation or for policing it effectively. Effective laws are most often those that co-ordinate rather than coerce, and that build on the wisdom of experience rather than impose novelty for its own sake. Changes in direction and policy do often have to be made, and prior research can often assist this transition by highlighting key problems.

At a national level, accurate assessment is often impeded by simple lack of information. There is a place for the collation and dissemination of legislation on key issues, as well as for further research on the actual operation of legal rules and institutions on property rights, on the establishment of wider systems of public land settlement, on possible approaches to effective controls of numerous forms of pollution as well as of greenhouse gases, and on the effective, possibly community-based, protection of crucial life-supporting ecosystems.

At the interface between national and international law, work is needed on the implementation of international rules at national level and systematic information is needed of the existence and scope of relevant treaty regimes. Information is also needed of the way in which participation in such regimes may assist Bangladesh to achieve internally desirable objectives as well as to ensure co-ordination of its internal polices with international norms.

ENDNOTES

[1] Notably, the presence of bays, deeply indented coast-lines, fringes of islands, etc.

[2] i.e., territorial sea, Art. 3 LOSC; contiguous zone, Art. 33(2) LOSC; exclusive economic zone, Art. 57 LOSC, and in some situations even the continental shelf, Art. 76(1) LOSC.

[3] The 1947 Award by Lord Radcliffe located the national boundary along the previous district boundaries of Khulna and 24-Parganas, Prescott, 1985, p. 176.

[4] "Guarantee Clause" of 1977 Agreement on Sharing of the Ganges Waters at Farakka states that if during a particular 10-day period, the Ganges flows come down to such a level that the share of Bangladesh is lower than 80 per cent of the quantum of water allocated for Bangladesh, the release of water to Bangladesh during that 10-day period shall not fall below 80 per cent of the value agreed in the Agreement.

REFERENCES

Churchill, R.R. and Lowe, A.V. (1988) *The Law of the Sea*. Manchester University Press, Manchester, 2nd edition.

Dolzer, R. (1992) The legal status of the commitments in the Convention on Climate Change and the need for future revisions, Second CICERO Seminar: *The Rio Convention on Climate Change - The Agenda for Research*. Oslo, Norway.

Farooque, Mohiuddin (1990) Water rights and management of water resources under conflicting demands in Bangladesh context. In *Workshop on Methodo-logy for Sustainable Water Resources Management and Development in Bangladesh*. Stockholm, SUWaR Report No. B1, 1990.

Farooque, Mohiuddin (1991) The state of environmental laws in perspective: the case of Bangladesh (Draft), UNDP/MOEF project on NEMAP, Dhaka.

Farooque, Mohiuddin, (1992) In defence of environmental order: the security of survival. In Bangladesh Institute of International and Strategic Studies (BIISS) (ed.) *Challenges and Prospects in South Asia in the 1990s*. BIISS, Dhaka.

Farooque, Mohiuddin, (1990) International law on shared rivers: Bangladesh perspectives. *Bangladesh Quest*, 3.

Farooqui, M. T. (1975) *The Law of Abandoned Property*. Dhaka.

Freestone, D. (1991a) The precautionary principle. In R. Churchill, and D. Freestone (eds.) *International Law and Global Climate Change*. Graham and Trotman/Martinus Nijhoff, London/Dordrecht, pp. 21-39.

Freestone, D. (1991b) International law and sea-level rise. In R. Churchill and D. Freestone (eds.) *International Law and Global Climate Change*. Graham and Trotman/ Martinus Nijhoff, London/Dordrecht, pp. 109-125.

Freestone, D. and Pethick, J. (1990) International legal implications of sea-level rise. In J.G. Titus (ed.) *Changing Climate and the Coast, Volume 1*. (Report of Miami Conference on Adaptive Responses to Sea Level Rise), May 1990, pp. 237-256.

Gehring, T. and Oberthur, S. (1993) The Copenhagen meeting [of the parties to the Montreal protocol]. *Environmental Law and Policy*, 23, 6-12.

Government of Bangladesh and IUCN (1991) National Conservation Strategy of Bangladesh (Draft). GOB and IUCN, Dhaka.

Habibur Rahman, M. (1991) Delimitation of maritime boundaries. Rajshahi University, Bangladesh.

Intergovernmental Panel on Climate Change (IPCC) (1990) *Strategies for Adaption to Sea-Level Rise*. Coastal Zone Management Sub Group to Response Strategies Working Group, IPCC.

Intergovernmental Panel on Climate Change (IPCC) (1992) *Global Climate Change and the Rising Challenge of the Sea*. Coastal Zone Management Sub Group (to Response Strategies Working Group), IPCC.

Ministry of Environment and Forest, Government of Bangladesh (GOB) (1991) *Bangladesh Country Report for United Nations Conference on Environment and Development* (UNCED). MOEF, Dhaka.

Ministry of Land (1990) *Government of Bangladesh Land Management Manual*. Ministry of Land, Dhaka (in Bengali).

Prescott, J.V.R. (1985) *Maritime Political Boundaries of the World*. Methuen, London .

Prescott, J.V.R. and Bird, E.C. (1989) The influence of rising sea levels on baselines from which national maritime claims are measured. In C. Grundy-Warr (eds.) *International Boundaries and Boundary Conflict Resolution*. International Boundaries Research Unit, Durham, pp. 279-300.

Soons, A. H. A. (1990), The effects of sea-level rise on maritime limits and boundaries. *Netherlands International Law Review*, 37, 207ff.

Wooster, W.S. (1985) Research in troubled waters - US research vessel clearance experience, 1972-1978. *Ocean Development and International Law*, 9, 219-239.

LEGAL SOURCES

BANGLADESH NATIONAL LAW

1905, Smoke Nuisance Act

1920, Agricultural and Sanitary Improvement Act

1940, Co-operative Societies Act

1950, East Bengal State Acquisition and Tenancy Act (SAT)

1952, Embankment and Drainage Act

1958, Displaced Persons (Land Settlement) Act

1958, Displaced Persons (Compensation and Rehabilitation) Act

1964, Disturbed Persons (rehabilitation) Order

1965, Law of Enemy Property

1965, Factories Act

1971, Law of Abandoned Property

1972, Constitution of the People's Republic of Bangladesh (subsequently amended)

1972, State Acquisition and Tenancy (Amendment) Order (Presidential Order 72/72)

1972, Abandoned Property (Control, Management and Disposal) Order (Presidential Order 16/72)

1972, Taking over Control and Management of Industrial and Commercial Concerns Order

1974, Wildlife(Preservation)(Amendment) Act

1974, Territorial Waters and Maritime Zones Act

1974, Maritime Zones Baseline Declaration

1976, Alienation of Land (Distressed Circumstances) (Restoration) Ordinance

1977, Environment Pollution Control Ordinance

1977, Territorial Waters and Maritime Zones Rules

1982, Acquisition and Requisition of Immovable Property Ordinance

1983, Motor Vehicles Ordinance

1983, Marine Fisheries Ordinance

1984, Land Reforms Ordinance and Rules

1989, Brick Burning (Control) Act

INTERNATIONAL TREATIES AND INSTRUMENTS

1951, International Plant Protection Convention.

1963, Treaty Banning Nuclear Weapon Tests in the Atmosphere, in Outer Space and Under Water

1972, Stockholm Declaration on the Human Environment

1972, Convention on the Prohibition of the Development, Production and Stockpiling of Bacteriological (Biological) and Toxic Weapons and on their Destruction

1972, Convention concerning the Protection of the World Cultural and Natural Heritage

1977, Bangladesh/India Agreement on sharing of the Ganges' water at Farakka and on augmenting its flows.

1977, Convention on the Prohibition of Military or Any other Hostile use of Environmental Modification Techniques.

1982, Law of the Sea Convention

1985, Vienna Convention for the Protection of the Ozone Layer.

1987, Montreal Protocol on Substances that Deplete the Ozone Layer.

1991, Espoo Convention on Environmental Impact in a Transboundary Context.

1992, Helsinki Convention on Protection and Use of Transboundary Watercourses and Lakes.

1992, UN Framework Convention on Climate Change

1992, Rio Declaration

CLIMATE CHANGE AND SEA-LEVEL RISE: THE CASE OF THE COAST

A. Kausher, R.C. Kay, M. Asaduzzaman and S. Paul*

7.1 INTRODUCTION

The future impact of sea-level rise and climate change on the coastal zone of Bangladesh will depend on the vulnerability and resilience of its physical, biological, social and economic systems. Previous studies have estimated future coastal impacts by calculating the areas of low-lying land and assuming that these areas will be permanently lost under rising seas[1]. Such studies have failed to address the considerable natural resilience of the Bangladesh coastal zone. For example, the supply of river sediments to the coastal zone may well provide a buffer against sea-level rise. Furthermore, there is considerable human resilience to environmental change; adapting to constantly shifting lands, floods and cyclones is part of the way of life of coastal people. Understanding of such factors is necessary in order to fully assess the future effects of climate and sea-level change on the coast.

The purpose of this chapter is to examine the relevant aspects of the human, physical and biological systems of the coastal zone of Bangladesh as they pertain to the climate change issue. Each system is considered in terms of its vulnerability and resilience to present and future stresses. Elements of each coastal zone system that require urgent management attention are identified, as are important areas of uncertainty and future research directions. In so doing, the chapter aims to support Bangladesh's commitment under the Framework Convention on Climate Change to "...develop and elaborate appropriate and integrated plans for coastal zone management ... " (Article 4 Section 1e) by summarising the key scientific and management issues relating to the potential future impacts of climate change and sea-level rise.

* The authors are respectively: Visiting Scholar, BUP, Dhaka; Coordinator of Coastal Planning, Ministry for Planning, State Government of Western Australia, Perth, Australia; former visiting scholar, BUP, Dhaka; and the late S. Paul was a PhD student at CEARS and Associate of BUP.

7.2 THE COAST OF BANGLADESH

The coast of Bangladesh is approximately 710km long (Snead, 1985), as estimated by measuring the distance around the Bay of Bengal between the Indian and Myanmar (Burma) borders. If the length of the tidal-estuarine coast is considered, the total shoreline of Bangladesh may be thousands of kilometres long.

The definition of the coastal zone used in this chapter is the area of the "Greater coastal districts," which is the most commonly used definition in Bangladesh. The coastal districts are the five Greater districts which adjoin the coast, namely: Chittagong, Noakhali, Patuakhali, Barisal and Khulna (Figure 7.1). The coastal districts, and hence the coastal zone as defined in this chapter, cover 23 per cent of the area of Bangladesh.

The transition between land and sea in Bangladesh is so gradual that small changes to the definition of coastal zone can produce very large changes in the extent and location of "coastal" land. Figure 7.1 shows the landward limits of the dry season salinity and tidal limits, both of which can also be used to define the area of the coastal zone. The comparison of coastal studies is often hampered by the use of different definitions of the coastal zone.

Waves, tides, river flow, sediment movement, plants and animals interact to shape coastlines. This shaping occurs each minute, each day of each year from decade to decade, century to century. The coastline of Bangladesh varies because the relative importance of these processes is not uniform around the Bay of Bengal.

Humans have traditionally altered the natural character of the Bangladesh coastal zone through their exploitation of its natural resources. The type and scale of human modifications have changed dramatically during the past 40 years as traditional Bengali practices have given way to modern technologies. These traditional and modern human modifications may have a profound effect on the vulnerability and resilience of the coastal zone to climate change and sea-level rise.

7.2.1 Three Major Zones

The coastline of Bangladesh can be divided into three main zones — the west, central and east coastal zones (see Figure 7.1). Each contains distinct coastal landforms and specific geological, physical and biological characteristics (Barua, 1991a, b). The west and central coasts are part of the Ganges-Brahmaputra-Meghna delta system, whereas the east coast is non-deltaic. In general, each of the three zones has one dominant, or controlling, factor. In the west, mangrove forests dominate the coastal fringe. Mangroves also control the interaction of inland non-mangrove areas with the sea. In the central zone, the sediment-rich freshwater flows and their interaction with tides are important factors. In the east, wave energy dominates the open-ocean coast, although mangroves are also important within sheltered estuaries.

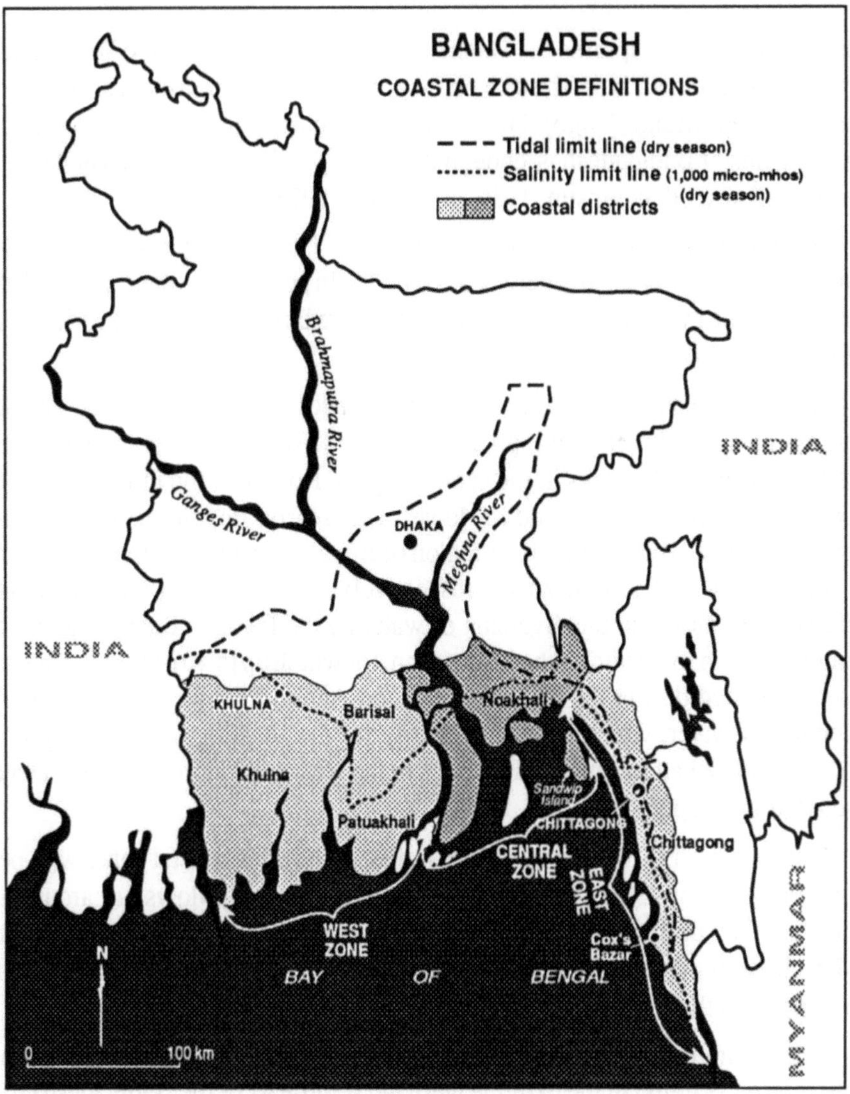

FIGURE 7.1: The coastal zone of Bangladesh

The present characteristics of the coastal zone will play a crucial role in determining the ultimate impact of any future climate change and sea-level rise. The relevant factors to consider include: rivers, tides and waves; soils and groundwater; population and settlement; uses of land and water (agriculture, forestry and fisheries); and extreme events. Each is discussed in a separate section below.

7.2.2 Rivers, Tides and Waves

The Meghna River captures the waters of both the Ganges and the Brahmaputra rivers and (combined) is the third largest source of fresh water to the world's oceans (after the Amazon and Congo). The combined annual discharge of the three mighty river systems of Bangladesh is enough to submerge all of Bangladesh under more than 9m of water (Thompson, 1989, after Khan, 1978)

River flow has a strong seasonal component. The ratio of peak monsoon flow to lean season flow is usually between 10 and 15. In peak flood years, such as 1988, the ratio can be even higher (Figure 7.2). There are large inter-annual variations in the discharge (see Chapter 4).

There are also large geographical differences in river flow. The absolute amounts of water flowing through the rivers of the west region average ten times less than the monthly flows in the central region. However, the ratio between the highest peak monsoon flow and the lowest lean season flow in the west is much more pronounced than in any other region of Bangladesh (39, compared to 14 and 17 for the east and central zones, respectively, in 1988; see Figure 7.2). This difference is caused by the diversion of water by the Farakka Barrage across the Ganges River in India during the lean season, for which up to 40 per cent of the total flow is diverted during the dry season (Flood Action Plan [FAP] 4, 1992). The diversion of water, which is used primarily for the flushing of Calcutta port, is thought to have had a number of impacts in the western region of Bangladesh (as discussed in Chapter 4).

The total river flow in the east is the lowest of the three coastal zones with peak monsoon flows of 4370 m^3s^{-1} during 1988. Mean monthly flows are around ten times less than in the central zone. The peak monthly flow occurs 1 to 2 months before those in the west and central zones due to the small catchments very close to the coast in the Chittagong Hill Tracts.

The minimum spring tidal range (the difference between high and low tide) of 2.6m to 2.7m is found in the extreme west and south-east of the Bangladesh coast. The tidal range increases progressively towards the north-east of the Bay of Bengal

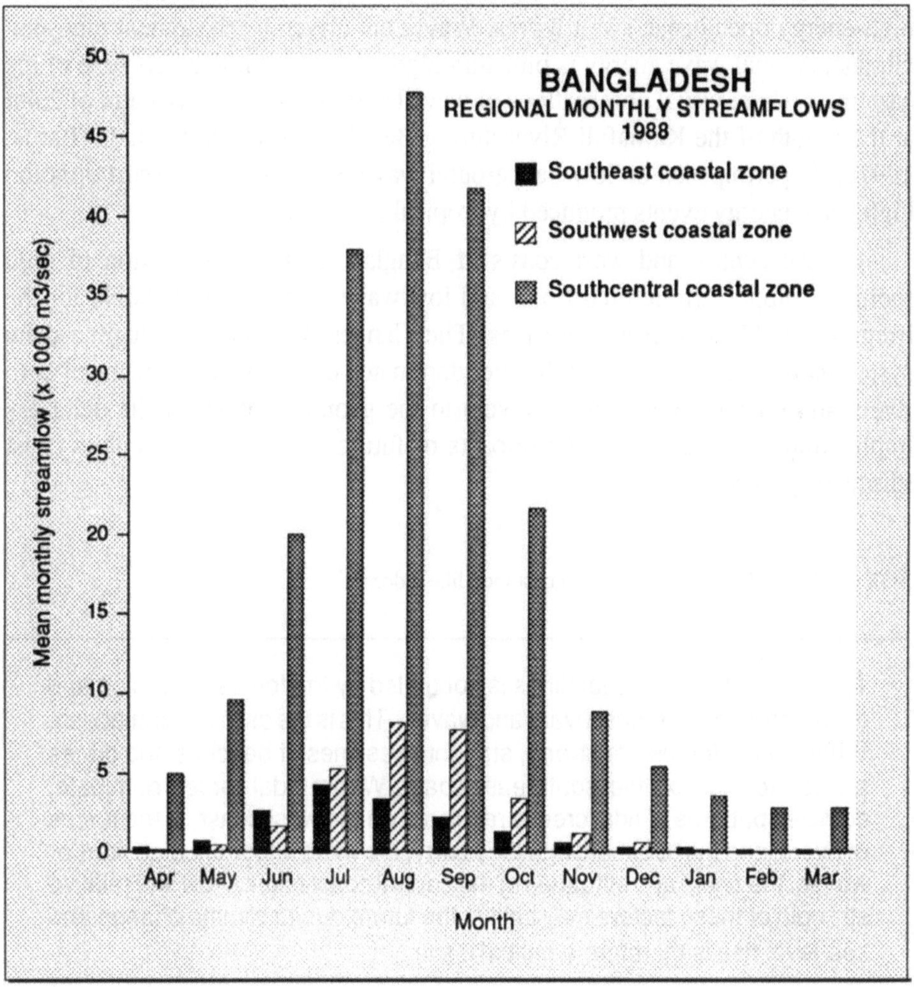

FIGURE 7.2: Regional monthly streamflows for 1988

Source: Bangladesh Bureau of Statistics, 1991, 22.

around the mouth of the Karnafuli River, Sandwip Island and north of Hatia Island. Here, the tidal range is over 4 metres. The maximum tidal range is 5.5m to 6.0m at the mouth of the Feni River.

The wave energy on the coast of Bangladesh is generally very low, although very few systematic wave measurements have been attempted. The position of

Bangladesh in the north-east corner of the Bay of Bengal shelters the coast from high-energy open ocean swell waves. Waves are driven by the annual monsoon climate. Annual wave energy is probably highest in the southern sub-zone of the east coast where visual observations of wave height indicate a maximum of 2.4m at the mouth of the Karnafuli River during Beaufort force 4 to 6 winds (Barua, 1991a). Superimposed on low background wave energies are the short duration high wave energy events produced by tropical cyclones.

On the central and west coasts of Bangladesh, the combination of high sediment supply, strong tidal flows and low wave energy is responsible for the irregular, highly embayed coastlines. The Ganges-Brahmaputra-Meghna delta system is, as a result, called a tide/river dominated delta. The importance of both rivers and tides (as opposed to waves) to the geomorphology of the delta has implications with respect to the impacts of future sea-level rise on delta coast (discussed below).

BOX 7.1: Why does the Bangladesh coast look like it does?

The natural shape of coastlines is controlled by the forces of erosion and accretion involving tides, rivers and waves. This is the case in Bangladesh. Where wave forces are strong, straight coastlines of beaches and dunes are formed, as on the south-east coast. Where tidal forces dominate, complex patterns of tidal creeks result, as on the west coast. If the forces of rivers, and the sediment that they carry, are more important than tides or waves, the *chars* and estuaries of the central coast result. How the relative strength of these factors may shift in the future due to climate change and sea-level rise is therefore a critical issue.

7.2.3 Soil and Groundwater

The types of soil in the coastal zone depend closely on the type, age and history of the land on which the soils have formed. These geomorphic factors interact with climatic, river flow and groundwater components to form distinctive suites of coastal soils rich in silt and clay (FAO, 1988). Five major types of coastal soil have been classified, which closely reflect the five physio-geographic units of the coastal zone (Figure 7.3) (FAO 1971, 1988; Chapter 4). Within the coastal floodplain areas of the deltaic coastal zone (physiography units 2a,b,c and 3b — Figure 7.3) soil types vary depending on local floodplain topography. The exposure to tidal flooding, in the unembanked condition, is the main factor which differentiates coastal soils from inland soils in Bangladesh.

West of the Meghna estuary, coastal soils are formed in Gangetic tidal floodplain deposits (Figure 7.3). Acid sulphate soils have been identified in the Sundarban where the forests have been cleared (physiography unit 2c — Figure 7.3). Acid sulphate soils are very poor soils for cultivation as they become toxic if allowed to dry out.

Non-saline grey tidal floodplain soils (physiography unit 2a — Figure 7.3) dominate the northern half of Patuakhali, central Barisal and north-east Khulna regions. These soils are very productive agriculturally.

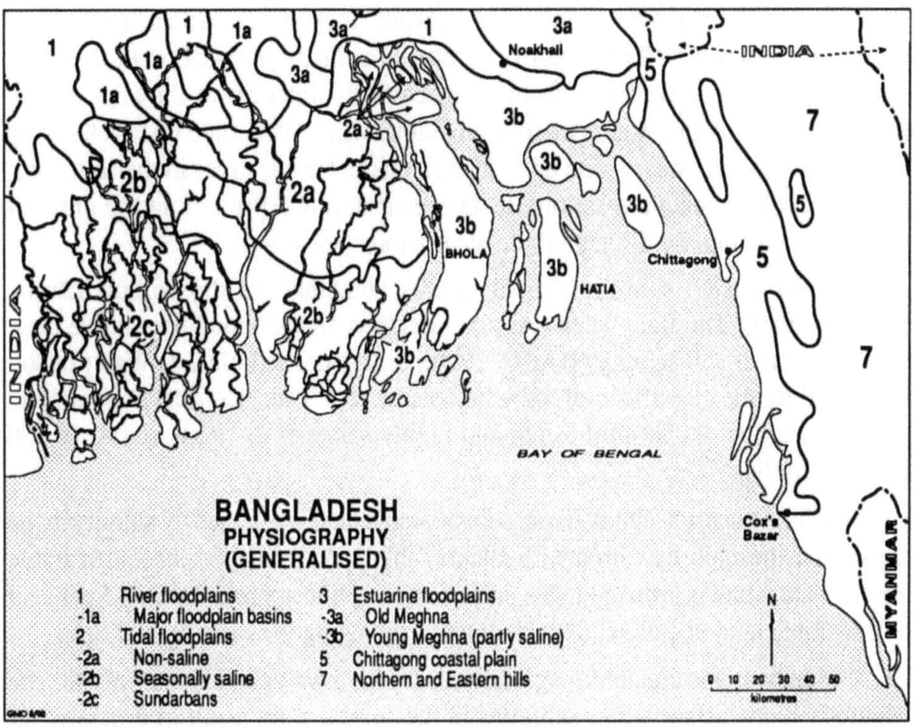

FIGURE 7.3: Bangladesh coastal physiography (generalised)

Source: Adapted from FAO, 1988 (see also Chapter 4).

The soils of southern Patuakhali, southern Barisal and central/north-west Khulna are grey floodplain soils, saline in the dry season (physiography unit 2b). Dry season salinity restricts dry season agricultural productivity, although the salinity of soil and surface waters is beneficial for shrimp farming.

In the central coastal zone, soils are classified as calcareous alluvium and are formed in silty young Meghna estuarine deposits (physiography unit 3b). Most become, to varying degrees, saline in the dry season.

In Chittagong District, the soils of the Chittagong Coastal Plain are formed on piedmont and river deposits, with characteristics similar to those occurring on piedmont and river floodplains within inland districts (physiography unit 5 — Figure 7.3). Specifically "coastal" soils are confined to areas of young tidal deposits, together with small areas of relic beach sand in the south and coral debris on St. Martin's Island (Jinjiradwip).

The salinity level in topsoils begins to increase as soon as monsoon rain ends. This occurs as a result of evaporation of moisture drawn to the surface by capillary action in dry weather. The majority of coastal soils are non-saline in the rainy season. From December onwards the reduced fresh-water flows allow increasing levels of estuarine salt water penetration. Saline waters penetrate up to 130km up the lower Meghna estuary, and up to 290km up the Passur River in the southwest (Nishat, 1988). Maximum salinity levels occur during April before the onset of monsoon rains rapidly dilutes salt concentrations (see Figure 7.4).

An estimated 1.4 million ha of land is vulnerable to dry season salinity (Mahtab,1989). The Bangladesh Agricultural Research Council uses a five band classification of soil salinity (BARC, 1990), as shown in Table 7.1. Water or soil is defined by the classification to be non-saline (S_0) if its electrical conductivity (EC) is less than 2,000 mmhos/cm and highly saline if its EC is above 16,000 mmhos/cm (S_4).

In Khulna district, about 70 per cent of arable land is affected by salinity in the dry season, the majority within the S_2 band (Table 7.1). Around 60 per cent of arable land in Patuakhali is affected in the same season, while approximately 15 per cent of the arable land of Barisal, Noakhali and Chittagong is saline (S_0 and S_2).

Groundwater is crucial for agriculture, drinking water and industrial use. The generally high surface water salinities in the coastal zone result in groundwater being the main source of potable water. There are both shallow and deep aquifers beneath the coast, separated by thick clay layers (Nishat, 1988). Shallow aquifer salinity levels and depths are extremely variable within short distances. In contrast, the deep aquifer is regionally continuous and its salinity levels relatively uniform.

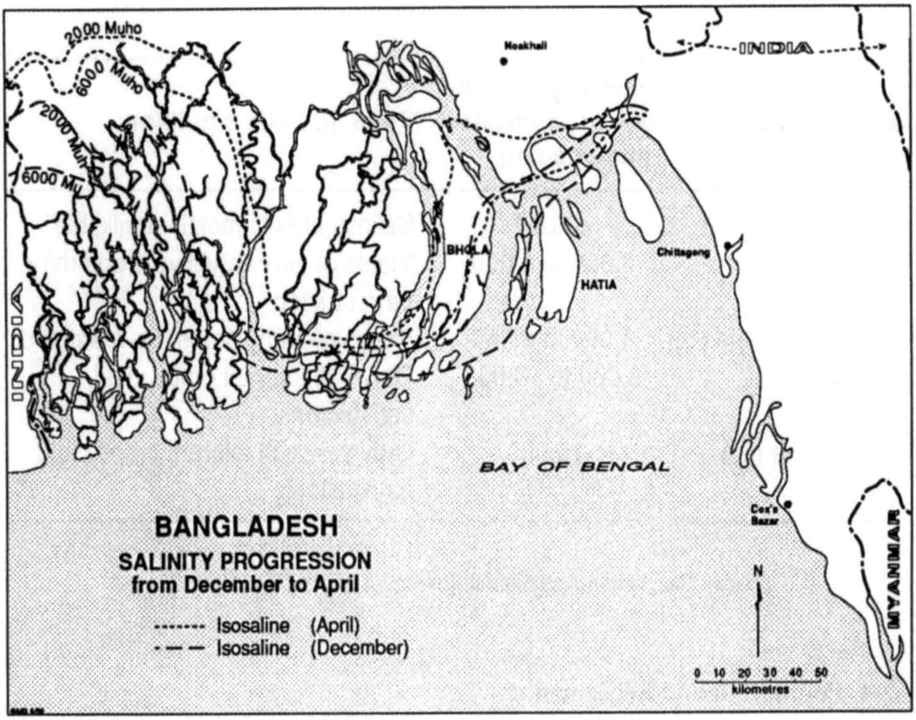

FIGURE 7.4: Seasonal fluctuation in the distribution of coastal soil salinity

Source: MPO, 1986.

The saline/freshwater boundary of the deep aquifer is close to the coast, except in the far west where the boundary is 120km to 160km inland.

There has been a rapid growth in the number of shallow tubewells in the west coastal zone, from approximately 20,000 in 1985 to 55,000 in 1991. This has been due mainly to government de-regulation policy and the reduction in subsidies paid on deep tubewells (FAP 4, 1992). Recent estimates suggest that in the southwestern portion of the west coastal zone, current water demands during the dry season exceed available resources (FAP 4, 1992). The depletion of coastal aquifers highlights the sensitivity of the groundwater system to possible future changes to the balance of fresh and salt waters due to sea-level rise and climate change.

TABLE 7.1: Classification of coastal area salinity

Salinity class	Conductivity of near-surface groundwater (mmhos/cm)	Plant growth condition
Non saline (S_0)	<2,000	Salinity effects mostly negligible
Slightly saline (S_1)	2,000 to 4,000	Yields of very sensitive crops may be restricted
Moderately saline (S_2)	4,000 to 8,000	Yields of many crops are restricted
Saline (S_3)	8,000 to 16,000	Only salt tolerant crops yield satisfactorily
Highly saline (S_4)	>16,000	Only very salt tolerant crops yield satisfactorily

Source: Bangladesh Agricultural Research Council (BARC), 1990.

7.2.4 Population and Settlements

Approximately 24 million people were recorded in the coastal districts by the 1991 census (BBS, 1991). The population in the coastal districts has grown from 7.2 million in the last 90 years (Figure 7.5). In the inhabitable coastal areas, population densities range from 613 people/km^2 (in Khulna) to 1257 people/km^2 (in Chittagong). The average population density in the coastal districts is 959 people/km^2, compared to the national average of 741 people/km^2.

The proportion of people living in the coastal districts between 1901 and 1991 has consistently been between 20 to 22 per cent of the Bangladesh population. This suggests that the factors which have caused the rapid growth of population throughout Bangladesh (described in Chapter 5) have affected the coastal districts similarly. It is therefore assumed that future coastal population growth will occur at rates similar to those of Bangladesh as a whole. On this basis, the coastal population may reach (45 to 50 million people by the year 2050[2] (see Figure 7.5)). But if the medium-term population projection of the United Nations is considered the coastal population may reach 60 to 67 million by 2050 (see Chapter 5).

In addition to the permanent coastal population, there are significant numbers of seasonal migrants to the coast, specifically to the chars of the central coast. Here, cattle grazing after each monsoon season has long been practised. Migrant

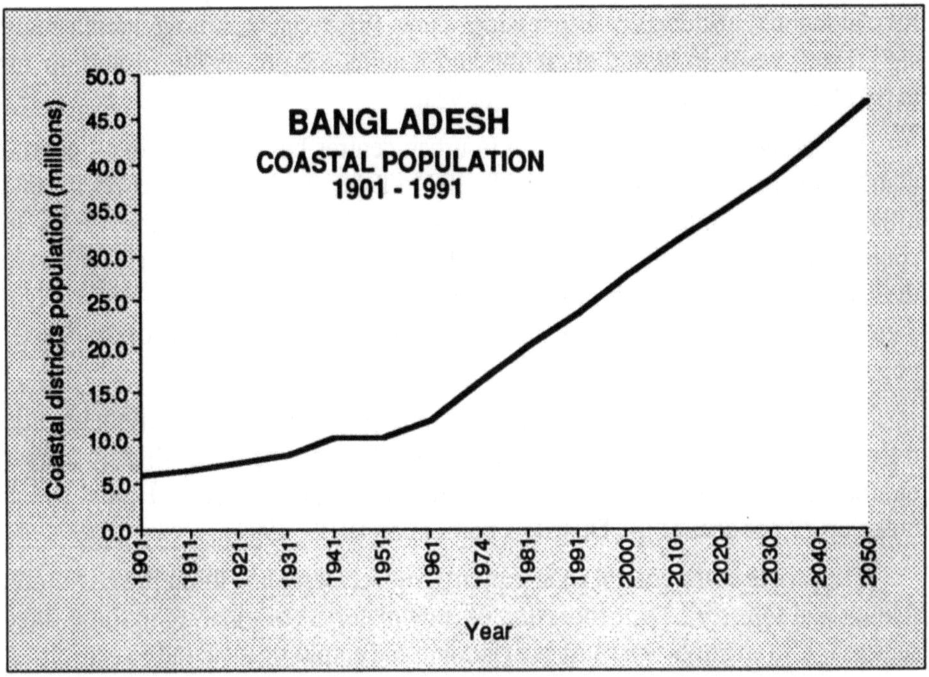

FIGURE 7.5: Coastal population growth 1901 to 1991, with a projection to the year 2050

Source: Bangladesh Bureau of Statistics (BBS), 1991 (1901-1991 data).

fishermen would also take advantage of favourable seasonal conditions to fish offshore during the cyclone-free months.

The numerous rural settlements in the coastal districts are located on any naturally high land. Where such land is unavailable and population pressure is strong, artificial mounds have been built on which individual homesteads are constructed. Their elevated location allows homes to remain flood-free while river and tidal flooding occurs on the surrounding fields. However, the raised mounds are not high enough to protect houses from storm surges produced by cyclones.

The two main urban centres in the coastal districts are Khulna, population 601,000, and Chittagong, population 1.57 million. Most coastal industry is located in these centres. Both are important centres of raw material processing. Mongla (located in Greater Khulna) and Chittagong are two important Sea Ports that handle the bulk of export and import commodities[3]. Important to Khulna are the paper and wood industries, which rely on the Sundarban mangrove forest for raw materials. In Chittagong, the Export Processing Zone (EPZ) is becoming increasingly significant, while in both Chittagong and Khulna the processing of shrimp for export has gained prominence in recent years.

The proportion of people living in urban centres has increased in all coastal districts over the last 100 years (BBS, 1991: Table 2.45). There has been rapid urbanisation in Chittagong since the 1950s and in Khulna since the 1960s. In contrast, Barisal, Noakhali and Patuakhali have not experienced urban growth to the same extent, although the 1981 statistics suggest that urbanisation is starting in these districts. Despite the drift to urban areas, more than four-fifths of the coastal population still live in rural areas.

The proportions of different religious groups in the coastal districts broadly reflect national averages[4]. In Chittagong and Khulna there are slightly higher numbers of Buddhist, whereas in Khulna there are large numbers of Hindu, which make up 27 per cent of the District's population (BBS, 1991).

The people of the coastal zone support themselves mainly through agriculture, fishing and forestry[5]. The processing of raw materials originating in these three sectors is also an important sources of income and employment in the coastal cities.

7.2.5 Use of Land and Water

Agriculture

Crop types and cropping patterns in the coastal zone, like inland agriculture, closely follow seasonal flood and climatic cycles. The different soil, flood and salinity conditions that occur in the coastal zone strongly influence crop suitability and crop rotations, both regionally and locally. The lighter textured soils and shallower seasonal flooding on ridges favour coconut and betelnut plantation. In contrast, the heavier depression soils are subject to deeper, more prolonged seasonal flooding favourable to rice cultivation.

Rice is the main crop in the coastal zone. Approximately 20 per cent of the country's rice acreage is within the coastal districts (BBS, 1991). The monsoon rice crop, locally called *aman*, produces the largest tonnage. The *aman* crop is transplanted from seedling beds during July-August and harvested from November to January. This rice cropping technique is abbreviated to *T. aman*. Broadcast sowing techniques, used elsewhere in some deeply-flooded parts of Bangladesh, are rarely used in the coastal districts.

During the pre-monsoon season *aus* rice can be planted, which matures within

80 to 120 days and is harvested within the monsoon season. *Rabi* can be grown if saline conditions are not too severe and if irrigation water is available. Increases in soil salinity during the dry season occur over much of the coastal zone and severely restrict winter crops. Rice, at mature growth stages, can be grown without any reduction in yield with a salinity of 2,000 mmhos/cm, although seedlings are more sensitive. Yields of potato and most vegetables are affected when salinity is more than 1000 mmhos/cm (BARC, 1990).

In Khulna, Barisal and Patuakhali districts, the *T. aman*-fallow cropping pattern is the most common (ESCAP, 1987). This pattern is one in which transplanted *aman* rice is grown in the monsoon season and the land is left fallow during the dry season (due mainly to salinity, but also to constraints imposed by absentee landlords).

Double and triple cropping is possible in Noakhali and Chittagong, with *T. aman*-fallow, *Aus-T.aman* and *Aus-T. aman-rabi* of approximately equal importance. This is due to the relatively low salinity levels during winter in the section of Noakhali District founded on old Meghna estuarine soils and in the estimated 134,865 ha of irrigated land in Chittagong district on which mainly *boro* rice is grown (BBS, 1991).

In Chittagong there is a 10:1 ratio in the production of High Yielding Varieties (HYV) to Local Varieties of rice, in comparison to a national average of 1.36:1. In contrast, in Barisal, Khulna and Patuakhali total production from local varieties is three times higher than from HYV (BBS, 1991: Table 4.55).

Forestry

The distribution of coastal mangrove forests in Bangladesh is shown in Figure 7.6. The natural coastal forest of the Sundarban[6] is the largest natural single area of mangroves in the world, covering 587,000ha. The formerly extensive mangroves in the east, Chakaria Sundarban, cover less than 1000ha due to the felling to make way for salt pans and shrimp ponds. There is a further 104,000ha to 109,000ha of coastal mangrove forest planted under the "Coastal Afforestation Project". These mangroves were planted in coastal areas without natural mangrove forests (mostly on the central coast) in order to protect coastal embankments and newly accreted lands from the impacts of cyclones.

Mangrove forests are critical to the people and economy of Bangladesh (Table 7.2). They supply fuelwood for domestic and industrial use, timber for industry and a range of other products. The forest is an integral part of the life cycle of many fish and shrimp species. The forest also contains many endangered species of global importance, including the Royal Bengal Tiger (*Panthera tigris*). The future health

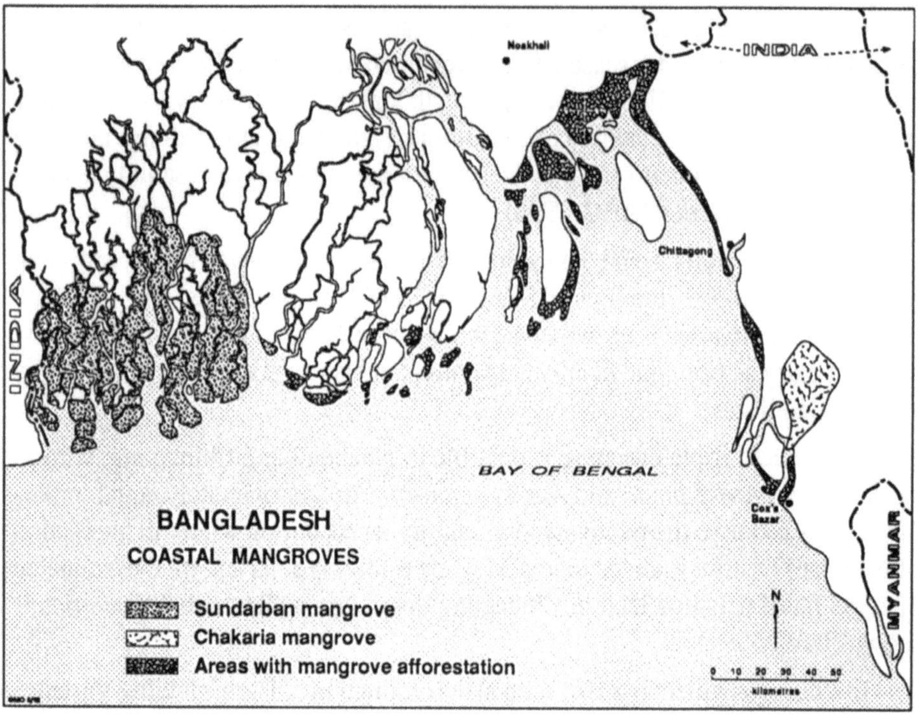

FIGURE 7.6: Coastal mangroves of Bangladesh

Source: ESCAP, 1987.

of the mangrove forests is a critical issue in the light of additional stresses that may result from climate and sea-level change.

Fisheries

Fish is the most significant source of animal protein in rural Bangladesh and is also one of the fastest growing export products in the country. The fisheries resource in the coastal zone grades from fully freshwater fisheries inland, through shrimp and other mixed freshwater/saltwater fisheries to fully marine fisheries in the Bay of Bengal.

Freshwater Fisheries, unlike brackish or marine fisheries, are found throughout Bangladesh. Freshwater fish production occurs in the open waters of rivers, canals and the river floodplain, and in the closed waters of lakes and ponds. Monsoon floods join these water systems into large interconnected units which act as fish breeding and spawning grounds. Fish breeding cycles are closely linked to the annual flood cycle; changes to the annual flood pattern, whatever the cause, can have serious implications for fishery resources.

TABLE 7.2: Ecological zonation, characteristic species and main economic uses of the Sundarban mangrove forest

Zonation	Area (km^2)	% of forest area	Bangladeshi name	Latin name	Main economic uses
Fresh water Zone	1,124	28%	Sundri	*Heritiera fomes*	Timber, fuelwood
Moderately Salt-water zone	1,823	46%	Gewa	*Excoecaria agallocha*	Pulpwood, matchwood baling board
Salt water zone	1,008	25%	Goran	*Ceriops decandra*	Fuelwood

Source: ESCAP, 1987, after Chaffey, *et al.*, 1985.

Brackish Water Fisheries. "Brackish water" is the term used for water that is more saline than freshwater, but is less saline than seawater. Brackish water is commonly found in the coastal zone where tide and river waters mix. The importance of *brackish water fisheries* has increased in recent years with the intensification of shrimp farming. There are an estimated 108,000ha of shrimp farms in the coastal districts (DOF, 1989). Shrimp farming is one of the fastest growing export industries in Bangladesh, with a turnover of 4,500 million Taka ($US 118.4 million) in 1990-1991 (Figure 7.7).

Traditional brackish water fishery combines the open-water fishery, using gill nets and set-bag nets, with the artificial pond-based *gher* fishery. The open-water fishery contains shrimp, hilsa and over 30 species of fin-fish, including mullet, catfish, anchovies, flat fish and giant perch.

Ghers, or artificial fish pens enclosed by low earth mounds, were built on the banks of tidal creeks to exploit the annual monsoon and fish breeding cycles. Water

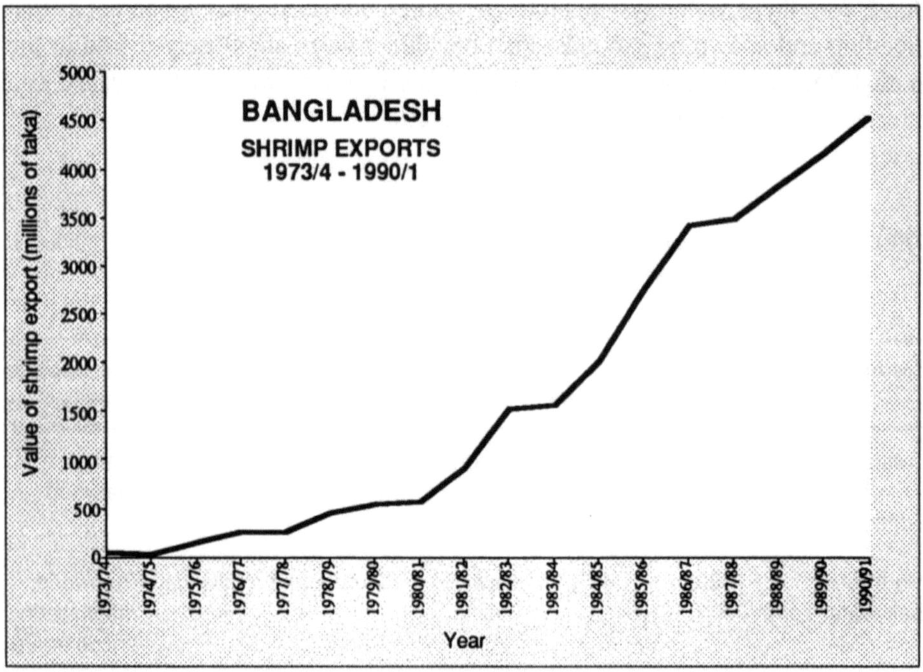

FIGURE 7.7: Annual value of shrimp exported from Bangladesh, 1973/1974 - 1990/1991

Source: EPB, 1993.

exchange between *gher* and creek was managed using wooden sluices. In spring, the tidal flows were allowed into the ponds through the sluice gates, carrying with them juvenile fish and shrimp larvae. Mature fish and shrimp were harvested from the *ghers* at the end of the dry season while the first monsoon rains began to flush the ponds of saline water. As the monsoon continued, rainwater flushing eventually led to fully freshwater conditions suitable for rice cultivation. Traditional *ghers* were therefore used for both fish/shrimp and rice harvesting.

The Coastal Embankments Project (CEP) stopped the traditional rice and fish/shrimp *gher* dual cropping practices in the project areas of Khulna and Chittagong districts. Traditional skills have not been lost, but were adapted to the new conditions imposed by the CEP. As with many other facets of the project, it has been difficult to evaluate the advantages and disadvantages of the change from traditional brackish water fisheries (Clark *et al.*, 1990). Nevertheless, what is certain is that the experience of traditional *gher* fishery has been successfully transferred to the brackish water ponds within the CEP, most notably within the Khulna area and the Chakaria Sundarban. Farmers in these two areas deliberately breach polder embankments to allow the exchange of salt water with tidal creeks.

The feature common to all areas is that shrimp produced specifically for export has increased, while fish production has decreased as a consequence. In Khulna district, shrimp harvesting is practised in rotation with rice cropping. In Cox's Bazar district, shrimp is grown in rotation with salt production. In other locations throughout the coastal zone, shrimp is produced all year round through multiple cropping using shrimp species tolerant to the different salinity levels experienced through the year (ESCAP, 1987).

In the western districts, the polders are prepared for shrimp immediately after the harvest of the *aman* rice crop in December and January. Salt water is first taken into the polders through sluice gates between early February and mid-March. The shrimp are left to grow naturally, and feed on algae contained in the tidal waters which are regularly exchanged during high tides. There is no loss of shrimp, because the shrimp are not in the migratory stage of their life cycle. When this stage is reached, bamboo slats are placed in the entrance to the polder to stop the escape of shrimp. Shrimp are caught in late May or early June by draining the fields and placing nets around the sluice gates. *Aman* rice crops can be grown in the same paddy field after the first monsoon rains flush the salt from the soil.

The rotation of intra-polder shrimping and *aman* rice crops remains controversial in Bangladesh. It is argued that since embankments were constructed to keep saline water out of rice fields, deliberately letting saline water into polders will cause salinization of neighbouring fields. Such contamination could reduce the production of rice crops. Saline water may also contaminate freshwater ponds within the polders used for drinking water and fisheries. Studies of the impact of shrimp farming and rice production in Khulna district found, however, that there is no significant impact on *aman* rice yields from dry season shrimp production. The impact of shrimp farming was found to be in the reduction of dry season grazing land. There were also significant social impacts of increased shrimp farming (Clark *et al.*, 1990; Delta Development Project, 1983). The studies found that shrimp cultivation favours rich landlords, to the disadvantage of the poor, mainly landless, people. However, it may be argued that even if shrimp cultivation was to completely inhibit paddy cultivation, shrimp farming would be preferable in economic terms, mainly because it forms an important source of foreign currency (excluding consideration of the broader environmental and social costs).

Marine fisheries form a key part of both the national and coastal economies and employ a significant proportion of coastal inhabitants. Over-exploitation of marine fish stocks off the Bangladesh coast may increase the need to intensify brackish water shrimp and fish farming within the coastal zone, with possible negative impacts on the coastal environment.

The high volumes of freshwater flowing into the Bay of Bengal reduces the sea's salinity close to shore. Consequently, marine fish species are generally located more than 60km offshore where salinity levels are higher. The need to travel such large distances offshore to catch marine fish species has traditionally restricted the fishing season to October-March, when the threat from cyclones is low. The use of motor-powered small fishing craft and large trawlers has extended this season by three months. Despite recent increases in large trawls, small fishing boats (with or without motors) still dominate marine fishing. An estimated 85 per cent of all fish landed in 1985 was by a fleet of up to 15,000 small craft (ESCAP, 1987).

Fisheries in the Bay of Bengal are extremely rich and varied, with 450 fish species recorded, 65 of which are commercially exploitable (ESCAP,1987). Data on marine fish stocks are sparse. Estimates of the standing stock of demersal and pelagic fish range between 360,000 and 450,000 tonnes. The maximum annual harvestable yield is estimated to be approximately 175,000 tonnes (Shahidullah, 1986).

The officially recorded fish catch has been rapidly increasing though the 1970s and 1980s. In 1988-1989, the estimated catch was 233,000 tonnes. The marine fish catch is estimated to rise by between 2 to 4 per cent to 270,000 tonnes by 1995 (Fisheries Resources Survey System, 1990). If the estimate of maximum sustainable yields is correct, the annual fish catch may already be seriously depleting fish stocks. This could have potentially catastrophic consequences for the long-term future of the marine fishing industry in Bangladesh.

7.2.6 Cyclone Impacts in the Coastal Zone

The Cyclone Risk

The Bangladesh coastal zone could be termed a geographical "death trap" due to its extreme vulnerability to cyclones and storm surges. The massive loss of life from cyclones is due to the large number of coastal people living in poverty within poorly constructed houses, the inadequate number of cyclone shelters, the poor cyclone forecasting and warning systems, and the extremely low-lying land of the coastal zone. Approximately 5.2 million people live within coastal areas at areas of high risk (High Risk Area) from cyclone and storm surge flooding within an area of

9,000 km^2 (MCSP, 1993).

As many as 10 per cent of the world's cyclones develop in the Indian ocean (Gray, 1968). The warm waters favour cyclone development during the pre-monsoon (March to June) and post- monsoon (October to December) seasons, with an average of just under two (1.77) cyclones each year (Talukder *et al.*, 1992). Once the cyclones are formed they generally move in a north-west to north-east direction and can cross the coast either in India, Bangladesh or Myanmar (Burma). Cyclones tend to cross the central and east Bangladesh coasts. The West coast has relatively few cyclones (Figure 7.8).

Bangladesh coastal people have always been aware of the risks of cyclones:

> ...the people have been living in the coastal areas and off-shore islands vulnerable to cyclones for centuries. Some of these remote areas have very old settlements. People used to build houses on raised land constructed by earth cutting...Although the infrastructure was far from adequate, people had been living with cyclones, being aware of all the risks (Talukder *et al.*, 1992).

The intense pressure on the land and resources inland from coastal areas forces people to remain in areas vulnerable to cyclones. These pressures are likely to remain in the future, exposing people again and again to the dangers of cyclone and flooding.

Past Disasters: Impacts and Response

Very scanty information is available about the number of deaths caused by cyclones in the 17th and 18th century (Figure 7.9). Recent records reveal that over 600,000 people have been killed by cyclones since 1960 (Haider, Rahman, and Huq, 1991). Amongst the cyclones that have hit the coastal area, those of 1970 and 1991 are considered the worst in recent history in terms of severity and damage.

The 1970 cyclone hit a vast area of the coastal zone, severely affecting Patuakhali, part of Barisal, Noakhali and Chittagong and particularly the coastal islands. The total population of Monpura and Char Kukrimukri islands was washed out overnight. The surge heights ranged from 6 to 9 metres and the total number of deaths was within a range of 220,000 to 400,000 (Haider, Rahman, and Huq, 1991).

On 29 April 1991, the worst cyclone to strike Bangladesh since the 1970 storm devastated the south-east coast.[7] With wind speeds of 240km/hour and surges heights between 6m and 7.5m, this storm also took an immense toll on human life. According to available statistics, the total number of deaths was 138,868 (Haider,

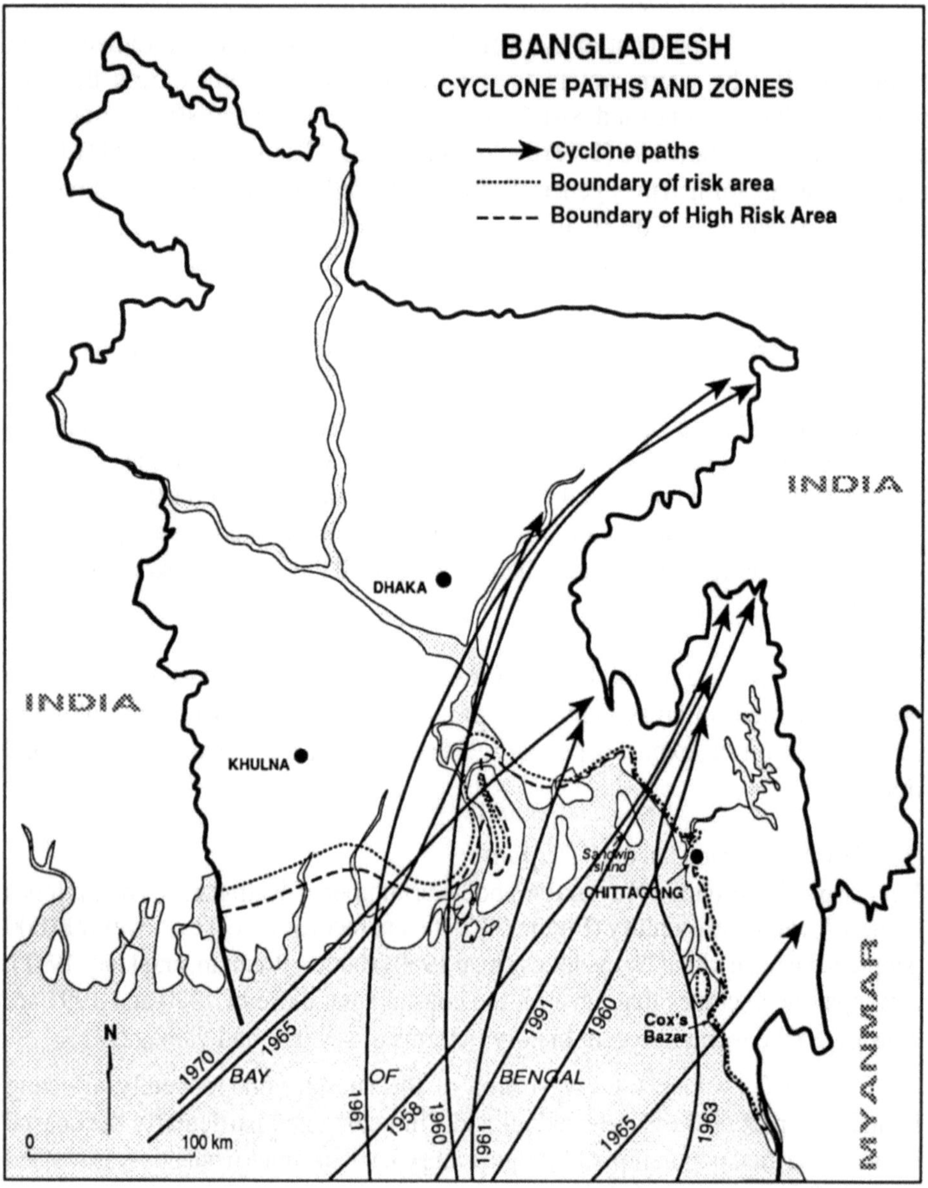

FIGURE 7.8: High risk cyclone areas of Bangladesh and cyclone paths, 1958-1991

Source: FAP 7, 1992 and Multipurpose Cyclone Shelter Project, 1993.

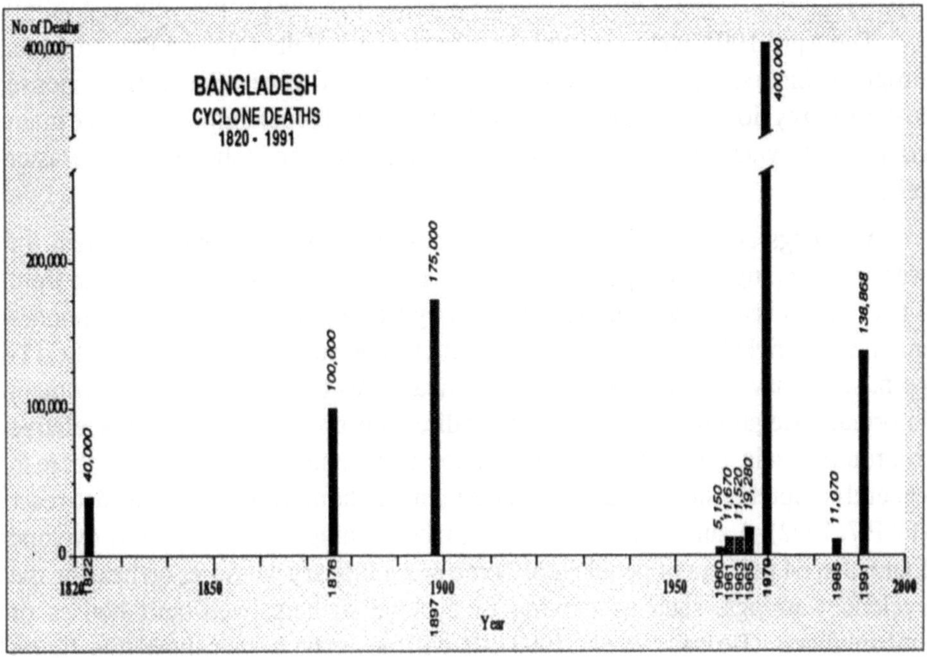

FIGURE 7.9: Number of cyclone deaths in the Bay of Bengal, 1820-1991

Source: Talduker, *et al.*, 1992.

Rahman, and Huq, 1991). The majority of deaths were below the age of ten and one-third of the fatalities occurred below the age of five; disproportionately more women died than men (Hossain *et al.*, 1992). The damage caused by the cyclone was enormous. An estimated one million homes were completely destroyed and a further one million damaged. Up to 60 per cent of cattle, 80 per cent of poultry stocks and up to 113,360ha of standing crops were destroyed (FAP 7, 1992)[8,9]. A Government of Bangladesh and United Nations Development Programme (UNDP) Task Force evaluated the damages at US$2.4 billion (GOB, 1992a), while the United Nations Centre for Regional Development (UNCRD) estimated the damage at US$3.98 billion (Hoque, 1991).

The strength of the 1991 cyclone was approximately the same as the 1970 cyclone. However, it claimed less than half the number of lives even though a larger number of people was exposed to the storm due to population growth since 1970. The number of lives saved has been attributed to a combination of better cyclone warnings, more cyclone shelters and the greater efficiency of post-disaster relief.

The April 1991 cyclone tested the effectiveness of the Cyclone Preparedness Program (CPP) set up in 1973 by the Bangladesh Red Crescent Society in response to the 1970 cyclone. It is a grass-roots volunteer based programme with three main components: warning, shelter construction and evacuation to shelters, and disaster relief [10].

Warnings are issued by the Meteorological Department using a system for warning shipping of an impending cyclone. The warning system uses a scale from 1 to11 and is issued over public radio and by CPP volunteers using megaphones mounted on rickshaws. The differences in cyclone severity indicated by the 11-point scale have limited use for coastal inhabitants because they do not contain information regarding the wind speed and direction, possible surge heights relative to ground levels, possibilities of inundation and damage, etc. However, there is general comprehension that the higher the number, the more urgent the need to react (FAP 7, 1992). A survey of the response to the warning of the April 1991 cyclone found that 64 per cent of people took some precautionary measures on hearing the cyclone warnings, such as moving to cyclone shelters, well-built houses or embankments (Hossain *et al.*, 1993). It will never be known how this figure compares to the total response of coastal residents to the warning, as this figure applies to the survivors only. Did those who died in the tragedy take heed of the warning in the same proportion as those who survived?

Reinforced concrete buildings designed to save human life from cyclones are the second part of the CPP program. A five-year shelter construction program was initiated in 1972, but was abandoned after only 234 shelters were built — an estimated 10 per cent of actual demand (Talukder and Ahmad, 1992). The program was re-started after the 1985 cyclone and a further 62 shelters were constructed.

Only 13 per cent of the cyclone survivors interviewed by a research team indicated that cyclone shelters were available to them. Reasons cited for non-evacuation to shelters included: the distance to the shelter; a lack of security of household assets; a perception that their own house was more secure; and a lack of drinking water and toilet facilities in shelters (Talukder *et al.* , 1992). Also, the cyclone shelters built after the 1985 cyclone have only one large communal room instead of the three-storey shelters built after the 1970 event. The 1985 shelters were found to be less effective partly because "Crowding with men is considered a violation of purdah for women. Cultural values and prejudices are often stronger than the need for survival" (Talukder *et al.*, 1992). Such cultural values must be considered in any future CPP schemes.

The post-cyclone effects on the survivors can be as severe as the immediate impacts of the cyclone itself. The risk of disease is enhanced, and rotting human and animal bodies increase water pollution. The most immediate requirement after a cyclone is to supply a huge quantity of fresh water. Due to disruption of communication, it is difficult to send fresh water, food and house-building materials to the whole cyclone affected area (most of which is remote or becomes remote). The post-1991 cyclone situation was not different from other previous severe cyclones. Local voluntary organisations started relief operations the day after the disaster. Supplies were limited, and the immediate relief operation was seen by local people as inadequate (Talukder and Ahmad, 1992). After the massive relief operation, the restoration of communication and transportation links was a high priority task.

As mentioned earlier, coastal embankments were severely damaged in many locations. The cost to repair the estimated 400 km of damaged embankments is 2,038 million *Taka* ($US 54.8 million[11]). Renewed coastal afforestation has been suggested for the 700 km of embankments in order to dissipate future cyclone wave energy; this would cost 200 million *Taka* (US$ 5.26 million) (FAP 7, 1992). A further 1,554 million *Taka* (US$ 40.9 million) is thought necessary for the long-term reconstruction of the entire embankment network.

The role of the coastal embankment is not to protect land completely from storm surge flooding. Instead, they are designed to dissipate the energy of cyclone waves and to allow overtopping. However, "for the protection of human lives, a coastal embankment that is breached is worse than no coastal embankment" (FAP 7, 1992). Embankments can give residents a false sense of security. As a result, upgrading coastal embankments and planting coastal forests to help protect the embankments is only a partial solution. However, it is thought that if embankments are constructed following proper engineering design criteria, and post-construction maintenance is done adequately, the chance of embankment failure is minimal for design (or below design) level events. In Potenga, during the 1991 cyclone, the embankment was washed out due to poor design and construction; the embankment was unable to resist the force exerted by the turbulent storm surge, resulting in a huge loss of life in the country-side villages.

The need for a holistic approach to cyclone impact reduction has been recently recommended (FAP 7, 1992). Such an approach would include shelters, suitable evacuation roads and improvements of the early warning system. Although such an approach is clearly warranted, there may be a requirement for an even broader approach to the cyclone problem:

... In spite of the many disadvantages associated with coastal occupancy,

> most residents expressed a willingness to continue living wherever settled. This indicates an indifference to the cyclone hazard despite their recognition of it. Fatalistic attitudes towards the environment combined with the absence of any alternative choice, particularly, better economic opportunities elsewhere, have been the main factors in determining their continued habitation in the hazardous zone (FAP 7, 1992)

A broader approach to the problem would consider cyclones from the viewpoint of those living with the cyclone threat — the occupants of the coastal zone. This would place their perception of cyclone danger within the context of their poverty, their lack of alternative (and safer) homes, and their close relationship to the coastal environment. These issues require urgent attention, or the lessons from the 1991 cyclone will not be learnt before the next cyclone strikes.

BOX 7.2: Coastal population at risk

> The 24 million people of the coastal zone are closely tied to the resources that land and water provide, either directly through fishing and agriculture, or in the processing of coastal fish, shrimp and timber. People work within the seasonal moods of the coast: monsoon floods and rains, dry season salinity and the seasonal threat of cyclones. However, rapid population growth has pushed many people to the margin of survival in the coastal zone, where there may be as many as 45 to 50 million people by the year 2050. Climate change and sea-level rise could then be putting a much higher number of people at risk — people with more tenuous means of livelihood than today.

The recent survey of cyclone shelter capacity by the Multi-purpose Cyclone Shelter Project (MCSP) found that, even with an expanded-shelter building programme proposed by donors, there will not be enough shelters for the projected 6.4 million people who will be living in the High Risk Area by 2002. To rectify this situation, the construction of 2,500 new cyclone shelters would be required (Figure 7.10). Considering the projected national population growth rate for the period 2000-2025 and 2025-2050, the population of the High Risk Area may reach 8.75 million and 12 million, respectively. For this increased population, 1300 and 3200 cyclone shelters would be required in addition to those proposed by MCSP.

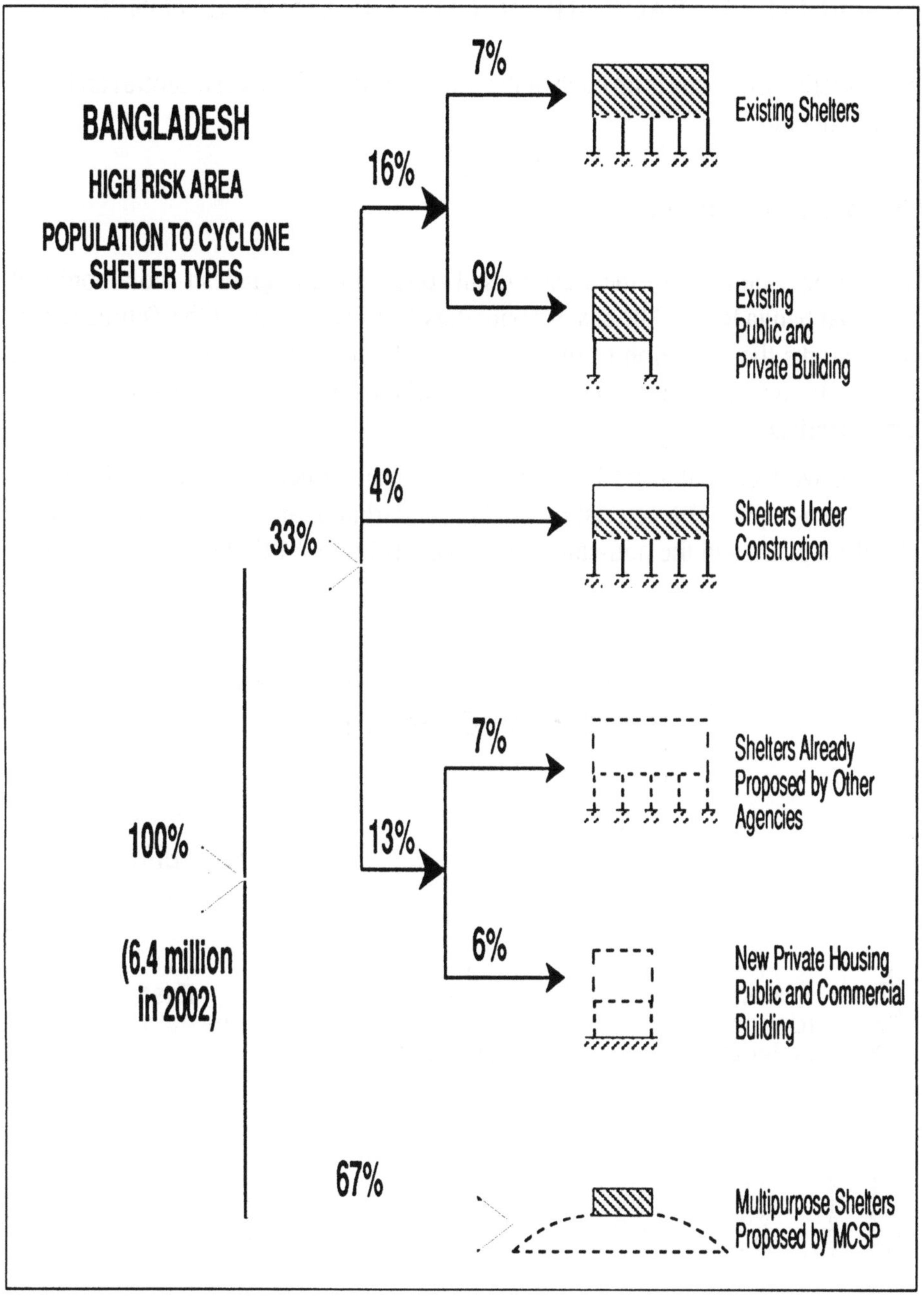

FIGURE 7.10: Distribution of possible future movement of High Risk Area population to different types of cyclone shelter accommodation

Source: Multi-purpose Cyclone Shelter Project, 1993.

7.3 SPECIAL CHARACTERISTICS OF THE COASTAL ZONES

In this section the focus is on each of the three coastal zones (west, central and east) in more detail.

7.3.1 West Coastal Zone

The western boundary of the west coastal zone is the Bangladesh-India border at the Hariabhanga River. The eastern boundary is the west bank of the Tetulia River, which marks the transition to the increased influence of the Meghna River. The northern limit is taken as the landward limit of the coastal districts of Khulna, and part of Barisal.

The west coastal zone has three distinct sub-zones defined by distinctive physio-geographic characteristics — the Sundarban mangrove forest, the saline tidal floodplain, and the non-saline floodplain (Figure 7.11). The far western part

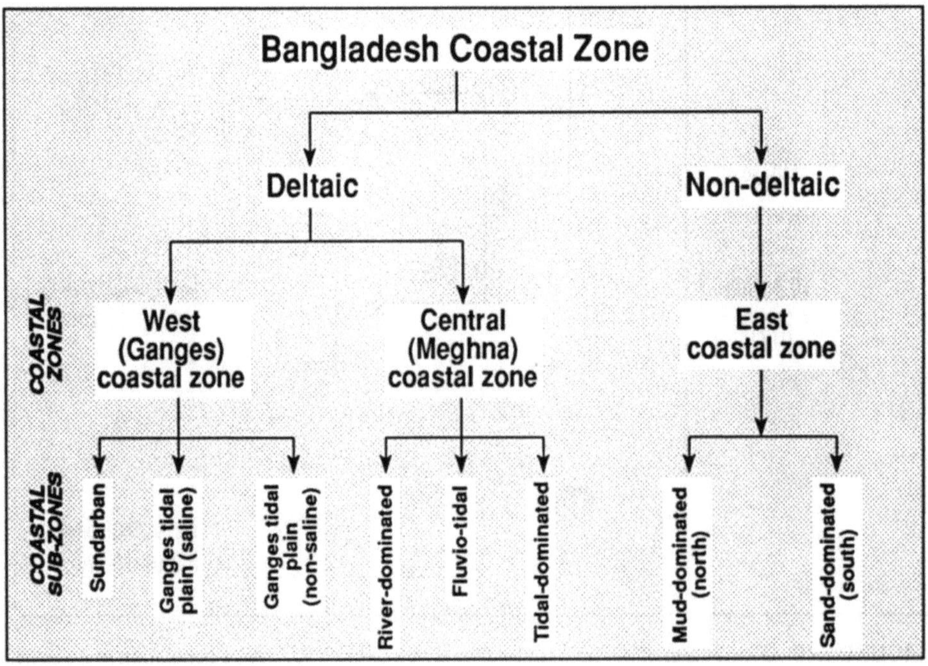

FIGURE 7.11: Bangladesh coastal zone sub-divisions

of the zone is the relatively inactive area of the Gangetic delta plain. Termed the 'moribund delta', it was until very recently (in geological terms) the main outlet for the Ganges River system. The reduction in the supply of sediment, and the increased interaction with saline water that resulted, favoured the colonisation of the Sundarban mangrove forest. The tidal floodplain areas are low-lying, with an intricate network of river channels. Salinity affects the soils and groundwater of the north-west and south-east of the zone. Extensive areas in the south and west have been embanked to create polders which protect land against tidal flooding with saline water (Figure 7.12).

Vegetation and Fauna

Mangrove forest (the Sundarban) occurs in the south-west of the zone. Its growth is favoured by sheltered, low energy conditions (away from high waves) where

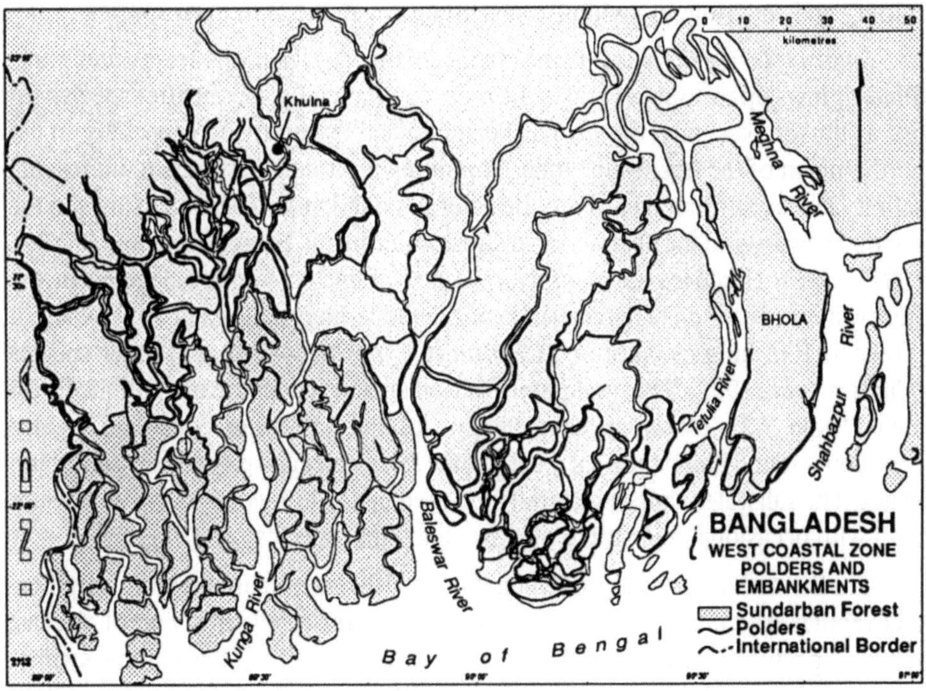

FIGURE 7.12: Coastal polders and embankments of the west coastal zone

Source: GOB, 1992b.

there is sufficient interchange of water and sediment to bring nutrients. Too much sediment, as well as too little sediment, can inhibit mangrove growth. Mangrove species are specially adapted to survive in saline waters. Different species thrive in different salinities through their own specialised adaptations to ensure growth and reproduction.

Mangroves have a strong interrelationship with tidal and river flows. The mangrove trees slow down these flows, allowing sediment to drop out of suspension, building up the mud flats and supplying nutrients to support mangrove growth. The interplay between tree growth and the flow of fresh and saline waters through the forest is extremely subtle. Any small disturbances to the forest, particularly in terms of water and sediment flows can upset this interplay and affect the forest.

The salinity distribution and degree of waterlogging within the forest determines the overall ecological zonation of the Sundarban into freshwater, moderately salt-water and salt-water zones (Table 7.2; Chaffey, *et al.*, 1985). Differences in salinity and waterlogging within zones produce variations in local species.

Sundri (*Heritiera fomes*) is the characteristic species of the freshwater zone. *Sundri* grows up to 15m tall and is the most commercially sought after species for use as timber. Recently, *Sundri* has been affected by top-dying especially in the northern part of the freshwater zone. Top-dying is thought to be due to increasing salinity levels, attributed partly to the natural reduction in freshwater flows to the region and partly to the abstraction of freshwater by the Farakka Barrage (FAP 4, 1992). *Gewa* (*Excoecaria agallocha*) is also found to varying degrees in the freshwater zone, and is characteristic of the moderately salt-water zone where it grows to 10m in association with *Goran* (*Ceriops decandra*) and other species (Chaffey *et al.*, 1985). The moderately salt-water zone covers nearly half the forest area (Table 7.2). *Goran* is characteristic of the salt-water zone and is also commonly found lining the tidal creeks of the freshwater and moderately salt-water zones. The saltwater zone is generally a closed understorey of *Goran* 4 to 5 metre high with occasional overstorey of *Gewa* and other species.

The Sundarban forest supports numerous and diverse mammals, birds, amphibia and reptiles of commercial and conservation importance[12]. It is one of the last remaining refuges of the Royal Bengal Tiger (*Panthera tigris*), the national symbol of Bangladesh, and the estuarine crocodile. Both are endangered species (GOB, 1991).

The forest is a multi-use natural resource of significance to the entire economy of Bangladesh. An estimated 500,000 to 600,000 people are directly employed in the commercial exploitation of mangrove timber, fuelwood resources, fish, honey, beeswax and mollusc shells (used to produce lime) of the forest (ESCAP, 1987).

The indirect economic value of the Sundarban is also vitally important, and has recently been estimated to be seven times greater than the direct economic value of the forest (FAP 4, 1992). The indirect value of the forest is a result of the protection the forest gives to the breeding of shrimp and fish larvae, thereby protecting the marine, brackish and freshwater coastal fishing industries. The forest also provides protection for Mongla port and Khulna city from storm surges.

The Sundarban has been systematically managed since 1887, when a system of forestry and fishery permits was established. Since then the forest has been a considerable source of revenue and now contributes approximately 2 per cent to the gross domestic product of Bangladesh (Task Force, 1991). Royalties are paid to the Department of Forestry in advance for the felling of sections of the forest. The royalty system also covers grass, *golpatta* (*Nypa fruticans* — used for thatching and fencing), *hental* (*Phoenix paludosa* — used for underthatching), honey, beeswax and shells collected from the forest. The Department of Fisheries collect royalties for fishing and shrimping rights in the forest, which can be considered an indirect economic forest resource.

The Sundarban is in a state of decline, attributed to the combination of the effects of empoldering, reduction in freshwater flows from human abstraction, increased agriculture, heightened silt deposition, and non-sustainable forestry management. Unless checked, the decline of the Sundarban will reduce its economic usefulness, with potentially serious implications for the coastal economy. In addition, depletion of the Sundarban could expose Khulna city and Mongla port to increased risks from storm surges.

Human Modification

Traditionally, small embankments were built by landlords to protect land from tidal flooding. Landlords used to pay for the construction themselves, on the understanding that they would collect a percentage of the crops produced within the protected polders.

Initiated in the 1960s, the Coastal Embankment Project (CEP) created a vast network of earthen embankments built to improve production within traditional embanked areas (Figure 7.12). Also, embankments were built in areas of the coastal zone previously unprotected. There are presently 80 polders[13] in the west coastal zone covering an area of 8,600km^2. These polders were designed to increase agricultural production through tidal flood protection and the enhancement of land drainage. Due to inadequate maintenance the coastal embankments cannot serve the desired purpose properly. Maintenance budgets have been too small to keep the infrastructures in a good state of repair.

The success of the CEP has been mixed[14]. Agricultural production has not been increased in many polders due to problems with saline intrusion, silting up of sluices, severe channel siltation (especially in the north-west of the zone), and deliberate breaching of embankments. In turn, this has impeded drainage of rainwater run-off from adjoining polders, with adverse impacts on crop production (though there may be benefits for shrimp and fish production[15]). Channel siltation has also created problems for river transport, especially for Mongla port and Khulna city.

Inter-polder channel siltation has seriously reduced the long-term sustainability of the polders. The recent Flood Action Plan study of the west coastal zone (FAP 5, 1992) has concluded that future human intervention will be required to maintain inter-polder channels and, as a result, to maintain agricultural production from polders on a sustainable basis. Suggested forms of human intervention include dry season river flow augmentation from the Padma River and the creation of large 'mega-polders'.

7.3.2 Central Coastal Zone

The western boundary of the central zone is drawn along the west bank of the Tetulia River (Barua, 1991a). The eastern boundary of the zone is drawn through the Sandwip channel and meets the coast at the mouth of the Feni River. The northern limit is taken as the landward limit of the coastal districts.

The central zone is the most dynamic of the three coastal zones. The sediment-laden rivers of the Ganges, Brahmaputra and Meghna deposit vast amounts of sediment into the Bay of Bengal at the mouth of the Lower Meghna estuary. These sediments have accreted to form the fertile and densely populated estuarine floodplains and offshore islands of the central coast. This sediment is moulded by strong tidal and fluvial processes, by moderate wave energy, and by high waves and surges produced during cyclones.

Tidal flows are strongest in the eastern part of the zone, whereas river flows are strongest in the western part. The relative strengths of tidal and river flows have been used to divide the zone into "river dominated", "fluvio-tidal dominated" and "tidal dominated" sub-zones (Figure 7.11; Barua, 1991a). Each sub-zone might be expected to respond differently to future sea-level rise and climate change.

Sediment Supply, Movement, and the Accretion/Erosion of Land
Total sediment volumes supplied to the central zone has been calculated to be 1,099

to 2,180 million tonnes per year depending on the method used and over what period of time the calculations are based (Barua, 1991a; Milliman and Meade, 1983). These river-derived sediment volumes are the second highest in the world, after the Yellow River in China.

The Brahmaputra River carries mainly silt-sized sediment, with a proportion of fine sand. In contrast, the Ganges River carries smaller mud-sized sediment (Siddique, 1989) with a proportion of fine sand. There is low seasonal variation in sediment concentrations in the coastal area (Eysink, 1983) because suspended sediment concentration is influenced more by tides than rivers. In the Sandwip tidal channel where tidal flows are strong, a measurement of 4,225 mg/l has been recorded, as compared to only 1,230 mg/l at the mouth of the Meghna River, where tidal flows are weaker (Barua, 1990). Suspended sediment concentrations can be twice as high during spring tides as in neap tides.

Freshwater river flows and saline marine flows interact to move the clay and silt sediment while it is still in suspension. Sediment 'falls-out' of suspension once suitable conditions are reached. Such conditions can be due to slack water, to particular chemical conditions of marine water, or to biological processes. Often sediment deposition is a result of a combination of physical, chemical and biological factors and is very difficult to predict as a result.

Newly-deposited sediment in the Bay of Bengal is unlikely to stay in the same position for long. Wave, tidal and river action can readily re-suspend sediment and move it to another part of the Bay. This could take place seconds, minutes, days, years or centuries after the original sediment deposition. Every new parcel of sediment that enters the Bay of Bengal from Bangladesh's rivers is therefore destined to many cycles of deposition, suspension and re-deposition before finally coming to rest.

The constant re-cycling of sediment, plus the annual additions from the monsoon floods, produce complex patterns of erosion and accretion in the lower Meghna estuary. These patterns can be analysed by comparing historical maps with modern satellite images. As shown in Figure 7.13, in the region west of the Tetulia River the present position of the coastline is relatively similar to the coast mapped by Rennell in 1760 and published in an atlas in 1779 (LaTouche, 1910). In contrast, in the region east of the Tetulia River there has been a general tendency for the coast to grow seaward, albeit with large fluctuations in the location and size of the offshore islands.

The shape of the central coast offshore islands and their erosion/accretion pattern are a function of the increased tidal energy from west to east. To the east, higher tidal ranges, volumes and speeds result in enhanced erosion and accretion.

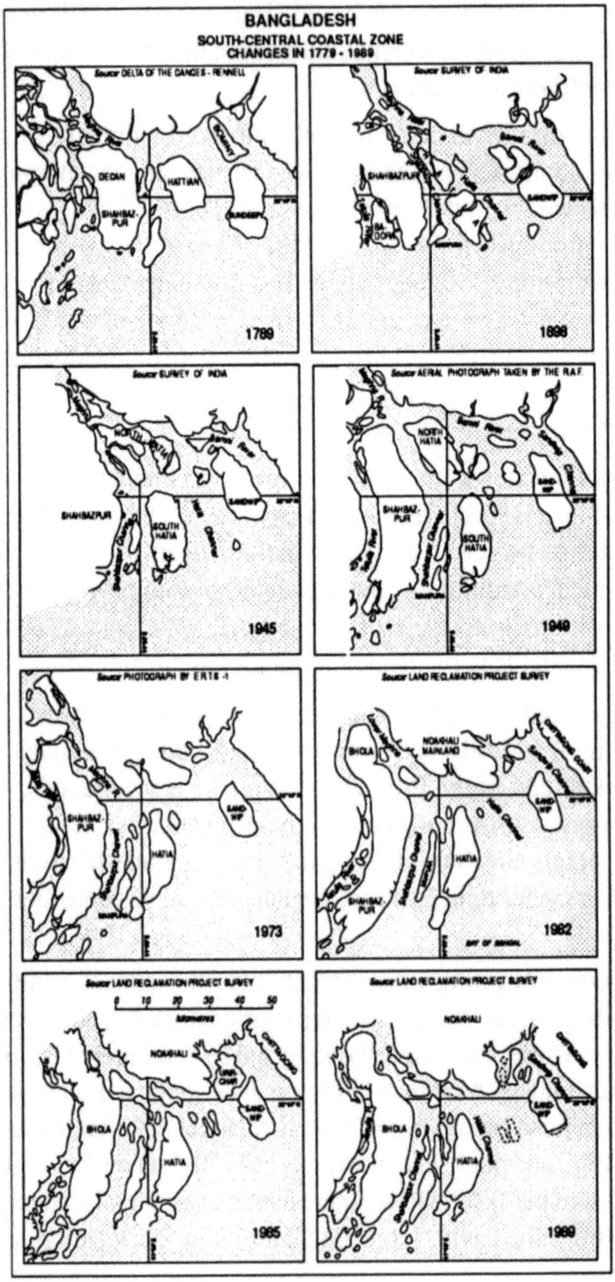

FIGURE 7.13: Historical changes in the south-central coastal zone, 1779-1989

Source: MPO, 1986 and LRP, 1990.

Island shapes are also different in the east with larger tidal flats produced by higher tidal ranges. West of the main Meghna channel, accretion of land generally occurs on the southern and western side of the coastal islands and erosion on the eastern side, due to the dominant westwards sediment transport in the area.

Analysis of the coastline changes between 1940 to 1963 reveals very large changes in the position of coastal land in the Lower Meghna estuary, but only small changes in the total amount of land (Eysink, 1983). The total area of accreted and eroded land is 1,077km^2 and 1,356km^2 respectively. Thus, the net loss of land area to erosion was 279km^2.

There are many uncertainties in measuring and predicting the rate and direction of sediment transport and the resultant land accretion and erosion. Computer modelling techniques are used extensively in this work. However, for a computer model to be accurate it must be calibrated and tested using data from the natural environment. Where such data are scarce or unavailable, as for large parts of Bangladesh's coastal zones, computer models can give unreliable or misleading results; this severely limits their predictive value. The paucity of data and modelling studies on sediment movement and land erosion/accretion is a major constraint to the assessment of potential sea-level rise impacts to the central coast.

Flora and Fauna

In contrast to the west coast, natural mangroves in the central zone are scarce. However, there have been an estimated 104,000ha to 109,000ha of mangroves planted by the Coastal Afforestation Project (CAP) since 1965 using mainly *Keora* (*Sonneratia apatala*), *Baen* (*Avicennia officinalis*) and *Sadda Baen* (*Avicennia alba*) (Katebi and Bari 1989, Khan, 1991).

During the first phase of the CAP, mangroves were planted on the seaward side of coastal embankments and in borrow pits. Such plantings were designed to protect the embankments from wave and tidal energy. An important side effect of the mangrove plantings was a significant increase in accretion of sediment. Later phases of the CAP were implemented specifically to enhance the natural land accretion and to expedite further the creation of coastal land expected from cross-dam construction.

The water of the central zone is rich with fish and shrimp. The mudflats also provide an important habitat for migratory waterbirds, of which 49 species have been recorded (GOB, 1991).

Human Influences

People have been reclaiming land from the Bay of Bengal for thousands of years.

The majority of reclamation has occurred in the central coastal zone. During the last century, and the first half of this century, feudal landlords used traditional embankment building techniques to reduce salt water incursion and reclaim small areas of land. The extent of this practice and its impact on the coast as a whole has not been determined.

Modern techniques of land reclamation overtook traditional techniques in the central coastal zone when the CEP (Coastal Embankment Project) was initiated. As part of the CEP, cross dams (embankments across tidal channels) were built to block the flow of water. The effect of cross dams is to fill up tidal channels with sediment. Cross dams succeed in areas where there are wide, shallow channels carrying water with high concentrations of suspended sediment. The construction of the dams is relatively easy under these conditions and the growth of new land is rapid because large volumes of sediment are quickly deposited.

The first major cross dam (Cross Dam No. 1) was built in 1957 between the Noakhali mainland and Ramgati Island in the Lower Meghna estuary. The 12km embankment enhanced the siltation of the Bamni Channel. A second embankment (Cross Dam No. 2) was built in 1964, 18km downstream of Cross Dam 1. The success of the Noakhali cross dams may also be due to the large pulse of sediment supplied to the central coast as a result of the 1950 Assam earthquake (Kingdon-Ward, 1951 and 1953) which induced landslides into the rivers of Tibet and north-eastern India. Approximately 900km^2 of land has been reclaimed by the direct action of the two cross dams (Tarafder, 1975). However, the alteration in the flow dynamics is thought to have enhanced erosion on the banks of other parts of the estuary. In comparison to the effects on accretion, the direct erosive effects of cross dams are much more difficult to ascertain. Consequently, calculating the net accretion of new land is difficult.

The social, economic and environmental success of cross dam construction is even more difficult to measure than the formation of new land. Questions regarding the overall benefit of such constructions to Bangladesh have stalled the construction of a cross dam designed to reclaim over 23,000ha of land from the sea north of Sandwip island.

7.3.3 East Coastal Zone

The southern boundary of the east zone is the border of Bangladesh with Myanmar (Burma). The north-western boundary is drawn through the Sandwip channel and meets the coast at the mouth of the Feni River. The inland limit is taken as the landward limit of the coastal districts.

The east coastal zone is the non-deltaic coast of Bangladesh and is characterised by a narrow coastal plain. The coastal plain is intensely cultivated, and the rural areas are densely populated as a result. The port city of Chittagong is Bangladesh's second largest city and has rapidly expanded. The coastal town of Cox's Bazar is the focus of the country's internal tourist industry and is being promoted to be a future centre of international, as well as local, tourism.

The zone can be divided into southern and northern sub-zones at the Karnafuli River. Each sub-zone is distinctive. The northern (mud-dominated) sub-zone contains a smooth, nearly level, estuarine mud coast in its northern-most reaches, whereas further south there are gently sloping sandy beaches with low (1m to 3m) dunes. The coastal plain varies in width from one to 12km (Siddique, 1967). The narrow estuarine plain is backed by loosely consolidated piedmont deposits which adjoin low hills. Although non-deltaic, this coastline is influenced by the dynamic forces operating in the adjacent Meghna estuary.

South from Chittagong to the Naaf River on the Bangladesh/Myanmar border, the shoreline of the southern sub-zone is a wave-dominated sandy coast with a narrow coastal plain and 145km of gently sloping sandy beaches backed by irregular sand dunes. There are ancient, raised beach-dune ridges behind the present-day beach. In addition there are 15km of rock cliffs. The coastal plain is backed by hills.

In the far south of the southern sub-zone, Teknaf peninsula, which is 75km long, separates the Naaf estuary from the sea. This peninsula has some hills reaching up to 300m high, but is generally a low sand spit with an average width of 160m at high tide (GOB, 1991), with 2km wide sand flats exposed at low tide containing rock boulders. Within the estuary are mudflats and mangroves that are important for resident and migrant birds. The area is recognised for its international conservation importance under the RAMSAR convention.

St. Martin's Island, or Jinjiradwip, is the only island with living coral in Bangladesh. It is 8km off the coast of Bangladesh. The island is approximately 8km^2 in area, with numerous surrounding small islands. The island is founded on sandstone and is generally 3m high, with a maximum elevation of 10m (GOB, 1991).

Sediment Supply, Movement, and the Accretion/Erosion of Land

The sediment supply to the southern sub-zone is thought to be limited. There are possible sediment sources from the rivers draining the Chittagong Hill Tracts, from the erosion of rock cliffs south of Cox's Bazar, from sand transported from the mouth of the Meghna River, or from sand derived from the continental shelf (Das,

1981). The exact source and volumes of sediment are uncertain.

In the northern sub-zone, the sediment transport is complicated by the action of strong tidal flows and wave action. Strong re-circulation of sediment is assumed to take place due to the tides. However, this has not been verified.

In the southern sub-zone, the dominant direction of littoral transport is thought to be to the north due to the southwesterly approach of swell waves (Siddique, 1967). However, at the extreme south of the area, at Teknaf and Moulavir Char, the spit is moving southwards, suggesting a southwards sediment transport (although the sediments supplied by the Naaf River may play a role). It is assumed that there is a re-circulation of sediment from the north to south in the offshore zone because there is no sand north of Chittagong and because a sand and shell island (Sonadia) exists off the coast of Cox's Bazar.

In the south, there is general coastal stability with localised areas of accretion, such as at Cox's Bazar. There is no significant shoreline erosion, which suggests a long-term balance between the volume of sediment supplied to the coast and the volume being removed, with large short-term fluctuations between erosion and accretion.

In the northern sub-zone, there is an overall trend of accretion, although erosion was reported during the 1930s (Siddique, 1967). In the central section of the coast there is localised erosion at Patenga.

Vegetation

Coastal sand dune vegetation has been modified at Cox's Bazar for tourist use. Some localised dune vegetation exists on the landward side of Moulavir Char spit. There is little vegetation on the rest of the southern section of the coast, although the coastal plain between the coast and the Chittagong Hill Tracts is cultivated with crops and trees.

Mangroves are mainly restricted to the delta of the Matamuhuri River, called the Chakaria Sundarban. There are poorly developed mangroves on the shore and along the creeks of the northern section of the east coast. The forest is under 12m high, and is dominated by various species of *Avicennia*, with *Avicennia tomentosa* (locally called *Baro baen*) being the most common (ESCAP, 1987).

Human Influences

The Chittagong industrial area had around 3km of sloping stone-faced sea defence wall. However, the 1991 cyclone destroyed this wall, and a new 7km to 10km long sea-wall is under construction (FAP 7, 1992).

The Port of Chittagong requires dredging to maintain its navigation channel. The dredge spoil is either used for raising land for new industry or housing, or is dumped at sea.

In 1975, there were 8,500ha of the Chakaria Sundarban, consisting of 20 mangrove species, reduced from an estimated natural area of over 20,000ha due to clearing for agriculture and salt pans (Choudhury, *et al.*, 1990; GOB, 1991). By 1988, only 973 ha remained, the result of clearing by cutting and burning for 8,000ha of shrimp ponds (Figure 7.14). Commercial shrimp farming in the area was started in 1977 with a 228ha site. Shrimp activity was accelerated by a Department of Fisheries pilot project in 1978 which covered 2,024ha. This large-scale development unwittingly encouraged clearing of the forest by local people, although the Department of Forests' policy at the time was for forest conservation. This emphasises the need for the coordination between government departments for the effective management of coastal resources.

The shrimp ponds in the Chakaria Sundarban produced an estimated 1.1 million kg of shrimp in 1989 valued at 165 million *Taka* (US$4.3 million) (Choudhury, *et al.*, 1990). Local shrimp farmers feel that the mangrove destruction

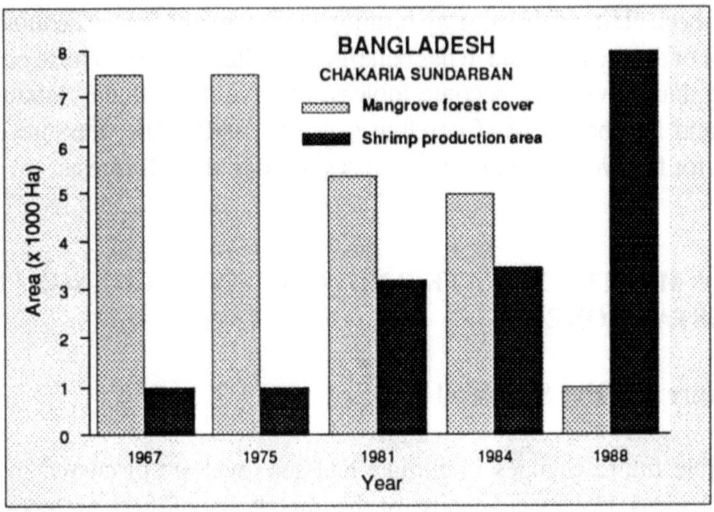

FIGURE 7.14: Changes in forest cover and shrimp farming area in Chakaria Sundarban 1967-1988

Source: Choudhury *et al.*, 1990.

has not affected their production, which is verified by estimates of annual production. It is unclear whether the addition of nutrients (or other measures) by farmers is acting to offset the impact of mangrove removal on the natural breeding cycle of the shrimp. In the long run, shrimp production may be reduced as a result of deforestation.

BOX 7.3: Vulnerability of coastal resource systems

> The current pressure on land resources (forest, agricultural and industrial land) and coastal fisheries is intense and unsustainable. Unchecked, this will lead to long-term resource depletion. The problem will be exacerbated by the rapidly increasing coastal population. These long-term trends of resource depletion are punctuated by the devastating effects of cyclones and storm surges. In future, climate change and sea-level rise threaten to contribute additional stresses to natural and human coastal resource systems that are already extremely vulnerable to environmental fluctuations and population pressure.

It has been claimed that shrimp farming has not increased the economic status of the poor of the area. Instead, it has benefited rich land owners. Those people who fished the mangrove have become jobless, or have become day labourers, and grazing land has been lost (Choudhury, *et al.*, 1990). The pressure on forest resources for fuelwood and other uses has shifted to the hill forests.

7.4 HOW MIGHT CLIMATE AND SEA LEVEL CHANGE IN THE COASTAL ZONE?

7.4.1 Future Relative Sea Level

The possible future changes in climate and sea level are discussed in detail in Chapters 2 and 3. However, because of the importance of these projections to the coastal zone, they are briefly summarised below.

Over the past 100 years, global-mean sea level has been slowly rising at a rate of 1.0mm to 2.0mm per year. Although the global temperature has also been rising over the same period, it is not yet possible to say conclusively whether the measured sea-level rise has been caused by global warming, but there is strong circumstantial evidence that this may be the case.

As a result of future global warming, the rate of sea-level rise is likely to increase. It is thought that this future sea-level rise will be 2 to 6 times faster than the present rate of rise. Estimates of future sea-level rise from the Intergovernmental Panel on Climate Change (IPCC; Warrick and Oerlemans, 1990) are shown in Table 7.3 (see Chapter 3).

Scenarios of changes in global-mean sea level must be combined with vertical changes in the land surface to produce estimates of 'relative' sea-level change for the Bangladesh coast. For past changes in sea level, the vertical land movement component is contained in measurements of relative sea-level change recorded by tide gauges, installed in ports to aid navigation. The available tide-gauge data for Bangladesh has been examined in Chapter 3. Because of the relatively poor data obtained by these tide gauges and the problems in comparing data from different tide gauges, the trends in vertical land movement cannot yet be clearly identified. The accurate estimation of such trends from tide-gauge and geological data is a priority area of future research.

In the absence of reliable information on relative sea-level change, it is assumed that there is a general trend of subsidence in the deltaic regions of the coastline, and a general trend of stability of the non-deltaic south-east coast. Therefore, it must be borne in mind that the projections of future sea-level rise listed in Table 7.3 may be applicable for the east coast of Bangladesh, but may be underestimates for the central and west coasts.

TABLE 7.3: Scenarios of future global mean sea-level rise

Global sea-level rise scenario by year (cm)				
	2010	2030	2050	2100
Low	4	8	15	31
Best-estimate	8	18	30	66
High	13	29	48	110

Source: Chapter 3, as derived from Warrick and Oerlemans, 1990.

7.4.2 Tidal Height and Range

Future changes in relative mean sea level may affect the strength and height of the tides in the Bay of Bengal. Because of the importance of tides in shaping the coast of Bangladesh, consideration of possible future changes to tides is critical.

The characteristics of tides are controlled by a number of local factors, the most important of which are relative mean sea level, the shape of the coastline, and the depth and shape of the continental shelf. Consequently, future changes to the tides of the Bay of Bengal will be a function of changes in mean sea level due to global warming, changes to the coastline of the Bay due to the erosion or accretion effects of sea-level rise and climate change, and changes to the sea bed as a result of erosion and accretion of the shoreline. However, change to the sea bed would have to be substantial to produce a measurable changes in future tides. In order to raise the sea bed sufficiently to influence tides, the required increase in sediment supply to the offshore zone would be massive. At present, an increase of this magnitude is thought to be unlikely. As a result, Chapter 3 concluded that it must be assumed that the present levels of high- and low-tides around the coast of Bangladesh will rise in the future by the same amount as the rise in future mean sea-level.

7.4.3 Tropical Cyclones

There has been great concern in Bangladesh that the number and intensity of tropical storms and severe tropical storms (cyclones) may increase in the future as a result of climate change. This concern arises from the argument that, since the development of tropical storms depends on a warm (warmer than 27°C) ocean surface for energy, and because the ocean surface is predicted to warm in the future, cyclones would become stronger and more frequent. However, there are a number of additional factors that must combine before a tropical storm can intensify into a tropical cyclone. These have to do with wind shear, wind direction and strength, and high mid-atmosphere humidity over the developing storm. The potential changes in such factors cannot yet be accurately predicted by climate models.

Consequently, Chapter 2 states that, on a global scale, the evidence for the potential future increase of tropical cyclones is 'weak', and that 'little can be said specifically regarding cyclone frequency and intensity in the Bay of Bengal'. Therefore, it is assumed that the present number and intensity of tropical cyclones that strike the Bangladesh coast may remain the same in the future.

7.4.3 Temperature and Rainfall

Chapter 2 summarised the possible future changes in the climate of Bangladesh and the south-Asia region. From an examination of global climate model results, it was concluded that a future warming is likely in the region. The climate models suggest that temperatures will increase in Bangladesh during all seasons by approximately 1.0°C to 1.5°C by 2030 for a "business-as-usual" scenario of greenhouse gas emissions. By 2050, Bangladesh is projected to be 1.5°C to 2.0°C warmer than today.

Climate models differ widely in their prediction of future changes in rainfall, and must be viewed as unreliable as a result. For 2030, the best-estimate scenario based on a number of climate models is for an increase of winter and summer rainfall of 5 to 10 per cent (with very large uncertainties) and an increase in monsoon rainfall of between 10 to 15 per cent (with a range of 0 to 20 per cent). By 2050, the best-estimate scenario shows an increase in winter and summer rainfall of 8 to 15 per cent (with very large uncertainties) and an increase in monsoon rainfall of between 15 to 23 per cent (with a range of 0 to 35 per cent) (Chapter 2).

7.4.4 River Flow and Sediment Supply

Future river flows may increase in Bangladesh due to climate change (Chapter 4). Higher future rainfalls during all three seasons in the Ganges-Brahmaputra-Meghna catchment may act to increase the supply of rainwater to rivers and groundwater aquifers in the coastal zone (Chapter 4). Predicted increases in monsoon rainfall may combine with enhanced snow melt in the Himalayas (due to higher winter snowfalls and higher summer temperatures) to increase peak river flows in the monsoon season. Also, if the number of extremely wet years increases in the future, the number of extremely high river levels may increase the probability of severe flooding (Chapters 2 and 4).

Increased future river flows may also increase the volumes of suspended sediment carried by rivers due to the increased erosion of soil in upland river catchments. In addition, increased river flood flows may increase channel instability, increasing river bank erosion and leading to higher volumes of sediment entering rivers. Consequently, the supply of sediment to the coastal zone may also increase. However, as a result of future sea-level rise, this increase in sediment to the coastal zone may be offset by the "back-water effect", whereby the rise in water level in the Bay of Bengal reduces the ability of rivers to carry sediment to the coastal zone.

A modelling study was carried out to simulate the changes in river-bed levels over the next 100 years due to an instantaneous 0.5m rise in sea level in the Bay of Bengal (Halcrow, 1992). Although sea level is predicted to rise slowly over the coming decades, this study gives an indication of the potential importance of the backwater effect. The study's results indicate that sedimentation in the lower Padma could raise the mean bed level by 0.22m after 100 years following an instantaneous rise in sea level of 0.5m (Figure 7.15). Measurable increases in river bed level are also predicted over 200km upstream from the sea.

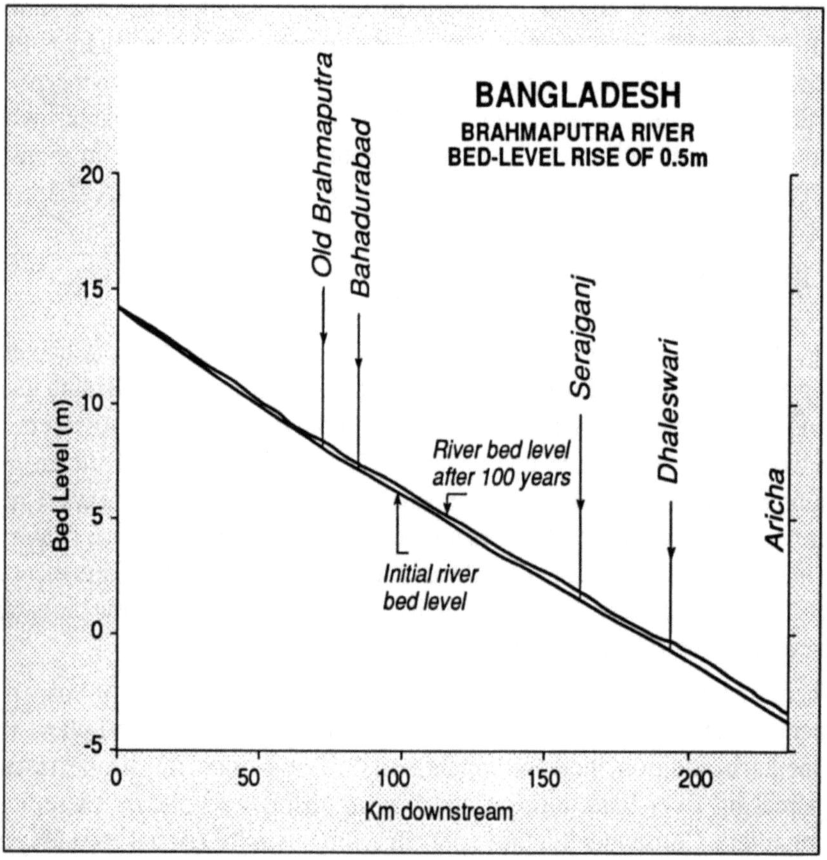

FIGURE 7.15: Modelled response after 100 years of the Brahmaputra River bed level due to an instantaneous sea-level rise of 0.5m

Source: Halcrow, 1992.

Climate changes are predicted to occur in the coming decades, whereas significant changes in river bed levels due to the backwater effect may be relatively minor during this time because of the considerable inertia in the river system. Consequently, an increase rather than a decrease in the supply of sediment to the coastal zone is likely to occur in the coming decades.

The combination of changes in temperature, rainfall and sediment supply are likely to have important implications for the future reaction of the coastal zone to sea-level rise. These implications are discussed in the next section.

BOX 7.4: How will climate change in the coastal zone?

Accurate predictions of future climate change for the coastal zone of Bangladesh are not yet possible. However, *scenarios* of change (that is, what *could* happen) based on climate models suggest a warmer, wetter future. Monsoon rainfall in particular may increase, perhaps by around 20 per cent by the middle of the next century, resulting in higher river flows and discharge. Sea level may rise by about 30cm. It is not clear whether the frequency and intensity of severe tropical storms (cyclones) in the Bay of Bengal would increase or decrease in a warmer world.

7.5 WHAT ARE THE POSSIBLE EFFECTS ON COASTAL WATER AND LAND?

This section examines the sensitivity, vulnerability and resiliency of the coastal systems to climate change and sea-level rise. The west, central and east coastal zones are considered separately.

7.5.1 West Coastal Zone

The Sundarban

Mangrove forests, in their natural condition, are extremely resilient to fluctuations in mean sea level. Mangroves have coped with sea-level changes during the past 10,000 years, which suggests that they can survive rates of sea-level rise or fall of up to 10mm to 15mm per year (Woodroffe, 1990). In general, future rates of sea-level rise, which are predicted to be a maximum of 10mm per year, are thought unlikely to endanger mangrove forest ecosystems (Woodroffe, 1990). However, the considerable natural resilience of mangrove forests to sea-level rise is seriously compromised in Bangladesh by two factors. First, the possible future landward migration of the Sundarban, as sea level rises, is likely to be initiated by landowners and farmers who will not wish to see their land reverting to mangrove. Second, the resilience of the Sundarban ecosystem depends on its health and, therefore, its ability to withstand the stresses imposed by climate change and sea-level rise. If the present trends of over-exploitation of the forest's resources continues, the natural resilience of the Sundarban will continue to decrease. The result will be a mangrove ecosystem that is increasingly sensitive and vulnerable to climate change and sea-level rise.

With or without landward migration of the Sundarban, future climate change and sea-level rise may modify the composition and location of mangrove species. Such modifications will be controlled by the relative changes in sea level and the input of freshwater as they affect the areas of the Sundarban characterised by freshwater, moderately salt-water and salt-water species. Scenarios of future climate change in Bangladesh (Chapter 2) suggest the possibilty of increased rainfall and freshwater flow. Such increases in freshwater input to the Sundarban, especially if they occur in the dry season, could offset the direct increase in salinity due to sea-level rise. However, without increased freshwater input there could be a gradual landward shift of salt-tolerant species, replacing species tolerant to less saline conditions. This could have serious implications for the freshwater mangrove species, including the *sundri*, the most valuable timber in the Sundarban. The *sundri*, if prevented from migrating inland, might not survive the increased salinity levels due to higher sea levels. *Sundri* and other freshwater mangrove species are presently affected by increasing salinity levels (e.g. the top-dying of *sundri*), thought to be due to a combination of natural causes and water abstraction at the Farakka Barrage.

Some erosion of forest on the seaward margin is also possible as sea level rises, assuming no significant increase in the supply of river-derived sediment due to increased rainfall.

Polders of the Southwest

The polders in the west coastal zone were built to increase agricultural productivity by isolating land from rivers, the majority of which are saline in the southwest. The polders generally only receive rainwater, except in some landward areas where river water is non-saline and can be used for supplementary irrigation. Excess freshwater is drained out through sluices at low tide. Future changes in rainfall, or changes in factors which control the efficiency of drainage of water from the polder, could have important consequences for agricultural production and plans for the maintenance of the polders.

Polders are particularly sensitive to two factors that may be affected by future climate change and sea-level rise: changes in the landward boundary and seasonal distribution of saline water; and changes to low tide levels. Both these factors may result from the combined effect of sea-level rise and climate change on the relative supplies of fresh and saline water, on intra-polder channel configuration, on the Sundarban and on future water management schemes, including possible river flow augmentation (FAP 4, 1992).

Enhanced rainfall in all seasons would directly increase the supply of freshwater to the polders. This is likely to be of increased importance if sea-level rise, or the deliberate breaching of embankments by shrimp farmers, causes enhanced dry-season soil salinity within the polder. Sea-level rise could heighten soil salinity through a gradual increase in the salinity of the groundwater.

It is uncertain how the landward limits of saline waters could change in the future as a result of possible increases in rainfall and sea level. Increased rainfall would tend to push the saline limits seawards, whereas the sea-level rise could tend to push the saline limits landwards. Therefore, it is uncertain how the availability of freshwater for the supplementary irrigation of polders would change in the future.

The configuration and depth of rivers in the southwest could be altered in the future due to changes in the volumes of freshwater and tidal waters from rainfall increases and sea-level rise, respectively. Changes to river channel form would affect the height and duration of the tide. This is important because freshwater from the polder can only be drained by gravity if the water level within the polder is higher than the level of the river. However, predicting the change in channel form is difficult due to the interaction of factors which affect channel form, including rainfall patterns, the Sundarban and mean sea level (which may affect tidal height and strength). For example, higher rainfall could increase the volumes of water and flow velocities in the rivers. This, in turn, could lead to erosion of the channel floors and produce an increase in tidal heights. Alternatively, depending on the flow velocity, sea-level rise could increase the amount of sediments washed into channels and increase sedimentation, thereby reducing tidal heights.

In addition, modification in channel form influences the location of areas of erosion and deposition along channel banks. At present, embankment erosion is not a major problem in the polders of the west. However, if both tidal and fluvial flow strengths increase in the future, an increase in bank erosion could be expected. Lessons may have to be learned from the polders of the central coast where embankment erosion is common.

Future changes to the ecology of the Sundarban due to climate change, sea-level rise and continued resource exploitation could affect the forest's ability to reduce the tidal and wave energy from the Bay of Bengal. Although it is clear that the sensitivity of the Sundarban to future environmental change has been increased through gradual increases in salinity and un-sustainable resource management, no accurate predictions of future change to the Sundarban have been made.

7.5.2 Central Coastal Zone

The constantly shifting sedimentary formations of the central coast do not lend themselves to simple analyses. The broad sensitivity of the central coast to climate change and sea-level rise are outlined here by examining the sediment budget. Below, the critical management and research factors in this dynamic section of the Bangladesh coastline are highlighted.

Figure 7.16 (a-b) makes a distinction between the constructive and destructive agents that move the sediment of the central coast. Constructive agents are those which act to accrete sediment, while destructive agents are forces which act to erode sediments. Depending on the circumstances, rivers, tides and waves can be either destructive or constructive. The relative strength of constructive and destructive agents varies within the central zone. In the east of the central zone, tidal forces (both constructive and destructive) dominate, whereas in the west of the zone constructive fluvial forces are strongest. This enables the division of the central coastal zone into tidal and fluvial-dominated sub-zones and an intermediary fluvio-tidal sub-zone (see Figure 7.11). These three sub-zones are likely to react differently in the future to climate change and sea-level rise because the relative importance of present-day tidal and fluvial forces will determine how future changes to tidal and fluvial forces will be manifested.

Constructive fluvial processes dominate the western sub-zone of the central zone. The huge volumes of sediment which are supplied to the coastal zone each year by the Ganges-Brahmaputra-Meghna river system form numerous new islands (*chars*) each year. However, the strong river flows and their interaction with the tides and waves of the Bay of Bengal also act as destructive agents, carrying much of this sediment into deep water. These flows can also erode previously existing land. The balance between deposition and erosion of river sediment controls the total amount and location of coastal land in this zone.

Numerous aspects of future sea-level rise and climate change could change this balance between deposition and erosion. Climate change may affect the supply of river sediment by changing the amount and geographical distribution of rainfall and its seasonal variation. Any future change to monsoon rainfall will be especially important. It is during the monsoon period that the majority of sediment is delivered to the central coast. With scenarios of increases in monsoon rainfall of 10 to 15 per cent and 15 to 23 per cent for the years 2030 and 2050, respectively, the future supply of fluvial sediment to the central coastal zone may well increase during the coming decades due to increased soil erosion in the upper river catchments and possibly greater channel instability and erosion of river banks. However, in the longer term the backwater effect due to sea-level rise may act to offset any increases in the supply of sediment to the coastal zone.

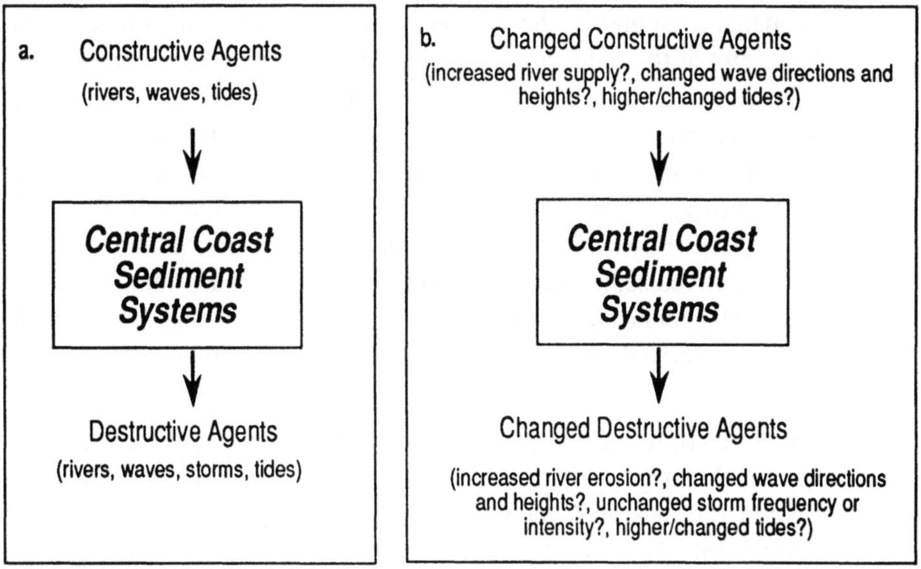

FIGURE 7.16: Schematic sediment budget diagrams for the central coast: a) Present situation and b) Possible future situation

Whether or not enhanced sediment supply causes an increase in the rate of formation of new land depends on future changes in the destructive agents. If sea-level rise is at the lower end of future predictions and wave and tidal energies remain constant, then land accumulation may increase in the future. Alternatively, higher sea-level and climate change may produce changes to wave and tidal energies which could accelerate the erosion of the central coast, even if fluvial sediment supply increases.

It is important to emphasise that it is currently not known whether wave climates and the frequency and intensity of cyclones in the Bay of Bengal will change in the future (see Chapters 2 and 3). Consequently, it is uncertain whether the cumulative effect of changes in sea level, waves, storms, river flow and sediment supply will cause widespread erosion or accretion. The critical factor in the fluvial-dominated central coast will be the balance between river-derived sediment supply and relative sea-level rise.

Toward the eastern side of the central coast, the tidal influence increases and gradually dominates over fluvial effects. Nonetheless, the sediment supplied to the Bay of Bengal by the Ganges-Brahmaputra-Meghna system still remains crucial because some proportion of the river-derived sediment enters the tidal system of the eastern central coast. The key difference between tidal-dominated sub-zone and the fluvial-dominated coast may be the time lag between source changes in sediment supply and receipt of the sediment at the coast. If increased future river-derived volumes of sediment are moved quickly to the tidal-dominated sub-zone, a new balance between erosion and accretion could result without a perceptible change from the present-day situation. If future increases in river-derived sediment supply are large, the erosive effects of sea-level rise could be minimal in the tidal-dominated sub-zone. However, if there is found to be a significant time lag in the supply of river-derived sediment to this sub-zone, erosion could dominate in the short term due to higher mean sea level. Widespread erosion of this sub-zone could also occur if river-derived sediment volumes do not significantly increase in the future.

The prediction of future change in the central coastal zone is further complicated by potential changes in sediment supply and tidal patterns resulting from water diversion schemes and embankment projects proposed under the Flood Action Plan (FAP 4, 1992). This coastal zone is already extremely sensitive to any future change in wave climate, tidal height and the frequency or intensity of cyclones. Consequently, any plans for further human interventions in this zone should consider the implications in the light of possible future climate and sea-level change.

7.5.3 East Coastal Zone

The northern mud-dominated sub-zone of the east coast lies south of the Ganges-Brahmaputra-Meghna delta complex. However, the form of the coast is controlled by the mud and silt supplied to the Bay of Bengal by the delta and moved by strong tidal flows.

The future evolution of this coastal section will be linked to future changes to the delta system and to the volumes of sediment that will be moved to, and deposited within, the northern sub-zone. Factors that will be critical in determining future coastline changes include: the future shape of Sandwip and, hence, the modification of incoming waves; the response of tidal flows between Sandwip and the mainland; and human-induced changes to the Noakhali shoreline, such as land reclamation projects or other engineering works.

The present coastline of the southern sub-zone appears to be very sensitive to changes in sediment supply, as evidenced by the fluctuations between erosion and accretion. Factors that contribute to this sensitivity may be the limited amount of sediment in the area brought by local streams and cliff erosion, and the conflicting north- and south-ward nearshore sand movement driven by summer and winter waves respectively.

If, after further study, the southern sub-zone is found to be sediment-limited, the future rate of sea-level rise may be assumed to be too fast for the available sediment supply and widespread erosion may be predicted. However, sediment supply to the coast could be enhanced by increased erosion in the adjoining river catchments from enhanced monsoon rainfall. Increased sediment supply could also result from higher rates of cliff erosion and future changes in wave energy and direction. If sediment supply from these two sources increases dramatically in the future, the potential impacts of sea-level rise may be offset.

BOX 7.5: Will the coast be drowned?

Sea-level rise does not necessarily imply permanent inundation of the present coastal zone of Bangladesh. The coastal zone exists in a dynamic balance between forces of erosion and accretion. Land subsidence and sea-level rise favour the loss of land to the sea, whereas supplies of freshwater and sediment to the coast promote the accretion of land. Whether climate change and sea-level rise will drown the coast depends critically on how the dynamic balance is maintained in the future.

Overall, it would appear unlikely that any increases in sediment supply to the east coast would be sufficient to offset the erosive effects of sea-level rise, at least in the short term (decades). As a result, a gradual shift away from the present situation of a dominantly accreting coastline (with local erosion) to a dominantly eroding coastline (with local accretion) may be expected with sea-level rise.

Predicting the onset of this change, and the possible magnitude of the shift to erosion, is impossible without further understanding of the processes that control the dynamic response of the coast. Critical questions that need to be addressed concern the sources, transport, and quantity of sediment in the area, and the coastline's reaction to changing sea levels during the past 10,000 years.

7.5.4 Implications for Resource Management

Fisheries

Higher ambient air and water temperatures and CO_2 levels may generally be expected to increase biological activity in all fish habitats. This could lead to increased fish stocks. But higher activity of plant and toxic algae could deplete oxygen levels, with detrimental effects on fish stocks.

Freshwater floodplain fisheries could also benefit from increases in rainfall, especially if accompanied by a lengthening of the rainy season. Increased rainfall could enlarge the area of dry-season water bodies in the coastal zone and reduce potential increases in salinity due to sea-level rise.

Offshore fisheries may be the least affected by future climate change and sea-level rise. The linkages between deep offshore fisheries and changes in flows of freshwater and sediment into the Bay of Bengal appear to be weak. The greatest impact may be on fish species which are dependent on the estuaries and creeks of the coastal zone for breeding or spawning. This may be particularly important for nearshore shrimp fisheries.

Possibly the most sensitive fishery sector to the future changes in climate and sea level is shrimp. Both natural open-water and farmed shrimp stocks are vulnerable. Shrimp life cycles are very closely tied to coastal mangrove forests. Deforestation, caused either by human pressure or degradation due to sea-level rise, could reduce shrimp production. Since detailed predictions of future changes to the Sundarban are not available, the future impacts on the shrimp fishery are very uncertain.

Shrimp farms are either on the banks of, or within, coastal polders. As mentioned above, climate and sea-level changes could affect such structures directly. It is difficult to say whether the future impacts on polders would affect fishery production beneficially or adversely.

The shrimp industry is Bangladesh's fastest growing foreign exchange earning sector. Gaining an understanding of the linkages between future climate and sea level impacts and shrimp production is a priority research area.

A secondary impact of sea-level rise on marine fisheries could result from changes in port and harbour facilities caused by coastal erosion, increased flood and tide heights, and changes in navigation channels.

Forestry

The mangroves of the coastal afforestation project were planted along the seaward side of coastal embankments to stimulate the accretion of land and to protect coastal embankments (see Figure 7.6). Mangrove forest planted on the seaward margin of embankments will be sensitive to future rise in sea level because they cannot naturally migrate inland.

The natural Chakaria Sundarban is rapidly being deforested to make way for shrimp farming (see Figure 7.14). Although this process is slowing, little natural forest remains. It seems reasonable to assume that any future changes to the remaining mangrove forest will be masked by human interventions related to shrimp farming.

The Sundarban of the west is already extensively modified by human use, as described above. It is believed that this current level of usage is not sustainable. Sea-level rise and climate change may be viewed as an additional threat which may exacerbate an already critical situation.

Sustainable Sundarban forestry management requires two initiatives to be carried out immediately: changed royalty fixing and enhanced policing (ESCAP, 1987). Ensuring that the terms of forest harvesting leases are being honoured is a difficult and dangerous job. Widespread overcutting often occurs, with officials unable to enforce regulations. Enhancing the ability of officials to do their job, possibly with assistance from the law enforcement agencies, is a priority (ESCAP, 1987).

BOX 7.6: Sundarban in a warmer world

> The species composition of Sundarban is likely to change in the future due to higher sea levels and changes in freshwater supply. Superimposed on these climate-induced changes will be modifications to the forest from human use. The pressure on forest resources will increase as the coastal population expands. Unsustainable forest use will heighten this pressure. The future of the forest under such multiple threats is uncertain. Sustainable management is vital to protect the forest resource now and to buffer the affects of climate change and sea-level rise in the future.

Effective policing is important if royalty pricing is changed to promote sustainable management. The ESCAP (1987) coastal management plan favours the increase in forest revenue by at least a factor of two, in order to increase revenue of Tk 400 to 500 million (US$10.5 to $13.1) annually. However, increased resource prices would increase the pressure for lawlessness and corruption, which could lead to *reduced* revenues in the long-term if adequate policing and public education programmes are not undertaken. Thus, pricing and policing must be integrated in order to ensure sustainable forest management.

Agriculture

Agriculture in the coastal zone may be affected by climate change and sea-level rise similarly to the rest of Bangladesh. The major difference is the sensitivity of coastal soils to variations in salinity.

Chapters 2 and 4 describe the potential future changes to climate and agriculture, respectively. Increased monsoon rainfall and temperature are likely to have both beneficial and detrimental effects on agriculture in Bangladesh. Potential increases in freshwater availability (surface and groundwater) could significantly benefit agriculture. However, greater rainfall could also lead to more frequent deep flooding and to waterlogging of inadequately drained polders, resulting in decreases in agricultural output.

The major limiting factor to agricultural productivity in the coastal districts is salinity, of both soil and sources of irrigation supply. As mentioned above, higher future sea level, if considered in isolation, may increase the extent and level of salinity, while increases in rainfall may reduce them. The net result cannot yet be predicted: average annual salinity limits may increase, decrease, or remain unchanged. Seasonal differences are possible. For example, if rainfall increases during the monsoon, but not during the dry season, there may be increased saline penetration in the dry season and reduced penetration in the monsoon season.

7.6 WHAT ARE THE POSSIBLE SOCIO-ECONOMIC CONSEQUENCES?

The potential secondary impacts of climate change and sea-level rise in the coastal zone include those on human migration, health and settlements.

7.6.1 Migration

Migration within, and to/from, the coastal districts must be viewed in the national context. There are two types of migration: seasonal and permanent. The seasonal migration of rural workers circulating within agricultural areas is widespread, as is the seasonal migration of fisherman to the coastal districts. Permanent migration tends to be from rural areas to the cities, especially towards the north and east of Bangladesh (see Chapter 5).

The eventual permanent migration of people out of, or into, the coastal zone is perhaps the most difficult of all potential impacts of climate change and sea-level rise to predict. The motivation for people to migrate is influenced by economic opportunities, the flooding, erosion, and accretion of land, and the structure of rural and urban society. As this (and other) chapter has shown, it is difficult to predict how any one of these factors, and the relationships between them, may change in the future.

In addition, the people of Bangladesh are very adaptable and resilient to changing environmental conditions. This resilience will play a crucial role in determining the numbers, location, and direction of those who may choose to leave. For example, on the west coast, saline intrusion could increase in the future and degrade the agricultural production of the area, forcing people to move inland. On the other hand, the same saline intrusion could enhance shrimp production, which in turn could attract migrants to the area. Similar examples can be drawn from all sections of the Bangladesh coast.

Although the resilience of coastal people is high in Bangladesh, it is not infinite. Presumably there are limits, or thresholds, of acceptable environmental change within which coastal people can still live and prosper in the coastal zone. Beyond this threshold, unacceptable environmental change could force migration. For example, future climate change and sea-level rise is likely to change present patterns of land accretion and erosion/submergence in the coastal zone. There will be many examples of migration decisions being made in the ever changing coastal zone of Bangladesh. These present day migration decisions require study in order to assess their role in modifying the potential impact of sea-level rise and climate change in the future.

7.6.2 Health

Three climate-related factors may interact to affect human health in the coastal zone:

- potential changes in the supply of potable groundwater;

- potential changes to the incidence of cholera, typhoid and other water-borne diseases; and
- potential changes to the incidence of malaria from alterations in the distribution of mosquitos in the coastal area.

Potable Groundwater

The shallow and deep aquifers are the main source of potable water in the coastal zone due to the generally high salinity levels of surface waters. Potable water is extracted from the aquifers by shallow and deep tubewells. Within a region, the salinity levels of shallow aquifers can change rapidly within short distances, while those of deep aquifers are generally uniform.

Increases in rainfall are likely to reduce salinity levels, whereas sea-level rise is likely to increase salinity levels. In addition, future saline penetration is likely to be affected by possible expanded use of shallow and deep tubewells by a growing coastal population (FAP 4, 1992). Thus, the impact of the combined effects of sea-level rise, climate change and an increased population on potable groundwater reserves is difficult to predict. If the salinity levels of coastal aquifers rise in the future, the coastal population may be increasingly forced to use surface water supplies. Coastal surface waters often harbour diseases including cholera, typhoid and diarrhoea. Consequently, increased use of surface water for domestic purposes could impair the health of coastal people.

Water-Borne Diseases

Up to 80 per cent of all illness in Bangladesh is linked to water-borne disease (Chapter 5), of which cholera and typhoid are the most important. The classic strain of *V cholera* is an indigenous micro-organism in brackish and saline estuarine waters. Water polluted by human faeces is thought to be the main transmitter of disease which can cause potentially fatal dehydrating diarrhoea, especially among adults.

Cholera incidence peaks in the dry season, when tubewells are dry and villagers turn to infected ponds for drinking water. Deep flooding events and cyclone surges can also precipitate outbreaks by limiting access to tubewells and disabling waste disposal, forcing people to use drinking water contaminated by waste carried by flood waters. In the year following the severe flooding of 1988, for example, the *V cholera* strain was concentrated in the western part of the central coastal zones which received very polluted flood waters from upstream (see Chapter 5). There was also a cholera outbreak after the 1985 cylcone, thought to be due to the increased dependence on ponds for drinking water. If more flood disasters occur in the future, the risk of water-borne diseases could increase.

Changes in the incidence of water-borne diseases produced directly as a result of sea-level rise and climate change will prove difficult to detect. This is because any future changes in population density, poverty and land use may also affect disease outbreaks. Moreover, these factors may, in turn, be influenced by natural resource depletion induced by sea-level rise or climate change — an interrelated set of factors.

Mosquito-Borne Diseases

There are over 100 breeds of mosquito in Bangladesh, each with specific habitats, breeding cycles and disease-carrying characteristics. The three most significant mosquito species are *Anopheles philippensis* and *Anopheles dirus*, both malaria carriers, and *Culex quinquefasciatus*, the main carrier for filariasis. All three species breed in freshwater; *Anopheles* prefers clean water, while *Culex* prefers polluted, stagnant water.

Anopheles mosquitos are found throughout Bangladesh: *dirus* in forested hills and *philippensis* on the floodplains. Because freshwater areas (ponds, water tanks, puddles, perennial streams etc.) are vital for breeding, the saline conditions of the coastal zone may inhibit breeding. If sea-level rise increases future salinity of streams and ponds, the incidence of malaria may be reduced in the coastal zone. On the other hand, an increase in monsoon rainfall could enlarge freshwater breeding areas.

Culex mosquitos are mainly restricted to the northwest region of Bangladesh. However, they are also found in Dhaka and Chittagong where they breed in stagnant and polluted freshwater ponds. If population increases in the coastal settlements, the areas of polluted freshwater ponds may also increase, creating potential breeding grounds for the *Culex*. Higher levels of salinity in the ponds would tend to create unfavourable conditions for the breeding of *Culex*.

7.6.3 Settlements

Coastal settlements are constantly evolving in response to the changing environment of the coastal zone. The environmental changes are both natural (e.g. the accretion or erosion of *chars* in the Meghna estuary) and human-induced (such as those arising from the Coastal Embankments Project, the introduction of high-yielding rice varieties, or the recent rapid expansion of shrimp farming). Adaptations by villagers to the threat of eroding land or rising waters may be to migrate or to raise individual homesteads. Such adaptations modify considerably the impacts of present-day environmental variations and change—a "resilience" in the face of environmental adversity. This resilience may be expected to reduce the potential impacts of future sea-level rise and climate change on the coastal population.

For example, it may be assumed that if the main two coastal cities, Chittagong and Khulna, become threatened by sea-level rise and climate change, their economic importance would warrant increased levels of protection by raised and strengthened embankments and walls. Additional arrangements may have to be made for the drainage of rainwater from behind embankments due to increased tidal channel levels from sea-level rise.

However, the resilience of the coastal inhabitants surely has limits. The thresholds of environmental change beyond which the adaptive capacity of coastal inhabitants is likely to fail are not clearly defined. Identification of the critical thresholds of environmental change in relation to the ability of the local population to adapt is a fundamental research issue that requires urgent attention.

BOX 7.7: Can coastal people adapt?

> In the past, the coastal people of Bangladesh have demonstrated a remarkable ability to adapt to changes in their environment. No doubt they will continue to do so in the future, although adaptation is likely to be more and more difficult. Environmental changes, including those arising from changes in climate and sea level, are expected to occur at the same time as the demands for food, fibre, and energy are increasing due to the rapidly growing population. The critical issue is the rates, or thresholds, of change that will exceed the ability of the coastal people to adapt.

7.6.4 Effects on Non-Coastal Districts

This chapter has concentrated on discussing the potential impacts of sea-level rise and climate change on the coastal districts. However, the greatest impacts of sea-level rise may, in fact, be felt in areas outside the coastal districts. As the coastal area contributes significantly to the national economy of Bangladesh, any future adverse impacts resulting from sea-level rise would possibly be felt nationwide.

Any "flow" between the coastal districts and the rest of Bangladesh may be viewed as a possible mechanism for shifting the future impacts of sea-level rise inland from the coastal zone. Such flows include natural resource flows (e.g. water, sediment), human-exploited natural resource flows, (e.g. fuelwood or shrimp) and human flows (migration).

For example, future sea-level rise could have the indirect effect of congesting drainage on the floodplain land behind the tidal fringe. During the monsoon, the depth of freshwater flooding could increase as a result. This could have disadvantages for agriculture, but advantages for natural fisheries. The net impacts on economic activities would thus be difficult to assess (see Chapter 4). Similarly, migration to non-coastal cities and the reduction in the supply of fuelwood, agricultural, and fishery resources to non-coastal districts could transfer the impacts to areas far beyond the direct reach of the sea. In short, vulnerability to sea-level rise is a combination of the direct impacts on coastal regions and their indirect effects on adjacent regions.

7.7 ENVIRONMENTAL POLICY FOR COASTAL AND MARINE AREAS

The Ministry of Environment and Forest of the Government of the People's Republic of Bangladesh has announced the *Environment Policy 1992 and Implementation Programme* in keeping with the broader objective of sustainable development through conservation of nature and protection of environment. The policy package contains four broad objectives for the coastal and marine environment, as follows:

1. to ensure the conservation and environmentally sound development of coastal and marine ecosystems and resources;

2. to prevent all local and external activities that lead to pollution in the coastal and marine areas;

3. to strengthen necessary research in order to preserve and develop coastal and marine environment and resources; and

4. to restrict exploitation of coastal and marine fisheries within sustainable limits.

The above four broad policy objectives are supplemented by eight specific "policy actions". The eight policy actions and the respective implementing agency(ies) of the Government are shown in Table 7.4. Policy Actions 2 to 8 deal with the specific coastal management issues relating to land stabilisation, water pollution (coastal and marine), and vigilance. Policy Action 2 strengthens the remit of the Ministry of Land and the Department of Forest to stabilise twenty accreted lands through planting mangrove species under the Coastal Afforestation Programme. Policy Actions 3, 4, 5, and 6 address coastal and marine water pollution issues, including the management of oil spills, waste collection, and transportation and management within the port area, and the creation of a special "cell" within the Ministry of Port Shipping and Inland Water Transport (IWT) to monitor coastal and marine waste disposal. Policy Action 7 identifies the Ministry of Port Shipping and IWT as the key agency for ensuring necessary security of staff and facilities involved in coastal environmental monitoring and conservation. (Piracy in coastal estuaries and rivers and in marine waters is a significant problem in Bangladesh [ESCAP, 1987].) Policy Action 8 specifies six agencies with responsibilities for the implementation of coastal environmental monitoring and conservation.

Broader issues of coastal natural resource management are dealt with in Policy Action 1 which identifies the Ministry of Environment and Forest as having a coordinating and supervisory role in the management process of coastal and marine areas. To facilitate this role, a special "cell" is proposed to be created within the Ministry that will act as the focal point for coastal zone management-related programmes and activities. Policies formulated by the proposed cell are to be implemented by the same Ministry through the Department of Forest, the Department of Environment, and the Forest Research Institute in cooperation with agencies of other ministries concerned.

It is expected that the policies related to the coastal and marine areas outlined in the *Environment Policy 1992 and Implementation Programme* could significantly improve the coastal zone management process, if proper attention is paid to them. However, it appears that the proposed cell within the Ministry of Environment and Forest may not be strong enough to act effectively, as the formal administrative linkages with various concerned implementing agencies are undefined. This cell has to be strengthened by high-level political backing for quicker and smoother functioning.

Previously, Jalal (1988) had proposed a "small but high level agency" that could act as an alternative to the proposed cell. Its structure could be like a "Board" that would be chaired by the Prime Minister. He proposed that the Board members should be drawn from the key government ministries, agencies and administrative heads of the two coastal divisions — Chittagong and Khulna. The day-to-day

TABLE 7.4: Plans of action relating to the coastal and marine environment of Bangladesh and implementing agencies

Action	Implementing Agencies
1. A special cell should be constituted in the Ministry of Environment and Forests to coordinate and supervise all activities related to the protection and development of the coastal and marine environment.	a. Ministry of Environment and Forest b. Department of Forest c. Department of Environment d. Forest Research Institute
2. Newly accreted lands should be handed over to the Department of Forest for afforestation in order to conserve and stabilise these lands.	a. Ministry of Land b. Department of Forest
3. Bangladesh Navy should police activities in their territorial waters in order to guard against any incidence of pollution. Department of Shipping will monitor this assignment.	a. Ministry of Defence b. Bangladesh Navy c. Ministry of Port Shipping and Inland Water Transport (IWT) d. Department of Shipping
4. Local and national contingency plans and financial provisions should be formulated in order to control marine pollution caused by water transport in marine waters.	a. Ministry of Port Shipping and IWT b. Ministry of Defence c. Bangladesh Navy d. Department of Shipping
5. Environmentally sound emergency systems should be developed for collection and transportation of wastes from Chittagong and Mongla ports and also for the collection and disposal of waste oil, billage, and other related wastes from ships.	a. Ministry of Port Shipping and IWT
6. A special cell should be created in the Ministry of Port Shipping and Inland Water Transport in order to determine the nature, composition, and characteristics of wastes and possible environmental hazards prior to giving permission for dumping wastesinto the sea.	a. Ministry of Port Shipping and IWT b. Department of Environment
7 An integrated Coast Guard system should be created and be deployed on an emergency basis under the Ministry of Port Shipping and Inland Water Transport to help manage and secure the coastal and marine environment and resources.	a. Ministry of Port Shipping and IWT
8 Appropriate measures should be taken in order to prevent pollution of the territorial waters, protection and conservation of the of coastal and marine environment, monitoring and conservation of newly accreted coastal lands, and proper utilisation of all coastal resources.	a. Ministry of Defence b. Bangladesh Navy c. Ministry of Port Shipping and IWT d. Department of Shipping e. Department of Forest f. Space Research and Remote Sensing Organisation (SPARRSO)

Source: Adapted from Government of of Bangladesh, 1992c.

functioning of the Board would be supported by a small group of permanent support staff. The Board could hire or retain the services of appropriate technical and managerial experts as and when required. According to Jalal (1988):

> ... the prime responsibility of the Board should be development of long-term strategies and policies on coastal resource planning and management, and not development and implementation of projects, which should rest with the line ministries /agencies.

Coordination and monitoring of implementation programme(s) should be vested with the Board.

The Environment Policy 1992 underscores the need to protect the country from natural disasters; this is a principal objectives of the policy package. However, no action programme has been formulated regarding the protection or mitigation of cyclonic hazard and storm surges. Also overlooked are the issues of climatic change and sea-level rise, which could have important ramifications in light of recent international agreements (see Chapters 2 and 6). For example, the Coastal Zone Management Subgroup (CZMS) of the Intergovernmental Panel on Climate Change (IPCC) has called for the integrated coastal zone management for all coastal and low-lying island nations by the year 2000. This call has been greatly strengthened by two of the outcomes of the United Nations Conference on Environment and Development (UNCED): the Framework Convention on Climate Change and the Agenda 21.

Under the Climate Convention, developed and developing countries must produce detailed reports on compliance of the various commitments of the Convention (Article 12) within six months and three years, respectively, from the date at which the Convention came into force (21 March 1994) or the date of the country's ratification, whichever is later. Bangladesh, having ratified the Convention in April 1994, will therefore have until May 1997 to produce an "appropriate" integrated coastal zone management plan by 1997 at the latest (extensions can be applied for). This plan is required to include an assessment of the adaptability of the coast to sea-level rise and to be set within a sustainable development framework.

7.8 KNOWLEDGE GAPS AND FUTURE RESEARCH NEEDS

This chapter has reviewed the present condition of the Bangladesh coast and its people. The review focussed on the sensitivity, vulnerability and resiliency of natural coastal systems, the use of natural resources by people, and the impact of tropical cyclones. It is concluded that population pressure and the dynamic, and often unpredictable, changes to the coastal zone have resulted in widespread depletion of coastal resources.

The assessment of potential future effects on the Bangladesh coast and its people due to changes in sea level and climate was conducted in the context of present-day population pressure and unsustainable resource use. It is concluded that there are many coastal systems that are both sensitive and vulnerable to future environmental change. There are several general areas in which research could improve the base of knowledge required to enhance adaptation and to guide policy.

First, there is need to examine the *sensitivity of the Sundarban mangrove species to changing salinity and other environmental conditions.* The Sundarban serves as a natural protective barrier to inland areas and has important ecological values for both fisheries and forestry. Maintaining the health of the Sundarban for sustainable use in the face of possible future changes in climate and sea level should be a high priority.

Second, for the reasons noted above, there is also an urgent need to investigate the *vulnerability of the Sundarban to increased future population pressure and resource use, within the context of potential changes in mangrove species and location due to climate change and sea-level rise.* Currently, the unsustainable use of the Sundarban is weakening the mangrove system and degrading its resource value, thus potentially increasing its vulnerability to future environmental changes.

Third, studies are required to determine the *effects of possible changes in rainfall, freshwater flows, and sea-level rise on polders.* The polders are crucial to the economic activities of the western coastal districts. Understanding of the potential disruptive effects of altered climatic and hydrologic regimes is fundamental to the consideration of policies and actions designed to reduce vulnerability.

Fourth, this chapter has emphasised the critical role of sediment in shaping coastal areas as climate and sea level change. Many land areas exist tenuously in a delicate dynamic balance between the forces of accretion and erosion. Whether or not coastal land will disapprear under the rising sea depends critically on how sediment supply will affect that balance. Especially for the *chars* and offshore islands of the central coast, studies are needed of the *sensitivity of coastal areas to potential changes in sediment supply and sea-level rise.*

This chapter also concluded that coastal systems are inherently quite resilient in the face of sea-level rise and climate change. There are two important features of this resilience that suggest areas of future research.

First, as mentioned above, the resilience of the coastal zone hinges on the supply of fresh water and sediment. River-derived fresh water and sediment may possibly increase in the future as a result of higher monsoon rainfall. Any future increase in freshwater flows and sediment supply to the coastal zone may offset the direct erosive effects of sea-level rise, especially in the west coastal zone (increased fresh-water supply) and the central coastal zone (increased sediment supply). Studies of the *relationships between climatic change, runoff and streamflow, and freshwater and sediment supply to the coast* are therefore critical in helping to discern the future impacts of climate and sea-level changes on the coastal zone.

Second, studies are also required of *the resourcefulness and adaptive capacity of coastal people.* The coastal inhabitants have extensive experience in adapting to the constantly changing environment of Bangladesh. However, the extent to which the coastal people, through their capacity for adaptation, can avoid the ill-effects of rapid sea-level rise and climate change in the future remains unclear. This is a crucial research need that has also been given high priority elsewhere (see Chapter 5).

7.9 WHAT SHOULD BE DONE?

At present, research into coastal resource use and management is overseen by numerous government departments located throughout the country. In the short term, coordination and integration of existing coastal zone research programmes may prove to be more important than carrying out large new projects.

The key to implementing policies for sustainable management of coastal resources is to concentrate on preserving the resilience of the natural and cultural systems. It is the people of the coast that may prove to be the most resilient to future changes. Traditionally, they have lived and worked with the natural cycles and variations of the coast. If this adaptability is preserved, future changes to the coastal zone may be absorbed to a large extent by its inhabitants. This resilience may have been underestimated in the past, especially by those accustomed to modern land-use practices.

Studies that examine the vulnerability, resiliency and adaptability of the bio-physical and human systems of the coastal zone to climate change and sea-level rise are required under the stipulations of the Framework Convention on Climate Change. The Climate Convention contains a number of specific references to coastal zone resources, to coastal zone management, and to the ability of the coast and its people to adapt to future coastal change. Perhaps the most significant coast-specific requirement of the Convention is that signatories "develop and elaborate appropriate and integrated plans for coastal zone management". The development and implementation of an integrated coastal zone management and development plan is urgently required, both for the sustainable management of today's coastal resources and to ensure that the future impact of climate change is reduced.

Many countries of the world are already planning for the potential future impacts of sea-level rise and climate change. Some of these countries, usually in the developed world, possess both comprehensive and effective coastal zone management policies and a clear understanding of how their coastlines work. Armed with these two essential weapons, such countries can start the battle to reduce the future impact of rising seas and a changing climate.

Bangladesh does not have integrated coastal zone management policies based on a comprehensive understanding of the biophysical system and its modification by human inhabitants. Such understanding is essential for effective sea-level rise impact mitigation. In Bangladesh, as in many developing countries, the problems of future coastal change due to sea-level rise are often perceived as insignificant compared to the everyday needs of combating poverty and feeding a rapidly expanding population. In Bangladesh, the future needs of the inhabitants of the coastal zone will be greatly compromised by the actions of today's poor and hungry people, with or without sea-level rise. Until coastal resources are managed with both the present and future generations in mind, the potential impacts of sea-level rise will remain obscure to those struggling for survival today. Policies to manage coastal resources sustainably are urgently required if the future impact of sea-level rise is to be reduced.

The coastal management policies of the *Environment Policy 1992* are a positive step towards sustainable coastal resource management in Bangladesh. However, the implementation of coastal management initiatives through existing administrative channels, as detailed in the *Environment Policy 1992*, will not significantly improve coastal resource management in Bangladesh without strong political commitment and high-level strategic policy directives and guidance. As a result, it may be appropriate to establish a central high-level coastal management coordination unit. Such a new coastal management unit could have responsibilities for strategic policy formulation, funding, policy implementation, and monitoring through existing government agencies. The new coastal management office should be closely linked to the Prime Minister's secretariat. It is concluded that a high-level central coastal zone management office will also be essential for coordinating coastal management policies and plans required by the Framework Convention on Climate Change and recommended by Agenda 21.

ENDNOTES

1 Many previous studies have used this approach including: BCAS (1990); Choudhury (1988); Choudhury (1989); Mahtab, (1989); Milliman *et al.*, (1989).

2 The projection of the population of the coastal districts is based on the following assumptions: Growth rate 1990-2000 same as 1980-1990 rate; growth rate 2000-2025 is 1.25 per cent; growth rate 2025-2050 is 1.00 per cent (assumed by the Bangladesh Planning Commission)

3 Ninety-two per cent of imports and 91 per cent of exports travel to and from Bangladesh by sea (BBS, 1991, Table 8.07)

4 The national religious composition is: 87 per cent Muslim, 12 per cent Hindu, 1 per cent Buddhist and less than 1 per cent Christian (BBS , 1991).

5 National averages of employed people aged 10 and above show that over 70 per cent of people work in the agriculture, forestry and fisheries sectors; 11 per cent in production and transport; and 7 per cent in sales. All other forms of employment make up less than 10 per cent of jobs.

6 The Sundarban forest is often, mistakenly, referred to as the "Sundarbans". The plural "Sundarbans" arises from the translation into English (Rainey, 1891). The singular, Sundarban, is used in this report.

7 The cyclone affected 180km of coast, and an area of 12,50 km^2, of which half was inaccessible to relief directly after the disaster (Hossain *et al.*, 1992).

8 Up to 470km of embankments were destroyed or badly damaged, exposing 72,000ha of *aman* paddy to salt water intrusion. The industries at Patenga, and salt and shrimp fields of Cox's Bazaar were also badly damaged. In the months prior to the cyclone, a feasibility study for the improvement of the coastal embankments in the central and eastern coasts, which had been damaged during the 1985 cyclone, was completed. Subsequent repair work had been delayed by attention being diverted to the inland floods of 1987 and 1988, and by contractual delays. The study found that the embankments were in a poor state of maintenance (FAP 7, 1992), and concluded that "Had the coastal embankments been in a better condition to withstand such a cyclonic surge, the damage would have been much less". The severity of the damage was concluded to be due to poor maintenance and sub-standard construction.

9 There is considerable information available on the April 1991 cyclone, due mainly to the publication efforts of NGOs (Haider *et al.*, 1992; Kafi and Alam, 1992; Khan, 1991; Talukder and Ahmad, 1992; Talukder *et al.*, 1992) and disaster experts (Hossain *et al.*, 1992).

10 The CPP (Cyclone Preparedness Program) is organised into 2,089 units which contain 10 volunteers each. Each volunteer has a specific task, such as cyclone warning, shelter management, rescue, first aid and relief (Quoreshi, 1992).

11 Bangladesh *Taka* converted to US$ at an exchange rate of 38 *Taka* per US$.

12 There are120 commercially-important fish species, 270 species of bird (including 95 waterfowl), 50 species of reptile, 8 amphibian species and 42 species of mammal including an estimated 350 Royal Bengal tigers, 40 to 70,000 rhesus macaques monkeys, 50 to 80,000 spotted deer, 20,000 smooth-coasted otters and 20,000 wild boars.

13 The present polders include 4,000km of embankments and 2,000 sluices completed by 1970 (Sarker, 1991).

14 A recent survey of polder performance in the West coastal zone has shown that 62 per cent, 27 per cent and 11 per cent of the CEP (Coastal Embankment Project) area is satisfactory, marginal and unsatisfactory respectively (FAP 4, 1992).

15 Shrimp farming was not one of the original planned uses of the polders. However, high saline levels and the natural adaptability of farmers has resulted in extensive shrimp farming within polders.

REFERENCES

Akonda, A. W. (1988) Bangladesh. In International Union for the Conservation of Nature and Natural Resources (IUCN) *A Directory of Asian Wetlands.* IUCN, Gland, Switzerland, pp.541-581

Bangladesh Agricultural Research Council (BARC)(1990) *Salinity of Coastal Soils.* BARC, Dhaka.

Barua, D. K. (1990) Suspended sediment movement in the estuary of the Ganges-Brahmaputra-Meghna river system. *Marine Geology,* 91, 243-253.

Barua, D. K. (1991a) The coastline of Bangladesh - an overview of processes and forms. In O. Magoon (ed.) *Coastal Zone '91.* American Shore and Beach Preservation Society, pp..2284-2301.

Barua, D. K. (1991b) Global warming and sea-level rise. *Holiday.* Dhaka, 4-5.

Bangladesh Bureau of Statistics (BBS) (1991) *Statistical Yearbook of Bangladesh* (12th Edition). BBS, Dhaka.

Bangladesh Centre for Advanced Studies (BCAS) (1990) *International Sea-Level Rise: A National Assessment of Effects and Possible Responses for Bangladesh.* BCAS, Dhaka.

Bangladesh Centre for Advanced Studies (BCAS) (1992) *Assessment of Vulnerability to Sea-Level Rise: A Case of Bangladesh.* BCAS, Dhaka.

Chaffey, D.R., Miller, F.R. and Sandom, J.H. (1985) *A Forest Inventory of the Sundarban* (Main Report). UK Overseas Development Agency, London.

Choudhury, A. M. (1988) Rise of sea-level and its effect for Bangladesh coast. Paper presented in the Workshop on Rising Sea Level and Subsiding Coastal Areas, Bangkok.

Choudhury, A.M., Quadir, D.A. and Islam, J. (1990) *Study of the Chakaria Sundarban Using Remote Sensing Techniques.* UNESCO/UNDP Regional Project RAS/86/120. Bangladesh Space Research and Remote Sensing Organisation (SPARRSO), Dhaka.

Choudhury, J. K. (1989) Effect of global temperature rise on the coastlines of Bangladesh. In O. Magoon (ed.) *Coastal Zone '89.* American Shore and Beach Preservation Society, pp.1360-1369.

Clark, J. C. and others (1990) *Evaluation of the Khulna Coastal Embankment Rehabilitation Project.* Asian Development Bank and United Nations Development Programme, Dhaka.

Das, S.C. (1981) Aspects of coastal morphology and bottom topography of the Bay of Bengal. In K.M. Elahi (ed.) *Perpectives on Bangladesh Geography,* Bangladesh National Geographical Association, Dhaka.

Department of Fisheries (DOF)(1989) *Annual Statistics.* DOF, Dhaka.

Economic and Social Commission for Asia and the Pacific (ESCAP)(1987) *Coastal Environmental Management Plan for Bangladesh. Volume II - Main Report*, ESCAP, Bangkok.

Elahi, K. M., Sharif, A. H. M. R. *et al.*, (ed.) (1992) *Bangladesh: Geography, Environment and Development*. Bangladesh National Geographical Association, Dhaka.

Export Promotion Bureau (EPB) (1993) *Export from Bangladesh 1972-73 to 1991-1992*. EPB, Dhaka.

Eysink, W.D. (1983) *Basic Considerations on the Morphology and Land Accretion Potentials in the Estuary of the Lower Meghna River*. LRP Technical Report 15, Dhaka.

Food and Agriculture Organization (FAO) (1971) *Soil Survey Project: Bangladesh. Soil Resources, Technical Report 3:* AGL SF/PAK 6, FAO, Rome.

Food and Agriculture Organization (FAO) (1988) *Land Resources Appraisal of Bangladesh for Agricultural Development*. 7 Reports (in 11 volumes). FAO, Rome.

Flood Action Plan (FAP) 4 - Government of Bangladesh, Sir William Halcrow and Partners and Danish Hydraulics Institute (1992) *South West Region Water Resources Development Programme (Draft)*, Dhaka.

Flood Action Plan (FAP) 5 - Government of Bangladesh and Mott MacDonald (1992) *South East Region Water Resources Development Programme (Draft)*. Dhaka.

Flood Action Plan (FAP) 7 - Government of Bangladesh Ministry of Irrigation, Water Development and Flood Control (1992) *Cyclone Protection Project II: Feasibility and Design Studies (Main Report)*, Dhaka.

Government of Bangladesh (GOB) (1988) *A Selected Directory of Industries*. GOB, Dhaka.

Government of Bangladesh (GOB), Ministry of Environment and Forest and the International Union for the Conservation of Nature and Natural Resources (1991) *National Conservation Strategy of Bangladesh: Towards Sustainable Development*. First Draft, July 1991., Dhaka.

Government of Bangladesh (GOB)(1992a) *Bangladesh Country Report to UNCED*. Ministry of Environment and Forest, Dhaka.

Government of Bangladesh (GOB)(1992b) *Support to Comprehensive Disaster Management*.14 December 1992, GOB, Dhaka.

Government of Bangladesh (GOB)(1992c) *National Environment Policy and Implementation Programme*. Ministry of Environment and Forest, Dhaka.

Gray, W.M. (1968) Global review of the origin of tropical disturbances and storms. *Monthly Weather Review*, 96, 669-700.

Haider, R., Rahman, A. A., and Huq, S. (1991) *Cyclone '91: An Environmental and Perceptional Study*. Bangladesh Centre for Advanced Studies, Dhaka.

Halcrow (Sir William Halcrow & Partners Ltd.) (1992) *1-D Morphological Modelling of the Brahmaputra River*. Halcrow, Dhaka.

Hossain, H., Dodge, C. P, and Abed, F.H. (1992) *From Crisis to Development: Coping with Disasters in Bangladesh*. University Press, Dhaka.

Hoque, M.M. (1991) Field study and investigations on the damage caused by cyclones in Bangladesh: A report on April 1991 Cyclone. In UNCRD, *Cyclone Damage in Bangladesh* . UNCRD, Nagoya, Japan.

Jalal, K. F. (1988) Towards a national strategy on coastal resource management. In H.J. Moudud, H. Rashid and A.A. Rahman (eds.) *Bangladesh Coastal Area Resource Development and Management*. Bangladesh Coastal Area Resource Development and Management Association, Dhaka pp.67-78.

Kafi, S. A. and Alam, A. U. (1992) *A Guide to Disaster Response Management* , Bangladesh Development Partnership Centre, Dhaka.

Kafiluddin, A. K. M. (1991) *Disaster Preparedness for Bangladesh Floods and Other Natural Calamities*. Dhaka.

Katebi, M.N.A and Bari, A. (1989) Coastal afforestation in Bangladesh. In H.J. Moudud, H. Rashid, and A.A. Rahman (eds.) *National Workshop on Coastal Area Resource Development and Management*. Coastal Area and Resource Management Association, Dhaka, pp.101-113.

Khan, A (1991) *A Preliminary Report on Environmental Impact Assessment of Cyclonic Storm and Tidal Surges in the Greater Chittagong and Noakhali District of Bangladesh*. National Forum of NGO's, Dhaka.

Khan, H.R. (1978) *A Study of Water Resources Development Activities in Bangladesh*. Department of Water resources Engineering, bangladesh University of Engineering and Technology (BUET), Dhaka.

Kingdon-Ward, F. (1951) Notes on the Assam earthquake. *Nature*, 167, 130-131.

Kingdon-Ward, F. (1953) The Assam earthquake of 1950. *Geographical Journal*, 119, 169-182.

LRP Estuary Information System (1990) *Hydraulic and Morphological Conditions in the South-eastern Estuary of Bangladesh*. LRP, Dhaka.

LaTouche, T. H. D. (ed) (1910) *The Journals of Major James Rennell*. The Asiatic Society, Calcutta.

Mahtab, F.U (1989) Effect of climate change and sea-level rise on Bangladesh. A report submitted to the Commonwealth Group of Experts, Commonwealth Secretariat, London.

Milliman, J. D., Broadus, J. D. and Gable, F. (1989) Environmental and economic implications of rising sea level and subsiding deltas: The Nile and Bengal examples. *Ambio*, 18(6), 340-345.

Milliman, J. D. and R. H. Meade (1983) World-wide delivery of river sediment to the oceans. *The Journal of Geology*, 91(1), 1-21.

Multi-purpose Cyclone Shelter Project (MCSP) (1993) *Summary Report*. Bangladesh University of Engineering and Technology and Bangladesh Institute of Development Studies, Dhaka.

Master Plan Organisation (MPO)(1986) *National Water Plan (Volumes I, II and III)*. Government of Bangladesh, Ministry of Irrigation, Water Development and Flood Control, Dhaka.

Nishat, A. (1988) Review of present activities and state of the art of the coastal areas of Bangladesh. In H.J. Moudud, H. Rashid and A.A. Rahman (eds.) *National Workshop on Coastal Area Resource Development and Management*. Coastal Area and Resource Management Association, Dhaka, pp.23-25.

National Oceanographic and Maritime Institute (NOAMI) (1992) *Proceedings of the IOC-UNEP Workshop on Impacts of Sea Level Rise Due to Global Warming*. NOAMI, Dhaka.

Quoreshi, A. H. (1992) Cyclone preparedness programme of Bangladesh. In Bangladesh National Commission for UNESCO, *Proccedings of the Workshop on Coastal Zone Management in Bangladesh*, 27-31 December 1992, Bangladesh National Commission for UNESCO, Dhaka, pp.395-399.

Rainey, J. R. (1891) The Sundarban: its physical features and ruins. *Geographical Society of London* XIII, 273-287.

Rashid, H. (1991) *Geography of Bangladesh* (2nd Edition). University Press, Dhaka.

Rizvi, A.I.H. (1969) Morphological changes in the coast of Chittagong, *Oriental Geographer* XIII. p.25-41

Saqui, Q.M.A.H.(1988) *Landlessness in Two Villages of Bangladesh*. National Institute of Local Government, Dhaka.

Sarker, M.A.B. (1991) Living with cyclones: Adequacy of existing structures. In *Land Reclamation Workshop*, June 27, 1991, Dhaka.

Shahidullah, Md. (1968) Marine fisheries resource management in Bangladesh and current status of exploitation. *Marine Fisheries Bulletin*, No.3.

Shahjahan, M. (1992) Development and environmental problems in the coastal area of Bangladesh. In T. Shibayama (ed.) *Symposium Coastal Processes in the Asian Region: Survey of Coastal Environment under Development Project*. Department of Civil Engineering, Yokohama National University, Yokohama, pp.21-52.

Siddique, J. A. (1967) Geomorphic features of the northern Chittagong coast. *Oriental Geographer* XI(2), 1-12.

Siddique, N. A., Khan, M. A. S. and Islam, H.R. (1992) Underplanting - A means to ensure sustainable mangrove plantations in Bangladesh. In Bangladesh National Commission for UNESCO, *Proceedings of the Workshop on Coastal Zone Management in Bangladesh,* Bangladesh National Commission for UNESCO, Dhaka, pp.1-12.

Siddique, M. (1989) Land erosion and accretion in the coastal area. In H.J. Moudud, H. Rashid and A.A. Rahman (eds.) *National Workshop on Coastal Area Resource Development and Management,* Coastal Area and Resource Management Association, Dhaka, pp.139-153.

Smith, E.C (1981) Closure of tidal channels in Bangladesh (II). *Land and Water International,* 44, 8-18.

Snead, R.E. (1985) The coast of Bangladesh. In E.C.F. Bird and M.L. Schwartz (eds.) *The World's Coastline.* Van Nostrand, New York, pp.761-765.

Sohrab, M. and Sarker, U. (1986) Endangered and threatened wildlife of the coastal zone of Bangladesh. *The Journal of the National Oceanographic and Maritime Institute Dhaka, Bangladesh,* 3(2), 29-37.

Talukder, J. and Ahmad, M. (ed.) (1992) *The April Disaster: Study on Cylcone Affected Region in Bangladesh.* Community Development Library, Dhaka.

Talukder, J., Roy, G. D. and Amhad, M. (1992) *Living With Cylcone: Study on Storm Surge Prediction and Disaster Preparedness.* Community Development Library, Dhaka.

Tarafder, M.R. (1975) Reclamation of land from the sea of Bangladesh. In *Proceedings of the Seminar on Flood Control and Allied Problems in Bangladesh,* Dhaka, pp.1-19.

Task Forces (1991) *Managing the Development Process. Vol II and IV,* University Press, Dhaka.

Thompson, P.M. (1989) *One Impact of Global Control on Agricultural Development in Bangladesh.* Flood Hazard Research Centre, Middlesex Polytechnic, U.K.

UNDP (1990) *Evaluation of the Khulna Coastal Embankment Rehabilitation Project.* (20 January), Asian Development Bank and United Nations Development Programme (UNDP), Dhaka.

Warrick, R.A. and Oerlemans, J. (1990) Sea-level rise. In J.T. Houghton, G.J. Jenkins and J.J. Ephraums (eds.) *Climate Change: The IPCC Scientific Assessment.* Cambridge University Press, Cambridge, pp. 271-281.

Woodroffe, C. (1990) The impact of sea-level rise on mangrove shorelines. *Progress in Physical Geography,* 14 (4), 483-520.

World Resources Institute (1990) *Bangladesh: Environmental and Natural Resource Assessment.* Centre for International Development and Environment, World Resources Institute, Washington D.C.

Index

Accretion, 369
Active adaptation, 77
Adaptation, 77
Agriculture 27, 143, 308, 338, 346, 386
Alluvion 301
Aman 16, 145, 161, 169, 210, 346
Antarctic Ice Sheet 105, 106
Antarctica 9
Anthropogenic sources 88
Asian Development Bank 232, 247
Assam valley 117
Atmospheric concentrations 8
Aus 16, 168, 346

Bangladesh
 climate change 5
 coastal zone 7
 mean annual rainfall 145
 population 13
 rainy season 145
 socio-economic environment 12
 water resources 14

Back-water effect 375
Barind Tract 145, 151, 152, 162
Barsha 158, 214
Bay of Bengal 1, 6, 10, 117, 123, 125, 127, 338, 340, 374
Bazars 296
Bengal basin 116-117
Bio-resource 16
Bonna 158, 214

Boro paddy 16, 160-161, 168, 173
Brackish water fisheries 349
Brahmaputra right embankment 161
Business-as-usual (BAU) 4, 21, 70, 112, 197, 375

Calcutta port 151, 338
Carbon dioxide 2, 40, 45, 177, 187, 290
Chakaria Sundarbans 25, 184, 351, 371, 385
Chandpur Irrigation Project 160, 206, 259
Channel stability 166
Char 206, 294, 301, 389
Chittagong Hill Tracts 338, 340
Chlorofluorocarbons (CFCs) 40, 49
Cholera 20, 254, 388
Climate change 8, 18, 20, 28, 31, 35, 36, 38, 70, 156, 162, 292, 319, 335, 375
Climate extremes 230
Climate models 14, 162
Climate sensitivity 4, 102
Climatic hazards 211
Coastline 336
Coastal afforestation 367
Coastal Embankment Project 350, 389
Coastal fisheries 188
Coastal flooding 123
Coastal polders 167
Coastal population 358
Coastal resources 27
Coastal Zone 24, 31, 335, 336, 340, 368, 377, 380, 382
Compartmentalization 161
Constitution of Bangladesh 297
Controlled flooding 161
Cooling effect of sulphate aerosols 51
Cropping intensity 155
Crop production 14

Crop seasons 210
Cyclones 10, 17, 23, 214, 352
Cyclone movement 127
Cyclone Preparedness Program (CPP) 352, 356

Deep tubewells 161, 343
Depth of flooding 160
Diluvion 294, 301
Discharge 157
Displacement 300, 302, 304,
Drainage 166
Droughts 12, 17, 20, 162, 216

Earth Energy Budget 4
Economic growth 225
Eco-refugees 292
Education 18, 247
Embankment design 165
El Nino/Southern Oscillation (ENSO) 40, 67
Employment 18, 243
Energy budget 4
Enhancd greenhouse effect 4
Environment policy 391, 394
Equillibrium changes 57, 60
Equillibrium warming 4
Erosion 7, 294, 369
Estuarine 16
Evapo-transpiration 163
Evapo-transpiration rates 145

Farakka barrage 151, 317, 338
Fauna 361
FCD/ I 231, 232
Fertilisers 154, 169
Fisheries 12, 27, 143, 186, 198, 308, 338, 348, 384

Floods 12, 17, 162, 210
Flood Action Plan (FAP) 151, 161, 165, 194, 197, 232, 364, 382
Flood control 231
Flood losses 14
Flood plains 12, 16, 17, 143, 145, 153, 161, 167, 208, 235, 340, 341, 384
Flood plain fisheries 188
Flood protection 14, 155, 156, 160, 175
Flood types 159
Forestry 12, 143, 184, 385
Fossil fuel burning 5
Framework Convention on Climate Change (FCCC) 22, 24, 28-29, 80-81, 86, 291, 320, 324, 327, 335,
Freshwater fisheries 348

Ganges-Brahmaputra-Meghna delta 25, 296, 336, 340, 382
Ganges-Brahmaputra-Meghna River Systems 14, 317,
Ganges delta 315
General Circulation Models (GCMs) 60-61, 73-74, 162
Ghers 349
Glacial-interglacial cycles 8, 36
Glaciers 103, 104
Global climate change 1, 2
Global climate models 4, 6
Global Environment Facility (GEF) 81, 85, 327
Global sea level 111
Global warming 1, 2, 5, 8-9, 63, 97, 103, 155, 198, 270, 373
Greenhouse gases 1, 35, 50, 54,
Greenhouse effect 4
Greenland ice sheet 107-108, 114
Greenriver concept 161
Gross-domestic product 82, 225, 227
Ground-water 338, 340, 342

Hats 296
Halocarbons 48

Health care 19, 247
Highest peak flow 157
High tide 151
Himalayas 119
Hatiya Island 339
Hoogly river 151
Hydrology 150
HYV seed 154, 169

Ice caps 104
Ice sheets 2, 103-104
Ijaradars 296
Indian Ocean 162, 315
Infant mortality 247
Integrated Coastal Zone Management Subgroup (ICZM) 292, 320
Intergovernmental Negotiating Committee (IPCC) 81
Intergovernmental Panel on Climate Change (IPCC) 1, 4, 9, 63, 292, 307, 394
International co-operation 23
International Trade in Endangered Species 319
Irrigation 14, 154, 210, 215, 231
Inundation 7, 209, 294, 299

Karnafuli River 339
Khas property 295
Kharif crop 16

Land holdings 153
Landless 12, 224
Landlessness 294
Land-ownership 153
Land Tenure 22, 153, 293
Land use 155, 294
Life expectancy 247
Light-intensity 179

Livestock 12, 16, 143, 190, 198

Madhupur Tract 116, 145, 151-152, 162, 184
Major rivers 156
Malaria 21, 256
Maldives 316
Malnutrition 12, 19, 153, 249-251
Mangrove ecosystems 197
Mangrove forests 25, 185, 347, 361
Maritime zones 23, 311
Marine fisheries 351
Mass balance 8
Mean annual flow 157
Mean low flow 157
Mean peak flow 157
Meghna River 338
Methane 2, 40, 45-46
Migration 13, 18, 243, 246, 293-294, 386
Monsoon 6, 66, 144
Monsoon rainfall 6, 76, 164
Moisture balance 163
Montreal Protocol 24, 49,305,321-322
Mountain glaciers 114

Natural resources 1, 12
Navigation 308
Nitrous oxide 2, 40, 47
Normal flooding 14, 160
Northern Hemisphere 5, 39

Oceanic heat uptake 102
Oceanic thermal expansion 8, 100, 103
Ozone-depletion 51
Ozone-depletion feedback effect 51
Ozone layer 48, 290, 321-322

Padma river 157
Passive adaptation 77
Photosysnthesis 177
Polders 378-379
Population 18, 236, 242, 344, 346
Population growth 26, 236-237
Post-monsoon 144, 162
Poverty 12, 19, 153, 249
Pre-monsoon 144, 162
Precautionary approach 22, 291
Private property 296
Projected changes 64
Property rights 22, 293, 296, 300

Rabi crops 16
Rabi-season 181
Radiation budget 50
Radiative forcing 4, 8, 50, 53, 101
Rehabilitation 309
Relative sea-level change 99-100, 121, 123, 373
Resettlement 302, 304
Resiliency 219, 377
Rio-Declaration 290
River flow 198, 336, 375, 379
Rural-urban migration 13

Salinity 166, 342, 344
Salinisation 7
Salish 297
Sandwip Island 339
Sanitation 20
Sea-level change 1, 7, 27, 97, 271
Sea-level rise 12, 23, 31, 115, 198, 292, 335, 343, 375
Sea-surface temperature 10

Sediment movement 336, 369
Sediment supply 364, 369, 375
Seasonal flooding 12, 16, 156, 158, 161, 188
Settlement 18, 236, 242, 344, 389
Shallow tube-wells 161, 176, 343
Shrimp farming 154
Societal change 261
Socio-economic effects 1, 17
Soil-moisture 180
South Talpatty Island 314
South-west monsoon 66
St.Martin's Island 190, 369
Steric rise 8
Storm surges 10, 26, 128, 213
Sulphate aerosols 4, 51, 62
Sundarbans 25-26, 31, 197, 361-362, 377, 385
Sundari 185
Surface water 156
Sustainability 221
Sylhet basin 145, 160

Tectonic movement 10
Temperature impacts 181
Thanas 214, 227
Thermal expansion 9, 100-101, 114
Tidal flooding 167
Tidal floodplains 167, 361
Tidal-height 374
Tidal sluices 151
Tide-gauges 10, 121
Tropical cyclones 10, 76, 374
Tropical storms 10, 76

Unemployment 152-153
Urban population 152

United Nations Conference on Environment and Development (UNCED) 22, 29, 81, 290, 394

Variability of climate 74
Vector-borne diseases 20, 253, 256, 260, 388
Vegetation 361, 370
Vienna Convention 24, 290, 321-322
Vulnerable 230
Vulnerability 18, 24, 31, 220, 242, 246, 261, 372, 377

Water resources 194, 307
Water vapour 4
Winter 144
World Bank 85, 232

Zamindar 295

MIX
Papier aus verantwortungsvollen Quellen
Paper from responsible sources
FSC® C105338

If you have any concerns about our products,
you can contact us on
ProductSafety@springernature.com

In case Publisher is established outside the EU,
the EU authorized representative is:
Springer Nature Customer Service Center GmbH
Europaplatz 3, 69115 Heidelberg, Germany

Printed by Libri Plureos GmbH
in Hamburg, Germany